SUSTAINABILITY PERSPECTIVES *for* RESOURCES *and* BUSINESS

SUSTAINABILITY PERSPECTIVES *for* RESOURCES *and* BUSINESS

Orie L. Loucks • O. Homer Erekson • Jan Willem Bol
Raymond F. Gorman • Pamela C. Johnson • Timothy C. Krehbiel

LEWIS PUBLISHERS
Boca Raton London New York Washington, D.C.

Library of Congress Cataloging-in-Publication Data

Sustainability perspectives for resources and business / Orie L.
    Loucks ... [et al.].
        p.   cm.
      Includes index.
      ISBN 1-57444-058-6 (alk. paper)
      1. Sustainable development. 2. Natural resources. 3. Industrial
management--Environmental aspects. 4. Economic development-
-Environmental aspects. I. Loucks, Orie L.
HC79.E5S8666 1998
333.7—dc21   98-38870
                                                                                                         CIP

   This book contains information obtained from authentic and highly regarded sources. Reprinted material is quoted with permission, and sources are indicated. A wide variety of references are listed. Reasonable efforts have been made to publish reliable data and information, but the author and the publisher cannot assume responsibility for the validity of all materials or for the consequences of their use.
   Neither this book nor any part may be reproduced or transmitted in any form or by any means, electronic or mechanical, including photocopying, microfilming, and recording, or by any information storage or retrieval system, without prior permission in writing from the publisher.
   All rights reserved. Authorization to photocopy items for internal or personal use, or the personal or internal use of specific clients, may be granted by CRC Press LLC, provided that $.50 per page photocopied is paid directly to Copyright Clearance Center, 222 Rosewood Drive, Danvers, MA 01923 USA. The fee code for users of the Transactional Reporting Service is ISBN 1-57444-058-6/99/$0.00+$.50. The fee is subject to change without notice. For organizations that have been granted a photocopy license by the CCC, a separate system of payment has been arranged.
   The consent of CRC Press LLC does not extend to copying for general distribution, for promotion, for creating new works, or for resale. Specific permission must be obtained from CRC Press LLC for such copying.
   Direct all inquiries to CRC Press LLC, 2000 Corporate Blvd., N.W., Boca Raton, Florida 33431.

   Trademark Notice: Product or corporate names may be trademarks or registered trademarks, and are used only for identification and explanation, without intent to infringe.

© 1999 by CRC Press LLC
Lewis Publishers is an imprint of CRC Press LLC

No claim to original U.S. Government works
International Standard Book Number 1-57444-058-6
Library of Congress Card Number 98-38870
Printed in the United States of America 1 2 3 4 5 6 7 8 9 0
Printed on acid-free paper

Preface

The 1990s have seen important new approaches by business to sustaining both corporate development and protecting the environment. These steps present a common theme, best expressed in the declaration of the Business Charter for Sustainable Development (BCSD). That declaration states, in part, that "Sustainability demands that we pay attention to the entire life cycles of our products," but it, and a conference report by the International Chamber of Commerce in 1991, goes on to project new commitments to partnerships between government, business, and society, and to product pricing that reflects full environmental costs. These new perspectives represent a significant paradigm shift on the part of business, one in which corporations are coming to accept responsibilities for a healthy environment, just as they accept their responsibility for healthy employees and consumers.

However, no similar shift has taken place in undergraduate education for either business majors or science majors. This book is being written to facilitate the sea change in business and science education that has begun to take form in the workplace. Our goal is to describe the environmental science and business foundations of the changes taking place in a form suitable for college seniors in business, the sciences, engineering, and other disciplines as well as for graduate students. We assume that all users of this text have taken an introductory course in environmental science or ecology, as well as one or two semesters in economics.

Beyond that, what is critical is to understand how and why the changes have come to businesses, small and large. It is our hypothesis that businesses, particularly large, multi-national corporations, are utilizing longer-term planning horizons than in the past, and they accept that sustainable development requires long-term availability of resources and a healthy environment if customers are to be satisfied. Thus, businesses and educational institutions must approach environmental management from a perspective that intentionally links long-term views of science and environmental monitoring with near-term and long-term business decision making. The book, therefore, also can be viewed as a step toward understanding how the business and science communities are adapting to new information about risks, including risks to the environment, and to new expectations by consumers as to business' responsibility for the environmental, economic, and social systems that make up our society.

In this view, sustainable development envisages an economy and production system wherein both economic growth and quality-of-life improvements can occur in a coherent, unified system complementary with, rather than antagonistic to the maintenance of natural capital. Achieving such a goal requires truly major adoptions of benign production technologies. However, the surprise we found in doing the case studies included here is that much of this technology has been available, and the only question was when business would adopt it in their own self-interest. Sustainable development, therefore, involves implementation of new technology, and a means of convincing our political institutions and the general public that policies to limit externalities, thereby "closing systems and loops," is good for business, good for resources, and good for people.

The Miami University Sustainability Project (MUSP) that led to the writing of this book has been a six-year collaboration of faculty from both the natural science and business fields. The project's goal has been to support curriculum and scholarship related to sustainability as envisioned in the previous paragraphs. We see such scholarship, for undergraduates as well as graduate students, as having to focus at the intersection of business and science in order to understand the feasibility and role of low-impact building and production improvements, and the sustaining of all forms of capital: forests, clean air and water, biodiversity, infrastructure, and knowledge. In September 1997, Miami University established the Center for Sustainable Systems Studies to focus its activities in teaching and investigating the sustainability of development.

With the support of a major grant from The Cleveland Foundation, MUSP has pursued development of these instructional materials for undergraduate seniors and graduate students, featuring the eight chapters on principles of sustainability and ten case studies. The latter illustrate how some companies have moved ahead of the curve in pursuing low-impact technology, new corporate standards for environmental protection and safety, and sophisticated environmental reporting. Their experience is showing the feasibility of marketing environmentally-friendly products, the use of full-cost pricing (including environmental externalities), and the empowering of stakeholders to be part of a total quality process, proceeding to sustainability. The core faculty have been drawn from the Richard T. Farmer School of Business Administration (Economics, Finance, Management, Decision Sciences, and Marketing) and the College of Arts and Science (Geography, Geology, History, and Zoology); but at other institutions many other fields will be able to contribute as well.

A further goal of this book, then, is to bridge the disparities between business, economic, and natural science views of resource use at an advanced level. Such bridging requires students to embrace current scientific knowledge about efficient, "sustained yield" practices in resource use, waste minimization, and pollution prevention, and new economic valuation and environmental accounting practices. Also important is the potential that arises from new corporate approaches to planning, marketing, and management, approaches that consider diverse stakeholder ethics and values. In such collaborations there is clearly a role, as well, for the public and for the state and federal government in policy making.

We see this text as supporting two other objectives:

- Interdisciplinary integration and synthesis as part of learning and living: the book supports interdisciplinarity by exploring how patterns of resource consumption by business and consumers can be consistent with a sustainable relationship between society and the natural world. Both students and faculty alike will share the responsibility of integrating information and insights from the natural sciences, business, economics, and public policy fields.
- Reflection, informed action, thoughtful decisions, and personal and ethical understanding: because the business decisions that lead to sustainable policies require balancing of human, moral, and natural resource constituency interests, the book, and courses that use it, will necessarily require students to confront the broader goals of liberal learning. Especially important are principles that derive from the synthesis of environmental and business ethics, both of which are major subfields in their own right, both brought together here.

Integral to the book and courses that use it are interdisciplinary discussions or projects that can be undertaken by groups of students from the diverse disciplinary backgrounds the book seeks to engage. We suggest the use of 75-minute class periods where, for many topics, both the business perspective and the natural science (or ethical) perspective can be presented briefly by different faculty, and an immediate discussion can lead to the desired intellectual synthesis. These discussions and projects provide opportunities for sharing ideas, values, and assessments of the science, economics, and business practices that lead to sustainability.

A common thread throughout the text chapters and case studies is articulation of changing business and scientific points of view, and movement toward use of measurements and values that support sustainable relationships between society and the natural world. Uses of the text will change from year to year, but the perspective to be retained will be the common vision that sustainability is both broadly desirable and possible, and must be approached through knowledge of the intertwined natural, social, economic, and political systems.

At the heart of this vision lies the paradigm of closed systems and loops in the sense that pollution and resource exhaustion are the result of open systems and open loops, while pollution prevention, sustained yields, and reduced liabilities are examples of controlling externalities, thereby closing the systems and loops. These essential principles have emerged prominently in personal life styles during the past 20 years, and are likely to expand rapidly through technology and commercial enterprises in the near future. Our book will have succeeded if we facilitate this transition even a little.

August 1998

Acknowledgments

We are indebted to many individuals, companies, and organizations who have provided intellectual and financial support for this work. As mentioned before, we are especially grateful to The Cleveland Foundation who provided a major three-year grant to support the necessary research and writing. In addition, we received financial support from the Council for Ethics in Economics and several sources from Miami University (including the Richard T. Farmer School of Business, the Liberal Education Council, and the Office for the Advancement of Scholarship and Teaching). We appreciate the support of individuals from the companies and organizations that worked with us in preparing the case studies: General Electric Aircraft Engines (Mark Singleton); Ashland Chemical (Glenn Hammer); Grand Trunk Railroad (Gloria R. Combe); Walnut Acres (Paul Keene); Cincinnati Gas & Electric (Van Needham); David J. Joseph Company (Skip Rouster); Fetzer Vineyards (Pat Voss); Procter & Gamble (Deborah Anderson); DuPont (Richard Knowles); and the Chemical Manufacturers Association (Dick Doyle and Charles Aldag). We have benefited from numerous comments from scholars who have presented seminars at Miami University and/or participated in two national conferences we sponsored at Miami University in the last five years, along with many colleagues at Miami University. We would especially like to express our appreciation for the insights provided by William Renwick, Allan Springer, Gene Willeke, Jim Cashell, Gary Allen, and Stanley Kane. Finally, and most important, we wish to thank Anne Dickey for her faithful efforts in designing, typing, and providing other support activities related to the preparation of the manuscript.

Dedication

We wish to dedicate this volume to Professor Emeritus Norman Grant, geologist and renaissance scholar, whose insight and dedication to a vision has made this synthesis possible.

About the Authors

PRINCIPAL AUTHORS

Orie L. Loucks,
B.Sc.F., M.Sc.F., University of Toronto; Ph.D., University of Wisconsin at Madison
Orie L. Loucks is Ohio Eminent Scholar of Applied Ecosystems and Professor of Zoology at Miami University. He received his doctorate in botany (with a minor in soils and meteorology), and has a masters degree in forestry. Although trained as a biologist and forester, Dr. Loucks has collaborated with scholars in many other fields, particularly engineering, economics, history, and the social sciences. His research has focused on the dynamics of lakes and forests, but has included policy related-work on acidic deposition, biodiversity, conservation, and the sustainability of cities. He has served on Boards for the National Research Council and The Nature Conservancy.

O. Homer Erekson,
B.A., Texas Christian University; Ph.D., University of North Carolina at Chapel Hill
O. Homer Erekson is Associate Dean for Academic Affairs and Professor of Economics for the Richard T. Farmer School of Business at Miami University. He received his doctorate in economics with undergraduate degrees in economics and political science. His environmental research has focused on estimation of hedonic pricing models and organizational and ethics issues related to business decision-making. He is Vice-Chair of the Task Force on Business and the Environment for the Council for Ethics in Economics.

Jan Willem Bol,
B.A., Davidson College; M.Sc., Ph.D., Warwick University
Jan Willem Bol is Adjunct Professor at the Institute for Environmental Studies at the University of Wisconsin-Madison. Until 1995, he was Associate Professor of Marketing at Miami University. He developed and taught numerous advanced marketing courses in academe and industry, with special expertise in business ethics. He presently owns and operates Amber Waves, an organic farm in Wisconsin.

Raymond F. Gorman,
A.B., Brown University; M.B.A., Duke University; D.B.A., Indiana University
Raymond Gorman is Associate Dean for Undergraduate Studies and Professor of Finance for the Richard T. Farmer School of Business and Director of the Center for Sustainable Systems Studies at Miami University. His academic background is in corporate finance and mathematical economics. His research interests include analysis of the effects of the merger and acquisition movement on the rate of resource depletion in target companies and the feasibility and utility of mutual funds specializing in "green" investments. He served recently as Editor of the *Mid-American Journal of Business*.

Pamela C. Johnson,
B.A., Willamette College; Ph.D., Case Western Reserve University
Pamela C. Johnson was Assistant Professor of Management at Miami University until 1998. Her research interests have included systems theory, organizational behavior, and business ethics. She presently is an organizational and environmental consultant in Seattle.

Timothy C. Krehbiel,
B.A., McPherson College; M.S., Ph.D., University of Wyoming
Timothy C. Krehbiel is Associate Professor of Decision Sciences and Management Information Systems at Miami University. His research interests include experimental design, quality improvement, and statistical education. His environmental research has explored the application of Total Quality Environmental Management in applied business decision making.

CONTRIBUTING AUTHORS

James M. Childs, Jr.,
James M. Childs, Jr., is Academic Dean and Professor of Ethics at Trinity Lutheran Seminary in Columbus, OH. He is the author of *Ethics in Business: Faith at Work.*

Norman K. Grant,
B.S., Ph.D., Edinburg University (Scotland)
Norman Grant was Professor of Geology at Miami University and is now retired. His interests included the relationship of society, business, and the environment to the use of earth resources, including their possible limits. He has edited a companion volume to this book entitled *Anthology of Readings in Sustainability Perspectives.*

H. Gregory Hume,
B.S. University of Kentucky, M.En. (Environmental Science), Miami University
H. Gregory Hume is a Waste Reduction Specialist at the Center for Applied Environmental Technologies at the Institute for Advanced Manufacturing Sciences. While a student in Miami University's Institute for Environmental Sciences, he was a graduate assistant for the Miami University Sustainability Project.

Allison R. Leavitt,
B.S. Purdue University, M.En. (Environmental Science), Miami University
Allison R. Leavitt is a former landscape architect, now an instructor and doctoral student in the interdisciplinary doctoral option, Department of Botany, Miami University. She was a graduate assistant for the Miami University Sustainability Project.

Alison Ohl,
B.S., University of Arizona; M.A., Miami University
Alison Ohl received her masters degree in economics from Miami University. From 1993 to 1997, she was the Project Assistant for the Miami University Sustainability Project. Presently, she is a research associate for Scudders, Stevens, and Clark in Cincinnati, OH, studying criteria for ethical investing.

David W. Rosenthal,
B.S., Wharton School (Pennsylvania); MBA, Southern Illinois University; DBA, University of Virginia
David W. Rosenthal is Associate Professor of Marketing at Miami University. He is an active case writer with special expertise in business ethics.

Steven Skeels,
B.S., M.S., M.B.A., Miami University
Steven Skeels received his undergraduate and masters degrees in Finance from Miami University, serving as a graduate assistant for the Miami University Sustainability Project. He presently is a scrap metals broker for David J. Joseph Company in Cincinnati, OH.

Nigel C. Strafford,
B.A. Cornell University, M.A., Miami University
Nigel Strafford received his masters degree in economics from Miami University, serving as a graduate assistant for the Miami University Sustainability Project. He currently is a doctoral student in organizational management at Case Western Reserve University.

Esi Wurapa,
B.A. Wells College; M.En. (Environmental Science), Miami University
Esi Wurapa is a former graduate assistant in the Miami University Sustainability Project and is now an environmental manager for Ford Motor Company.

About the Center for Sustainable Systems Studies

The concept of interdisciplinary scholarship and teaching organized around sustainability started at Miami University in Oxford, OH in the early 1990s when faculty from the Richard T. Farmer School of Business and from the College of Arts and Science informally established the Miami University Sustainability Project. In September 1997, this initiative became established as the Center for Sustainable Systems Studies. The mission of the center is to provide a focus and operational leadership for research, teaching, and outreach at the intersection of business, natural and applied sciences, and the humanities implicit in society's expectation of sustainable development. The Center has three objectives:

- To support and facilitate the teaching of interdisciplinary courses on sustainability perspectives for students in business, natural science, design, technology, natural resources, and the humanities.
- To expand the opportunities for faculty research in the development and use of benign technologies and sustainable industrial development and investment.
- To provide a focal point for outreach initiatives through which industry leaders and faculty from other universities can be convened for information exchange, workshops, conferences, and seminars.

Table of Contents

SECTION I

PRINCIPLES OF SUSTAINABILITY

Chapter 1
The Context of Sustainability 3
O. Homer Erekson, Orie L. Loucks, and Nigel C. Strafford

Chapter 2
Natural Science Foundation of Sustainability: Health and Integrity of
Resources 23
Orie L. Loucks and Allison R. Leavitt

Chapter 3
Sustainability and Economic Well-Being 41
O. Homer Erekson

Chapter 4
Natural Resource Conflicts and Sustainability 63
Orie L. Loucks and Esi Wurapa

Chapter 5
Environmental Ethics and Corporate Decision Making for
Sustainable Performance 81
Jan Willem Bol, Pamela C. Johnson, and James M. Childs, Jr.

Chapter 6
Valuation and Reporting 105
Raymond F. Gorman

Chapter 7
Sustainability and Business Management Systems 139
O. Homer Erekson, Timothy C. Krehbiel, and Alison Ohl

Chapter 8
Decision-Making and the Environment: Integrating Scope
and Values 167
O. Homer Erekson, Orie L. Loucks, Jan Willem Bol, Raymond F. Gorman, Norman K. Grant, Timothy C. Krehbiel, Allison R. Leavitt, and Nigel C. Strafford

SECTION II

CASE STUDIES ON SUSTAINABILITY

Chapter 9
Ashland Chemical: Achieving Sustainability Through the Use of Responsible
Management Systems—Internal Environmental, Health, and Safety Auditing
Case Study Number 1 183
Raymond F. Gorman

Chapter 10
Maxxam Group Inc.'s Takeover of the Pacific Lumber Company
Case Study Number 2 213
Nigel C. Strafford, Raymond F. Gorman, Timothy C. Krehbiel, and Orie L. Loucks

Chapter 11
Walnut Acres, Organic Farms
Case Study Number 3 231
Jan Willem Bol, Allison R. Leavitt, and David W. Rosenthal

Chapter 12
From the Earth to the Table: Fetzer Vineyards and Its Bonterra Wines
Case Study Number 4 249
O. Homer Erekson and Allison R. Leavitt

Chapter 13
General Electric Aircraft Engines: Pollution Control Investment Analysis
Case Study Number 5 271
Raymond F. Gorman and H. Gregory Hume

Chapter 14
The David J. Joseph Company: Closing the Loop in the Automobile Industry
Case Study Number 6 279
Jan Willem Bol and Allison R. Leavitt

Chapter 15
Environmental Hazards in Transportation: The Response of Grand
Trunk Railroad to a Derailment
Case Study Number 7 291
O. Homer Erekson and Timothy C. Krehbiel

Chapter 16
The Procter & Gamble Company: Reducing Packaging Waste in the
United States and Germany, 1987–1994
Case Study Number 8 303
Jan Willem Bol, Orie L. Loucks, David W. Rosenthal, and Steven Skeels

Chapter 17
Electricity Load Management and Stakeholders:
Cincinnati Gas & Electric Company
Case Study Number 9 323
H. Gregory Hume, Alison Ohl, Raymond F. Gorman, and Orie L. Loucks

Chapter 18
Community-Industry Dialogue in Risk Management: Responsible Care® and
Worst-Case Scenarios in the Valley of the Shadow*
Case Study Number 10 337
O. Homer Erekson and Pamela C. Johnson

Permissions 355

Author Index 359

Subject Index 365

Section I

Principles of Sustainability

1 The Context of Sustainability

O. Homer Erekson, Orie L. Loucks, and Nigel C. Strafford

CONTENTS

1.1 Introduction 3
1.2 Foundations in Ecology and Economics 4
    1.2.1 Carrying Capacity and Sustainability 4
    1.2.2 Social and Economic Views of Sustainability and Sustainable Development 7
    1.2.3 Economic Growth and the Environment 8
    1.2.4 Growth and the Environment: A Contrary View 9
    1.2.5 Aggregated Determinants of Environmental Trends 10
    1.2.6 Global Welfare Curve 11
1.3 A Unified Approach to Sustainability 12
    1.3.1 Hierarchical Conceptualization of Sustainability 12
    1.3.2 Three Sustainability Principles from Systems Thinking 13
        1.3.2.1 Principle 1 14
        1.3.2.2 Principle 2 16
        1.3.2.3 Principle 3 17
1.4 Values and Process Principles 17
1.5 Conclusion 17
1.6 References 20

1.1 INTRODUCTION

Recent assessments of business, sustainable development, and the environment[1,2] present a common theme, best expressed in the declaration of the Business Charter for Sustainable Development (BCSD): "Sustainability demands that we pay attention to the entire life cycles of our products." The declaration goes on to prescribe new commitments to partnerships between government, business, and society, and to product pricing that reflects environmental costs.[3] In the chapters and case studies that follow, we describe a paradigm shift in business and technology systems, an unprecedented change in business approaches to resources, and the corporation's view of public health and environmental responsibility. We also describe a paradigm shift in our understanding of ecological systems as they are used by society. This sea change, for business and ecological science, as for business education and science education, follows a realization that consumer acceptance, competitiveness, and profitability will proceed together by "changing course," as outlined in the BCSD.[1,4]

For business, science, and engineering students to understand the context for such profound change, we need to consider how the business community sees its role in the environmental, economic, and social systems that make up our society, and how it adapts to new scientific findings. This book explores the idea that sustainability is central to understanding the paradigm shift. *Sustainability, as we shall see, covers many principles, but ultimately it envisages a system whereby economic growth and improvements in the quality of life occur in a coherent, unified way that is complementary with, rather than antagonistic to the maintenance of non-renewable resources such as oil, coal, and minerals, and renewable resources such as forests, farms, and fisheries.* The idea of sustainability also can be a vehicle for showing business and political leaders, and the general public, that policies designed to "close cycles and loops" scientifically are good business practices over the long term. Major innovation in benign technologies and in the reconciliation of manufacturing, consumption, and disposal of goods is required to achieve such sustainability, however, including the goal of maintaining healthy stocks of renewable resources and biological biodiversity.

Sustainable economic development, while having the potential to mean many things, should be understood as the outcome of private action by consumers, business, and other stakeholders pursuing individual goals while intentionally accounting for the long-term impact of their actions on the greater natural and human systems. Consensus for these actions requires dialogue among all these stakeholders. Therefore, sustainability concerns communication about shared expectations for the future as much as it concerns science and business fields. Sustainable development, thus, requires a unified system among people, their resources, and businesses, maintained in a dynamic equilibrium and able to sustain the viability of natural and social systems.

In this chapter, we introduce the many facets of sustainability as an integrating concept, and provide a context in which to understand its component principles.

1.2 FOUNDATIONS IN ECOLOGY AND ECONOMICS

> "If the present growth trends in world population, industrialization, pollution, food production, and resource depletion continue unchanged, the limits to growth on this planet will be reached sometime within the next 100 years. The most probable result will be a sudden and uncontrollable decline in both population and industrial capacity."[5]

This statement is the first conclusion from a highly controversial study published in 1972 as part of The Club of Rome's Project on the Predicament of Mankind. The study was published under the title *The Limits to Growth,* and it led to two decades of vigorous research attempting to confirm or refute the pessimistic conclusion. Critics of the study argue that the conclusion was inevitable given the structure of the study.[6] In particular, although the authors used a complex model based on systems dynamics, they coupled relatively fixed limits of resource availability with exponential growth in population, production, and waste. The results, these critics argued, were skewed by neglecting important economic and social feedback loops, such as the adjustment of resource prices to increasing scarcity.

In a sequel to their earlier study, Meadows, Meadows, and Randers[7] recognized these shortcomings and attempted to include the capacity of the earth and of society to absorb the consequences of growth. However, as illustrated in Figure 1.1, their first conclusion in 1992 is still very similar to the original:

> "Human use of many essential resources and generation of many kinds of pollutants have already surpassed rates that are physically sustainable. Without significant reductions in material and energy flows, there will be in the coming decades an uncontrolled decline in per capita food output, energy use, and industrial production."

Undoubtedly, this conclusion again will be subjected to debate and counter evidence. We will not attempt to mediate the debate here. However, to appreciate the extent to which we may be reaching "limits to sustainability," we will introduce a framework for long-term projections based on ecological and economic systems theory.

1.2.1 CARRYING CAPACITY AND SUSTAINABILITY

Human beings always have altered the earth by degrading ecosystems, at least locally. For most of the history of the human species, however, the stresses imposed over large areas have been minimal.

The Context of Sustainability

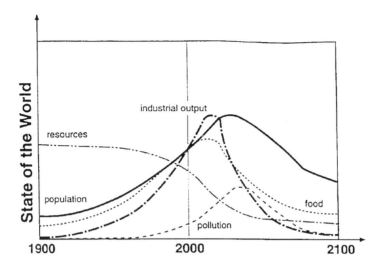

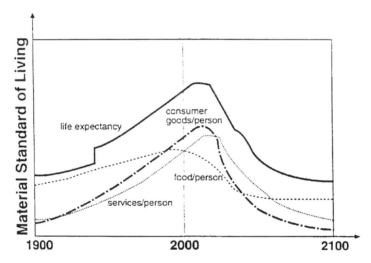

FIGURE 1.1 The "Standard Run" from *Beyond The Limits:* World society proceeds along its historical path as long as possible without major policy change. Population and industry grow until a combination of environmental and natural resource constraints eliminate the capacity of the capital sector to sustain investment. Industrial capital begins to depreciate faster than the new investment can rebuild it. As industrial output falls, food and health services also fall, decreasing life expectancy and raising the death rate. (*Source:* Reference 7)

But, since the middle of this century, population growth has been rapid, and scientific studies have shown the rate and scale of ecosystem change has increased dramatically, as described in Box 1.1. Such increases in population are part of the reason non-renewable resources have been depleted in some areas, forest and fishery stocks have been reduced, soils have been degraded, and coastal water quality impaired. People or their livestock either consume or degrade approximately 40 percent of the energy fixed world-wide by photosynthesis in terrestrial systems each year, leaving only 60 percent for all other land-based plants and animals.[8] And the magnitude of this impact continues to grow, probably doubling with each doubling of the world population. At the same time there has been a quintupling of global economic output since mid-century.

Ecologists are now asking whether there is a threshold in the diversion of photosynthetic projects beyond which the Earth's productivity and resource base cannot function as we have known it.

Box 1.1 Population Growth as a Consideration for Sustainability

How severe is our world population growth? Are we moving toward a situation where agriculture and fisheries can no longer provide adequate food?

The world continues to experience the most rapid growth rate in its history. As shown in Figure 1.2, the world population in 1950 was approximately 2.5 billion, more than double the 1 billion on earth in 1750. Total population in 1996 is near 6 billion. Improved health conditions worldwide and the compounding effect of a large population base have resulted in projections of an annual increase in world population of 90 million people, well into the next century.[10] These projections indicate that population will double in the next 30 years and reach 12 billion by 2100.

There is evidence, however, that the population growth rate has slowed somewhat. Since 1986, the average annual growth rate, world wide, has declined from 1.75 percent to 1.56 percent. This trend has resulted from a sharp decline in the fertility rate in China, as well as a smaller share of the world's women bearing children each year, and a decline in the average number of children per child-bearing woman.[11]

Another important dimension of the world population story is the distribution of population and growth rates across countries. As shown in Figure 1.2, developing regions of the world dominate the growth rate. In 1993, 94 percent of population growth occurred in developing nations and these nations now have 78 percent of the world's total population.[11]

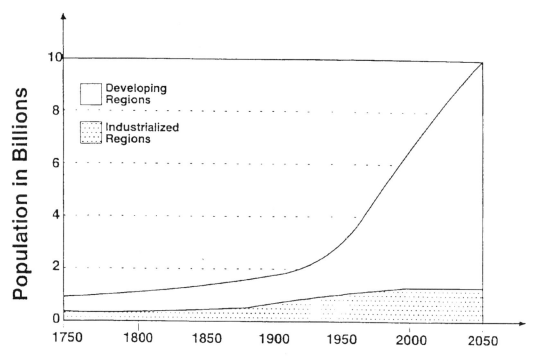

FIGURE 1.2 Trends and projections in world population growth, 1750–2050, for developing and industrialized regions of the world. (*Source:* Reference 10)

We do not know the answers, but imprudent use of resources has the potential to reduce, irreversibly, the capacity of the earth to sustain biological diversity and the cycling of water and carbon in the biosphere. To explore the consequences of human activity more formally, it is useful to think about the concept of *carrying capacity*.

The Context of Sustainability

Carrying capacity as a concept grew out of the biological foundations of population ecology some 70 or 80 years ago. Mathematical expressions for the growth of populations of insects and other animals recognized resource opportunities and constraints as follows: When a few individuals of a species are introduced in an area with abundant resources, simple reproduction increases the numbers slowly at first, but exponential growth follows (like a high compound interest rate). After a modest number of generations, the population is quite large, and the rate of further increase is always seen to level off, creating what is known as a sigmoid (or S-shaped) curve (see Figure 1.3). This pattern is so universal that the population, where it has leveled off, (identified as K in equations for this response) has come to be known as "carrying capacity." The concept corresponds with wildlife and waterfowl managers' ideas as to the natural limits they should expect for the wild game to be grown or harvested on public hunting grounds, but it also has some equivalency to the idea of a sustained harvest in forests, fisheries, and agriculture. Another way of conceiving of carrying capacity is "the largest number of any given species that a habitat can support indefinitely."[9]

Although the idea of "carrying capacity" is one aspect of defining the sustainable yield for a particular species, it falls far short of what we are defining here as "sustainability." First of all, the focus of carrying capacity is on one species singled out from its supporting ecosystem. If we applied carrying capacity to a social or economic system, we would be implying a limiting capacity for the human species, but our technology provides for more flexibility than that. Secondly, ecological systems are rarely able to maintain "stability" for a particular population at the so-called carrying capacity level for any extended period of time. There are too many interacting factors and competing populations. Thus, sustainability can be seen as a concept that seeks to represent populations and resources dynamically, and, in the case of the human species, to manage access to these resources such that carrying capacity is neither degraded nor fixed. The result is a scientific view of "sustainable development."

1.2.2 Social and Economic Views of Sustainability and Sustainable Development

For the reasons above, carrying capacity for the human species, utilizing the earth as a whole, must be seen as an immensely complex problem. Carrying capacity depends on a myriad of

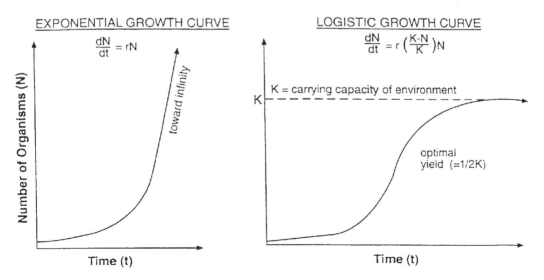

FIGURE 1.3 Exponential growth of a population in the absence of any resource limits (left), and logistic growth (sigmoid curve) in the presence of resource limits, or carrying capacity (right). Note the equation for logistic growth.

interdependent factors, including levels of human consumption and production, and a wide range of effects and benefits from technology.

As policy makers and scholars have considered economic growth and development internationally, the raising of consumption standards in developing countries to the levels of developed countries has come to be seen as potentially degrading of the natural resource base of the developing countries. Such economic development also would induce significant environmental stresses on ecosystems all around the world. These concerns were the context for the idea of "sustainable development" outlined by Caldwell.[12] The report of the World Commission on Environment and Development[13] (WCED), entitled *Our Common Future,* put the concept into words as follows:

> "Sustainable development is development that meets the needs of the present without compromising the ability of future generations to meet their own needs. It contains within it two key concepts: the concept of 'needs', in particular the essential needs of the world's poor, to which overriding priority should be given; and the idea of limitation imposed by the state of technology and social organization in the environment's ability to meet present and future needs."

The report also states that "sustainable development [is] a goal not just for the `developing' nations, but for the industrial ones as well." The view of sustainability in this text is consistent with the WCED statement, but we go further, arguing that for sustainable development to be approached meaningfully, consumers and businesses in developed countries will need to be able to reevaluate their current patterns of consumption and production and modify them. In fact, case studies later in the book show that when sustainable development is applied effectively, it structures social and economic activity in developed as well as developing countries.

1.2.3 ECONOMIC GROWTH AND THE ENVIRONMENT

Macroeconomists typically argue that there are only three major macroeconomic goals: price stability (limited inflation), full employment, and economic growth. The worthiness of economic growth as a goal, measured as the annual rate of change in gross domestic product (GDP), or gross domestic product per capita, is seldom questioned. In fact, some economists argue there is an important positive relationship between economic growth and measures of environmental quality.

The argument for economic growth to support the environment proceeds somewhat as follows: As income increases during the early stages of economic development, there is environmental degradation owing to resource exploitation. Beyond some point, however, environmental quality improves as society begins to support environmental legislation and pursues environmental remediation, both as individuals and through non-governmental organizations.[14] This relationship, as shown in Figure 1.4, seems to have validity for basic services such as sanitation and drinking water supplies, and for certain air pollutants such as suspended particulates, sulfur dioxide (SO_2), and carbon monoxide (CO).[15]

The most significant underlying premise for the beneficial outcome of economic growth, however, is that human-made capital (equipment and buildings), and people, can be substituted for the natural capital used in the early stages of development. In fact, technological advances, as evidenced by increases in human knowledge, methods of production, and social organization provide an impressive basis for substitution of resources. Economic growth, therefore, is seen to be "sustainable" and thus capable of long-term increases in per-capita consumption, so long as sufficient savings are set aside by current generations to finance investment (the creation of human-made capital) in line with growth in population.[16] People concerned with degradation of resources, loss of species diversity, or global warming, are directed to consider the potential for appropriate new technologies leading to compensatory investments in other human-made capital to be provided to future generations for alternative resources and welfare.[17]

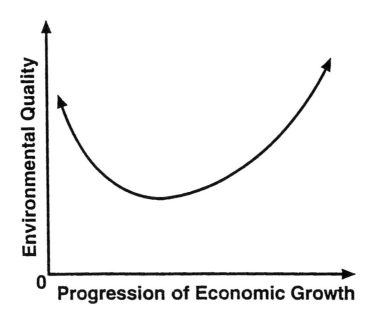

FIGURE 1.4 The long-term relationship between environmental quality and an economy's growth history. After a period of environmental degradation, increasing wealth enables citizens to demand and afford environmental protection.

1.2.4 GROWTH AND THE ENVIRONMENT: A CONTRARY VIEW

Other economists and many scientists offer a contrary view and dispute the model depicted in Figure 1.4. Their argument is conveyed colorfully by the claim that "sustainable growth is a bad oxymoron."[18] There are several dimensions to this viewpoint.

First, these scholars argue it is not possible to have *unlimited* substitution between natural and human-made capital. In fact, rather than accepting the notion that natural and human-made capital are substitutes, they maintain that both are complementary in production. Economic activity is seen as requiring both human-made capital and the support of natural resources whose productivity is driven by solar energy. Furthermore, the argument goes on to point out the increasing scarcity of stored natural capital (such as fisheries and old-growth forests), thus viewing the small stocks of natural capital as a limiting factor in present economic development.[19]

Moreover, physical laws are seen as limiting the extent to which human-made capital can be substituted for natural capital. The very creation of human-made capital intrinsically requires the use of natural capital and, furthermore, results in some energy becoming unusable or in a lower energy yield in the future.* Moreover, since the environment is of finite size, the waste created and disposed as part of production strains or exceeds the assimilative capacity of local water and air resources, and ultimately even of the earth. These views are evident in many scientific papers written about the effects of a growing human population on the long-term availability of natural resources and on the functioning of the biosphere. No consensus has been reached yet as to the long-term consequences for the earth from an expanding human population, but sustainable development is seen as part of the solution.

Another concern of scientists and others who reject the optimism of the U-shaped economic growth/environmental quality relationship is the limited applicability of the empirical evidence.

---

*The First Law of Thermodynamics states that matter and energy can neither be created nor destroyed. When natural capital is used in production, some of it must be returned to nature in the form of residuals. The Second Law of Thermodynamics states that nature's capacity to convert matter and energy is not unlimited. In the process of using natural capital, it is transformed into lower energy (higher entropy) forms.

Economic growth has been correlated with improvements in pollutant loads, despite short-term, local control costs (e.g., sulfur and particulates abatement), but this growth also has led to accumulations of other industrial wastes, and increasing impacts from pollutants that produce long-term dispersed damages (such as from acid deposition and carbon dioxide emissions). Few studies have incorporated these negative relationships into the growth/environment relationships. Moreover, economic growth continues to result in localized reduction in resource stocks in the case of forests and fisheries,[14] with occasionally severe negative impacts on local communities. Finally, the evidence offered in support of economic growth leading to improvements in environmental quality has tended to be studies of limited scope that could ignore delayed or inter-generational feedbacks, processes that are important for a large-area or long-term balance of the environment and the economy.

1.2.5 Aggregated Determinants of Environmental Trends

The interplay between population, production and the environment is described by Ehrlich et al.,[20] who are credited with formulating a basic equation about human impacts on the environment over time, now known as the Ehrlich-Holdren Identity. In this equation, lifestyle, population, and technology are brought together as

$$I = P \times C \times T,$$

where I denotes environmental impact at a point in time, P is population, C is consumption per person, and T is the environmental impact per unit of consumption. Our concern in a discussion of sustainability is to understand whether the impacts, I, are increasing over time.

The basis of this relationship can be appreciated more fully by understanding why the right-hand side is multiplicative. According to Ehrlich et al., the main point is that the outcomes from increasing consumption (and associated measures of environmental degradation) do correlate well with the factors that influence consumption (such as disposable income, packaging, and average distance from a workplace). The multiplicative form of the relationship indicates that these factors are interdependent in their ultimate effect.

An implication as to how we may project the outcome of "sustainable development" can be seen now in a modified expression of the Ehrlich-Holdren equation, rewritten to more clearly convey environmental impact:

Environmental Impact = population $\times$ consumption of goods per person

$\times$ environmental consequences per quantity of goods consumed.

However, sustainable development, by definition, means that the environmental impact term (EI) here should not increase over time. Since population is increasing, then for EI to remain the same, either consumption per person or the environmental consequences per unit of goods must decrease. The mass balance approach implicit in the Ehrlich-Holdren equation provides an explicit means by which accounting principles can measure consumption per capita, and impact per unit of goods consumed. The adoption of more benign technology reduces this impact, reducing aggregate environmental impact. Consumption of goods per person, seen frequently as a measure of affluence, may be reduced as more human needs are met from smaller quantities of virgin raw material, or with fewer residuals, as recycling and new technology are introduced. Thus, aspects of lifestyle, together with benign technology, have potential to reduce the impact of a growing population, thereby facilitating the sustainability of economic development.

The critical point to be considered can be seen as follows: population, including immigration, for many western countries increases at a rate of 1 to 2 percent per year. Therefore, long-term aggregate environmental impact can remain constant only if the technology improvement $\times$ consumption per person can be decreased by 1 to 2 percent per year. Although these percentages are small, they represent very significant decadal change (10 to 20 percent) and a substantial technical challenge, locally and globally. Greatly improved economic and natural resource accounting data are required if we are to monitor the outcome from using improved technologies, thereby measuring sustainability as opposed to violations (exceedences) over carrying capacity.

1.2.6 GLOBAL WELFARE CURVE

Other authors[21] have taken the Ehrlich relationship between economic measures and the environment and presented it in the form of a Global Welfare Curve (Figure 1.5). The horizontal axis shows *environmental impact* as *ET*, where *E* is aggregate economic metric, capturing entropic throughput, and *T* is the degree to which technology is environmentally malign (as opposed to benign). The welfare benefit from global economic activity is shown on the vertical axis (as population × average welfare). Thus, these authors have only slightly recast the Ehrlich-Holdren equation by letting $ET = P \times A$ be a series of points on the curve shown in Figure 1.5. The section of the curve from *A* to *C* corresponds with the path of human development, moving from the development of agriculture, through the development of technology and human-made capital. During these developments there was a corresponding expansion of population and affluence. Point *D* is the point of maximum welfare thought to be maintainable by sustained inputs of matter and energy from natural capital, without reducing this capital. At higher levels of malign technology (i.e., beyond point *D*), welfare declines because of excessive use of resources and diminishing (in fact, negative) returns from the technology being used, even to the point of irreversibility. These negative returns are seen by Wetzel and Wetzel[21] as due to an increasing proportion of society's activity and resources (i.e., throughput) being devoted to substitution of resources and repair of large-area ecosystem damage.

We will return to this discussion in later chapters and consider determining where we in the United States are on this Global Welfare Curve. For now, suffice it to say that many scientists and economists in our society are concerned that we have reached the point of negative returns, where "further growth beyond the present scale is overwhelmingly likely to increase costs more rapidly than it increases benefits, thus ushering in a new era of 'uneconomic growth' that impoverishes rather than enriches."[22]

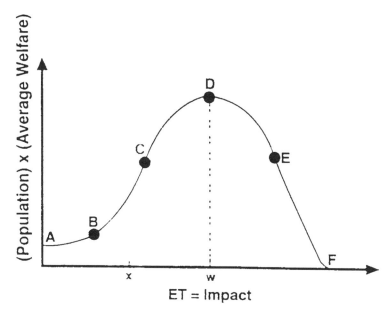

FIGURE 1.5 The Global Welfare Curve. (*Source:* Reference 21)

1.3 A UNIFIED APPROACH TO SUSTAINABILITY

The ideas presented here have been developed from work of many authors who advance the ideas of "sustainable development" and "sustainability" as alternatives to the conventional goals of growth discussed earlier. Some of their definitions go back many years,* but much modern usage begins with the simple statement in *Our Common Future*[13] quoted earlier, where "sustainable development" was seen as economic activity that *"meets the needs of the present generation without compromising the ability of future generations to meet their own needs."* Beyond this, a more elaborate view of sustainability is described by Liverman et al.[23] who see it as

> "the indefinite survival of the human species (with a quality of life beyond mere biological survival) through the maintenance of basic life support systems (air, water, land, biota) and the existence of infrastructure and institutions which distribute and protect the components of these systems."

Goodland and Ledec[24] highlight the social system components by defining sustainable development as "a pattern of social and structural economic transformations (i.e., development) which optimizes the economic and societal benefits available in the present, without jeopardizing the likely potential for similar benefits in the future." In this definition, as in others cited, we see that sustainable development, or the state of a system as being sustainable, can be both a scientific and a social concept, sometimes short-term, but more often inter-generational, and sometimes local, but more often regional, national, or international. A complete definition will have to consider all of these scales.

1.3.1 Hierarchical Conceptualization of Sustainability

We have seen that growth and carrying capacity were themselves complex concepts. The definitions above show that sustainability is even more complex, so as we seek a full conceptualization we should consider a hierarchical approach. The important elements of sustainability, as we have come to understand them, are shown hierarchically in Figure 1.6. The relationships can be understood from the top down, i.e., from sustainability as both a social and a scientific concept, as well as from the bottom up where specific principles, conditions, or measurements, as cited by many authors, can be met.

From the highest level, we can see *sustainability as a holistic concept,* and can distinguish three *Foundational Divisions* (the mid-size ellipses in Figure 1.6), together defining three sets of principles or metric types (like primary colors). Each is comprised of a few broad principles, or aggregated measures, which we will define and discuss in this and ensuing chapters. On close examination, we see that some of these principles can be quantified using a number of individual measures (or indicators). Experience exists already with aggregated metrics for economic activity, such as Gross Domestic Product, which clearly is related to some of the measures indicated in Figure 1.6. Similarly, the Index of Biotic Integrity[25] is a widely known aggregate measure conveying the condition of large biological systems.

The three foundational divisions outlined here are defined as:

1. *Systems Principles and Measures*—for renewable resources and cycling in systems where sustained productivity, resilience, low externalities, and closure of cycles or loops is sought.

---

*There are many definitions of sustainability development throughout the economics and scientific literatures. For a gallery of these definitions, see Pearce, D. W., Markandya, A. and Barbier, E. B., *Blueprint for a Green Economy,* Earthscan Publications, London, 1989.

The Context of Sustainability

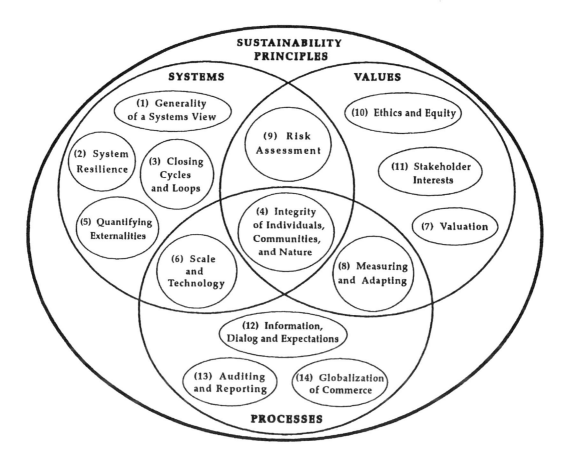

FIGURE 1.6 A synthesis of our research shows that sustainability (or sustainable development) can be thought of and measured in terms of fourteen interacting principles, aggregated into three foundational divisions, as shown here. These principles are identified here only by a word or phrase. They are clustered in the three divisions, one relating to natural and societal systems; one on expressions of ethics and valuation; and one concerning communications and interaction processes. Each principle is developed fully in this and later chapters.

2. Ethics and Values Principles and Measures—human and ethics oriented; the basis for the amenity valuation, audits, community integrity, and stakeholder interests.
3. Coordination Processes and Measures—the means by which dialogues, consensus or alienation as to sustainable expectations is expressed, locally to globally, and pursued or adapted to.

Further definition of these principles is provided in Box 1.2, and in individual chapters throughout this book.

1.3.2 Three Sustainability Principles from Systems Thinking

Here we develop more fully the ideas implicit in the first foundational division of the sustainability concept, *systems*. Our approach requires that the natural world and the elements of human, political, scientific, social, and economic spheres be viewed as interwoven, and for this we have the concept *systems*. Systems, like the environment or the economy, are seen often as healthy and functional, but sometimes dysfunction can be recognized (as, for example, when resources are depleted and the externalities leading to such outcomes are denied or ignored).

Early in this chapter, we referred to the complex systems models used by Meadows et al.[5,7] in their research. For them, as for us in this book, systems can be thought of as groups of interacting, interdependent parts linked together by exchanges of matter, energy, capital, values, or information. These interconnections are the essence of sustainability. For us, these linkages lead to the first of the 14 principles of sustainability from Figure 1.6, *the generality of a systems view.** We define this principle as follows:

1.3.2.1 Principle 1

> A general systems perspective holds where groups of interdependent and interacting parts are linked together by exchanges of energy, matter, capital, values, and information, thereby capturing a broad range of societal systems' behaviors within a unified and potentially sustainable model.

From this definition, we argue that much of the confusion inherent in earlier definitions of sustainable development can be overcome when that concept is broken down into its two components: sustainability and development. Thus, a further definition: What is to be *sustained* are the complex systems and subsystems that support society and the environment; what is to be *developed* are human beings and their individual potentials. Sustainability, then, is a condition or state of our industry, environment, economy, and society (and its values) that ensures sustained development of the human potential.

As we focus on sustaining systems, then, it is useful to consider the basic structure and properties of systems. A system is a group of interacting, interdependent parts or subsystems, often organized hierarchically into natural levels of increasing complexity.

To ensure that the interacting parts can be measured and evaluated quantitatively, we have to be able to define the boundaries around them. An example is shown in Figure 1.7, and others are illustrated in Chapter 2. Boundaries also can be established over time. These systems often also are characterized by the following features:

- *Nonlinear Interactions between the Parts*—In his book *The Fifth Discipline,* Senge[26] notes that, "Reality is made up of circles, but we see straight lines. Herein lie the beginnings of our limitations as systems thinkers." Consider the Linear-Sink Model as shown in Figure 1.7a. Here resources flow in a straight line through the ecosystem. This can be contrasted with the Unified Cycling Model shown in Figure 1.7b, where energy enters the model and is transformed, but remains within a closed system.
- *Complex Feedback Loops*—Systems scholars look for system boundaries that minimize interaction between the subsystem under study and other parts of a surrounding system or neighboring systems. However, systems are partially characterized by the nature of interactions across these boundaries. One subsystem's output becomes another subsystem's input. Using a spatial metaphor, one subsystem's output becomes a downstream subsystem's input. Moreover, outputs to another system can feed back as a new input to the original system, modifying it. These outputs and returns are called feedback loops and are important components of systems where information or modified material (or capital) flows are able to change the upstream system.
- *Limits and Thresholds*—The systems we will encounter have extensive dimensions in time and space. Often, environmental systems and subsystems must include switching points that arise when threshold levels or limits are reached and passed. For instance, ecologists are concerned that clear-cutting of old growth forests can affect spotted owl populations,

---

*Models of systems may be classified according to three criteria: (i) *realism*—simulating system behavior in a qualitatively realistic way; (ii) *precision*—simulating system behavior in a quantitatively precise way; and (iii) *generality*—representing a broad range of systems' behaviors with the same model.[28] While we will be interested, obviously, in both realism and precision in the models and approaches we use in this book, our emphasis here is on the generality of systems and the usefulness of approaching sustainability using a unified model.

The Context of Sustainability

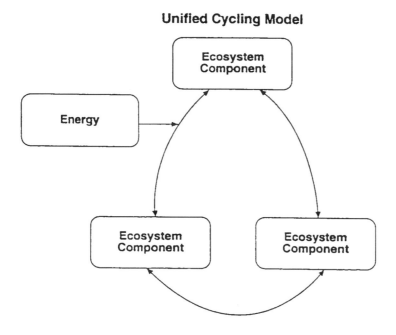

FIGURE 1.7 Two views of flows in complex systems: The Linear-Sink Model excludes feedback effects from wastes to the other components; the Unified Cycling Model shows complete feedbacks. (*Source:* Reference 27)

not directly, but by diminishing the food chain below a threshold that is able to support owls, thereby creating a switching point.
- *Emergent Properties*—When a system is considered as a whole, rather than as a separate listing of its various subsystems, unexpected (emergent) properties of the larger system often can be recognized. Because emergent properties are recognized in systems whose parts interact non-linearly and with feedback loops, it is not possible to understand such systems by examining only their constituent subsystems. Neither does it follow that the entire system is optimized when the functions of individual subsystems are optimized separately.

Another of the principles in the systems division (Figure 1.6) is *resilience,* a second principle and measurable property of sustainability which we will discuss here and refer to in later chapters. We define resilience in systems as follows:

1.3.2.2 Principle 2

> A system is healthy and sustainable when, following a significant stress or intervention, its variables show resilience, defined as the capacity of the system to accept a fluctuation and recover, returning the variables to characteristic levels or patterns.

For the business, science, and societal context of this book, it is useful to indicate the three components of resilience that are present when societal systems are sustainable [18, 29]:

- *Ecological Resilience*—Nutrient supplies and food energy to maintain a native species, community, or human population can fluctuate, returning quickly to a characteristic level. Moreover, the tolerance for waste (assimilative capacity) can fluctuate, but cannot be exceeded for long periods. An important corollary is that harvest rates of renewable resources should be kept in line with regeneration rates for those resources, and that depletion rates of non-renewable resources should not be greater than the rate at which substitutes can be created.
- *Economic Resilience*—Economic systems can be organized to promote efficient use of resources to meet the needs of people, maximizing the flow of income while maintaining the stock of natural and human-made capital necessary to provide these benefits. The productivity of the economic subsystem is dependent upon the stock of natural capital, human-made capital, and labor, as well as the state of technology and the management of these factors of production. Efficiency also requires that environmental costs arising from externalities are expressed, or otherwise accounted for in societal decision-making.
- *Social Resilience*—Political and social systems also experience change, but need to be structured to ensure that social change occurs in ways that preserve justice and equity among stakeholders.* One writer has conceived of this system as "the social scaffolding of peoples' organizations that empower self-control and self-policing in the peoples' management of natural resources."[30] Sustainable systems are ones which promote *intragenerational* or *socio-geographical* equity (elimination of poverty and adverse consequences of environmental externalities over a diverse geographic area), and *intergenerational equity* (preserving the rights of future generations to pursue welfare within the context of a resource base similar to that available to current generations). Sustainable systems also minimize destructive domestic and international conflicts, many of which are rooted in environmental inequities or exploitation strategies.

A third principle from the systems foundation division (Figure 1.6) concerns the need, with sustainability, to achieve fully closed systems. It is interesting to note that the word *environment* is derived from the Old French, *viron,* meaning "circle."[31] From a circle, we can envision fully closed systems describing the environment, with feedback processes linking all the components of our sustainable system. Thus, the third principle of sustainability to be considered in this chapter is the *closing of cycles and loops.* We define it as follows:

---

*Stakeholders is a term referring to all institutions and persons (and even other species) affected by the actions of a business enterprise. This includes shareholders, suppliers, buyers, consumers, employees, government agencies, communities, and world citizens. Obviously, it refers to all those who have a "stake" in the actions of a particular business or firm.

1.3.2.3 Principle 3

Boundaries of sustainable systems should allow measurement of stocks and flows among the components of subsystems and with adjacent systems, identifying externalities and feedbacks and determining whether cycles and loops are open or closed.

Closure or non-closure of loops relates to many other principles listed in Figure 1.6, from the quantitation of externalities, to auditing and reporting, to the integrity of health, communities, and nature. These linkages illustrate, again, the dynamic and evolutionary, but closed system nature of sustainability and sustainable development.

These three definitions of elements within sustainability are summarized along with eleven others in Box 1.2.

1.4 VALUES AND PROCESS PRINCIPLES

The previous section has introduced the basic elements of systems thinking, but Figure 1.6 shows that sustainability operates in a context of other major groups of principles making up a unified system. We cannot expect, at this point, to understand the intricate relations among all the components, and how each component helps define others, but by the end of the book we will. What is important is that we begin to understand sustainability as not one condition projected to exist at some future time, but a set of conditions that interact and affect one another like the contents of a living cell as shown in Figure 1.6.

The second cluster of principles (values division) relates to the methods of valuation and the processes inherent in values goals. In the upper right-hand corner, the principles center on six valuation challenges (such as identification of multiple stakeholder interests) that society faces in assessing the health and development of the overall system and its component parts. In the lower divisional ellipse, there are six process principles (such as the process of environmental reporting and auditing by business) that relate to the way society organizes itself and its communities to pursue sustainability.

Reflecting on all the principles in Figure 1.6 and Box 1.2, one sees inter-disciplinarity and connectedness. For instance, systems principles and values principles are connected through the principle of risk assessment. The definition of risk assessment (Box 1.2) shows both the identification of risks inherent in characterizing relationships within the system, and the valuation issues that arise upon quantifying those risks. Environmental reporting and auditing link values principles and dialogue or process principles, while scaling and technology connect systems principles and process principles.

Drawing all three types of principles together, and thus closing the loop in this taxonomy of the components in sustainability, is the principle of *integrity of health, communities, and nature*. We will return to the scope and interdisciplinary nature of this principle in the next chapter. Other principles will be explored in each of the following chapters, as the subject matter of sustainability is developed more fully.

1.5 CONCLUSION

Figure 1.6 helps to introduce a way of thinking about our natural environment and the relationship of human beings to it, as well as a set of processes that allow our economy and society to be a dynamic, resilient system. Thus, we conclude at a point where we can gather the ideas presented throughout this chapter into a cohesive whole that defines the broad context for sustainability. Sustainability envisages a system whereby economic growth and/or improvements in the quality of life occur in a unified system that is complementary with, rather than antagonistic to, natural capital. Sustainability involves an undertaking to build consensus among industry, political institutions,

Box 1.2 Overview of Sustainability Principles

Sustainability envisages an economic and social system whereby economic growth and improvements in the quality of life occur in a coherent, unified system that is complementary with, rather than antagonistic to the maintenance of renewable and non-renewable resources.

I. Systems Principles and Measures

1. **Generality of a Systems View**—A general systems perspective holds where groups of interdependent and interacting parts are linked together by exchanges of energy, matter, capital, values, and information, thereby capturing a broad range of societal systems behaviors within a unified and potentially sustainable model (Chapter 1).

2. **System Resilience**—A system is healthy and sustainable when, following a significant stress or intervention, its variables show resilience, defined as the capacity of the system to accept a fluctuation and recover, returning the variables to characteristic levels or patterns (Chapter 1).

3. **Closing of Cycles and Loops**—Boundaries of sustainable systems should allow measurement of stocks and flows among the components of subsystems and with adjacent systems, identifying externalities and feedbacks, and determining whether cycles and loops are open or closed (Chapter 1).

4. **Integrity of Individuals, Communities, and Nature**—Sustainability is achieved when all system elements contributing to overall integrity, understood as a state of being whole, are not allowed to compromise the ability of individuals, communities, or nature to function (Chapter 2).

5. **Quantifying Externalities**—Externalities occur when the actions of one person or company impacts the welfare of another person, company, species, or other systems component in a way that is not reflected in market prices. In a sustainable system, the impacts of all externalities are fully valued (Chapter 2).

6. **Scale and Technology**—Sustainable systems require minimizing the degradation of natural capital at scales from local to global, immediate to long term. Changes in technology can put natural capital at risk or facilitate its efficient use at all scales (Chapter 3).

II. Values Principles and Measures

7. **Valuation**—Equitable and efficient allocation of resources across space and time requires that the values of goods and services reflect all of the costs and benefits from their production, consumption, and disposal. Inherent in this valuation process is the internalizing of externalities, appropriate discounting of future costs and benefits, and consideration of other issues related to intergenerational and intragenerational equity (Chapter 3).

8. **Measuring and Adapting**—Sustainable processes require that system characteristics and outcomes be measured in quantitatively precise ways, and that mechanisms be included to allow the resulting information to modify human behavior so as to maintain system integrity (Chapter 4).

9. **Risk Assessment**—Human impacts on the environment involve uncertain consequences owing to the imprecision of measurements and scientific models. Sustainable systems must include identification and quantification of risks and determination of the level of acceptable risk to system components (Chapter 4).

10. **Ethics and Equity**—Sustainable systems include human appreciation of the intrinsic value of nature and the ethical positions assumed by business and its stakeholders. A sustainable

system provides for consideration of equity among humans, locally and globally, as well as other species, and between present and future generations so that system integrity can be maintained (Chapter 5).

11. **Stakeholder Interests**—All stakeholders, human and non-human, play an integral role in the functioning of a sustainable system. Therefore, these stakeholders must have an effective voice in decision-making processes affecting the system (Chapter 5).

III. Process Principles and Measures

12. **Information, Dialogue and Expectations**—Sustainable systems enable all participants to engage in effective dialogue, bringing their own traditions and perspectives to the decision-making process in an atmosphere of mutual respect (Chapter 5).

13. **Auditing and Reporting**—The information systems of individuals and organizations must be designed so that the financial consequences of an environmental audit are accounted for in the corporate reporting system. The system should reflect the true values of assets and liabilities, and be designed so that all stakeholders can utilize the information (Chapter 6).

14. **Globalization of Commerce**—Sustainable processes include explicit recognition of international interdependencies among consumers, businesses, and governments, with the result that solutions to environmental problems cross cultural and national boundaries (Chapter 8).

---

and the general populace supporting policies that will close the loop and reconcile manufacturing and disposal of wastes with the need to maintain the earth's capital, through waste minimization, recycling, adoption of benign technologies, and environmental conservation. Ultimately, sustainable economic development is driven, in part, by the self-interest of consumers, businesses, and other stakeholders pursuing individual goals, while intentionally considering the impact of their actions on the greater unified natural and human subsystems. Such a goal requires a *dialogue* among the various stakeholders representing these subsystems. The result should be a unified system able to maintain a dynamic equilibrium and able to regenerate itself to maintain its viability.

In the early part of this chapter, we introduced the *Limits to Growth* literature and noted some of the criticisms to it. The Meadows et al.[7] study still has an extremely pessimistic first conclusion in predicting the likelihood of an unsustainable future, given current conditions and behaviors. But typically, critics of this work often ignore the second and third conclusions of the report. In particular, the 1992 study goes on to conclude:

> "This decline is not inevitable. To avoid it two changes are necessary. The first is a comprehensive revision of policies and practices that perpetuate growth in material consumption and in population. The second is a rapid, drastic increase in the efficiency with which materials and energy are used."

Meadows et al.[7] also conclude, "A sustainable society is still technically and economically possible. It could be much more desirable than a society that tries to solve its problems by constant expansion. The transition to a sustainable society requires a careful balance between long-term and short-term goals and an emphasis on sufficiency, equity, and quality-of-life rather than on quantity of output. It requires more than productivity and more than technology; it also requires maturity, compassion, and wisdom."

We invite you to join us as we search for a sustainable system—one that serves both present and future generations.

1.6 REFERENCES

1. Schmidheiny, S., *Changing Course: A Global Business Perspective on Development and the Environment*, MIT Press, Cambridge, MA, 1992.
2. Smart, B., *Beyond Compliance: A New Industry View of the Environment*. World Resources Institute, Washington, D.C., 1992.
3. International Chamber of Commerce, *World Industry Conference on Environmental Management*, Vol. 2. *Conference Report and Background Papers*, ICC, Paris, 1991.
4. Erekson, O. H., Loucks, O. L., and Aldag, C., The dimensions of sustainability for business, *Mid-Am. J. Bus.*, 9(2), 3, 1994.
5. Meadows, D. H., Meadows, D. L., Randers, J., and Behrens, W. W., III, *The Limits to Growth*, Universe Books, New York, 1972.
6. Tietenberg, T., *Environmental Economics and Policy*, Harper Collins, New York, 1994.
7. Meadows, D. H., Meadows, D. L., and Randers, J., *Beyond the Limits*, Chelsea Green Publishing, Post Mills, VT, 1992.
8. Vitousek, P. M., Ehrlich, P. R., Ehrlich, A. H., and Matson, P. A., Human appropriation of the products of photosynthesis, *BioScience* 36(6), 373, 1986.
9. Postel, S., Carrying capacity: Earth's bottom line, *Challenge*, 37(2), 4, 1994.
10. World Resources Institute, *World Resources, 1994–1995: A Guide to the Global Environment*, Oxford University Press, New York, 1994.
11. Sachs, A., Population increase drops slightly, in *Vital Signs 1994: The Trends That Are Shaping Our Future*, W. W. Norton, New York, 1994, 98.
12. Caldwell, L. K., Sustainable development: viable concept—attainable goal?, paper presented at Linking Concepts of Sustainability, Santa Fe, NM, 1994.
13. World Commission on Environment and Development, *Our Common Future*, Oxford University Press, London, 1987.
14. Arrow, K., Bolin, B., Constanza, R., Dasgupta, P., Folke, C., Holling, C. S., Jansson, B., Levin, S., Maler, K., Perrings, C., and Pimentel, D., Economic growth, carrying capacity, and the environment, *Science*, 268, 520, 1995.
15. Grossman, G. M. and Krueger, A. B., Environmental impacts of North American Free Trade Agreement, in Garber, P., Ed., *The U.S. Mexico Free Trade Agreement*, MIT Press, Cambridge, MA, 1993, 137.
16. Branson, W. H., *Macroeconomic Theory and Policy*, Harper & Row, New York, 1972.
17. Toman, M. A., Economics and 'sustainability': balancing trade-offs and imperatives, *Land Econ.*, 70(4), 399, 1994.
18. Goodland, R., and Daly, H. E., Environmental sustainability: universal and non-negotiable, paper presented at Ecological Society of America, Knoxville, TN, 1994.
19. Folke, C., Hammer, M., Costanza, R., and Jansson, A., Investing in natural capital—why, what, and how?, in *Investing in Natural Capital: The Ecological Economics Approach to Sustainability*, Jansson, A., Hammer, M., Folke, C., and Costanza, R., Eds., Island Press, Washington, D.C., 1994, 1.
20. Ehrlich, P. R., Ehrlich, A. H., and Holdren, J. P., *Ecoscience: Population, Resources, Environment*, W. H. Freeman, San Francisco, 1970.
21. Wetzel, K. R. and Wetzel, J. F., Sizing the earth: recognition of economic carrying capacity, *Ecol. Econ.*, 12(1), 13, 1995.
22. Daly, H. E., and Cobb, J. B. Jr., *For the Common Good: Redirecting the Economy Toward Community, the Environment, and a Sustainable Future*, Beacon Press, Boston, 1989.
23. Liverman, D. M., Hanson, M. E., Brown, J. B., and Merideth, R. W., Jr., Global sustainability: toward measurement, *Environ. Manage.*, 12(2), 133, 1988.
24. Goodland, R. and Ledec, G., Neoclassical economics and principles of sustainable development, *Ecol. Model.*, 38(1/2), 19, 1987.
25. Karr, J. R., Assessment of biotic integrity using fish communities, *Fisheries*, 6(6), 21, 1981.
26. Senge, P., *The Fifth Discipline: The Art and Practice of the Learning Organization*, Doubleday, New York, 1990.
27. Allenby, B. R., Achieving sustainable development through industrial ecology, *Int. Environ. Aff.*, 4(1), 58, 1992.

28. Costanza, R., Waigner, L., Folke, C., and Mäler, K-G., Modeling complex ecological economic systems, *Bioscience,* 43, 545, 1993.
29. Munasinghe, M., *Environmental Economics and Sustainable Development,* The World Bank, Washington, D.C., 1993.
30. Cernea, M., The sociologist's approach to sustainable development, *Finan. Dev.,* 30(4), 11, 1993.
31. Romm, J. J., *Lean and Clean Management,* Kodansha International, New York, 1994.

2 Natural Science Foundation of Sustainability: Health and Integrity of Resources

Orie L. Loucks and Allison R. Leavitt

CONTENTS

2.1 Introduction 23
    2.1.1 Health and Integrity of Ecosystems and Economics 24
    2.1.2 System Properties and the Measurements of "Integrity" 26
        2.1.2.1 Principle 4 26
2.2 Natural Capital and Its Conversion to Human Use 28
    2.2.1 Naive Uses of Resources 28
    2.2.2 Liquidation of Natural Assets for Financial Gain 29
    2.2.3 The Sustainable Development Paradigm: Open System and Closed Loops. 30
2.3 Pollution Threats to Sustainability 32
    2.3.1 Principle 5 32
    2.3.2 Naive Disposal of Waste 32
    2.3.3 Sustainable Waste Disposal and Development 33
    2.3.4 Ecologically Persistent Chemicals 34
    2.3.5 Air, Water, and Biodiversity as Threatened Common-Property Assets 36
    2.3.6 Public-Private Partnerships as a Paradigm for Sustainability 37
2.4 References 39

2.1 INTRODUCTION

In this chapter we will look at the history of sustainability in human societies, from early times to the present. Our societies, whether hunters and gatherers or a world dominated by multi-national corporations, have all depended on low-cost natural resources for food, fiber, water, shelter, minerals, and other material goods. Many resources were available to early human societies simply for the taking, but as populations have grown, increasing demand has led, at least in developed nations, to tools and technology that make the entire earth's resources accessible to human needs. For the future, the vital question is this: Can new technology always assure sustainable supplies of needed resources (or appropriate substitutes), and can technology assure appropriate capacity for waste disposal?

To answer these questions, we will continue to develop the three principles outlined in Chapter 1 and add two more. To structure the discussion, however, we explore three models that describe how values and social conventions adopted by humans at different times in our history have influenced the balance between taking resources (and creating wealth for some in our society), and the apparent deprivations imposed on the majority by those actions. But in addition to benefits, there are two kinds of potential negative outcomes: *liquidation of resources for human use induces local scarcity or substitutions, and over-use of resources such as land, air and water to dispose of wastes creates local deprivations.* Neither problem need be serious, but the long-term disinclination to recognize either has led to problems whose solution now lies in what was introduced in Chapter 1 as sustainable development.[1]

Historically, we can recognize three different sets of human conventions for resource use at different times in our history.

1. In the pre-history period, naive conversion of natural resources for human use was constrained by limited energy and technology, but food, shelter, and fuel could be provided to a small number of human communities quite sustainably.
2. Within the period of recorded history, small-scale farming, fishing, and trading brought about conversion of natural assets for local gain through regional trading, creating local wealth. These historical conventions are being challenged now, however, under a principle of fairness to future generations and preventing impoverishment of people who must remain and live where resources have been depleted.
3. Sustainable use of natural resources is being envisioned now for a high-population, globally-integrated technological era wherein society seeks to convert the *sustainable productivity* of natural systems to wealth without impoverishing their long-term productivity.

For the *disposal of wastes,* on the other hand, we have used land, air, and water as sinks. Here we can distinguish two phases:

1. Naive disposal of settlement wastes and production residuals into natural systems, practices that became a problem for society as population densities grew and, later, as the wastes from new industrial technologies could not be assimilated in the air, soil, and water.
2. Waste disposal limited within the assimilative capacity of air, water, and land systems is a phase that is becoming possible now as new, low-impact technologies are becoming available along with society's and modern industry's awareness of an obligation to achieve sustainable use of all resources.

The sections that follow explore how and why the shifts from one human convention to another have taken place. It builds on the principles of sustainability developed in Chapter 1, i.e., the generality of a systems context, the resilience of those systems, and the necessity for closed cycles and loops in order to sustain those systems. The linkages that exist between economic, social and human systems, and the ecological systems that support human populations, will be emphasized.

2.1.1 Health and Integrity of Ecosystems and Economics

For a half-century or more, the term "healthy" has been used to describe a strong, fully-functioning economy. For less than 20 years the term also has been used to describe productive, fully-functioning ecosystems.[2] Scholars have noted the many similarities in definition and dynamics between healthy economic and ecological systems. Both approaches define the primary variables and processes of the system in terms of stocks, and the rates at which materials are exchanged in both economic and ecological systems.[3] These authors and others have noted similarities in the balancing of costs and gains in the two types of systems (Table 2.1), and among yield and internal couplings for both types of systems. The major point, of course from Chapter 1, is that both economic and ecological systems, while inherently complex, are also resilient, that is, they are capable of supporting a quasi-equilibrium condition around which observable variables in the systems can fluctuate, within certain limits. *Resilient systems are healthy and sustainable when, following a significant stress or intervention, their variables show a capacity to recover and return to their characteristic levels or patterns.*

TABLE 2.1
Measures of Integrity and Health in Ecological and Economic Systems

Variable	Ecological Systems	Economic Systems
1. Stock Balance	• Standing crop (as biomass or carbon) • Fish or forest stocks • Carbon density (c/m^2 across regions) • Biological diversity	• Natural capital • Financial/property assets • Timber stock/ore body • Property value (tax base) • Knowledge/patents/copyrights
2. Productivity (rates)	• Photosynthetic product (per unit of time) • Growth (per unit of stock/per unit of time) • Decomposition of carbohydrate (per unit of time)	• Aggregate manufacturing product (per unit of time) • Product per unit of input (per unit of time) • Waste product recovery/reuse (per unit of product per unit of time) • Annual return on assets (per unit of time)
3. Yield (rate)	• Harvestable product (units per year)	• Sales/assets (over time)
4. Production efficiency (ratio of rates to stocks) Performance efficiency	• Annual productivity (per unit of standing crop) • Net growth/assimilation (output/input)	• Sales/R&D spending
5. Coupling (connectedness)	• Diversity of species consumed and of consumers • Variety of (negative) feedbacks • Hierarchical framework (depth of food chain linkages)	• Pareto optimality • Competing suppliers • Constraints market • Matrix management
6. Stressors	• Severe stock depletion (over harvesting) • Toxic effects on productivity and efficiencies • Biodiversity or system function impoverishment	• Material scarcity • Monopolistic exploitation • Market failures for public goods • Inflation • Divergence of private vs. social cost
7. Health/integrity	• Resilience (as capacity to recover) • Meta-stability of stocks, productivity and efficiency • Optimal system coupling • IBI, Index of Biotic Integrity (composite of 1–6, above)	• Resilience (as capacity to recover) • Functional stability of stocks and flows • Gini coefficient • Optimal interlocking of investments and directors • Low unemployment (percent) and low inflation

Suggestions have been made that an approach similar to human health diagnosis could be applied to ecosystems,[2] but many ecologists have adopted the term *integrity,* rather than health, to describe the ideal condition of renewable natural resources. This term has not been applied to economic systems, but it is descriptive of them. In economics, instead, the varied indices provided each month from the Department of Commerce and other agencies, when viewed collectively, provide a general expression of economic health (whether poor or good). Unlike the descriptors of natural system integrity (measures that quantify long-term sustainable function, as shown in Table 2.1), the common measures of economic activity (such as Gross Domestic Product, see Chapter 3) do not include elements such as depreciation of land or other resources and, therefore, do not contribute to measuring economic sustainability.[4,5] Thus, for the public, questions of health, integrity, and sustainability lead to questions about how economic activity that leads to local depreciation or externalities is being measured or not measured. Externalities, or market failures, can be recognized in economic systems that appear to support an essential local livelihood (which may be unsustainable environmentally), but they need not be tolerated. Communities that exist to complete logging of the last old-growth forest in a state are an example of a non-sustainable element. The apparent policy differences implicit between environmental activism, and cavalier or indifferent business practice, have been, until recently, confrontational and mutually exclusive. Sustainable development, as we will see throughout this book, should frame the relationship in new ways, opening up the possibility that environmental concerns, the sustainability of communities, and good business practice do share common ground and common goals.

2.1.2 System Properties And Measurement Of Integrity

We have referred several times to the functioning of natural resource systems. By this we mean to convey that ecological systems are living, and within limits, are self-renewing, even when humans remove some of the resource stocks, or even reduce air or water quality. In using the term system we are referring also to the linkages and functions among the many elements making up a renewable natural resource: clean water with oxygen and appropriate nutrients to support the function of plant growth; animals (including birds, mammals, and insects) able to live by eating plants or their debris after they die (thereby converting waste to useful product); and fungi and bacteria that complete the decomposition of dead tissue and the function of returning nutrients again to the living system.[6] These relationships are illustrated in the example shown in Box 2.1. *While none of the variables in an ecosystem function like constants (indeed, they are highly sensitive to changes), a healthy system with a high level of integrity has sufficient natural resources and absorbable accumulation of waste.*

Thus, we articulate a fourth principle of sustainability shown in Figure 1.6:

2.1.2.1 Principle 4

> Sustainability is achieved when all system elements contributing to overall integrity, understood as a state of being whole, are not allowed to compromise the ability of individuals, communities, or nature to function.

What is often overlooked, however, is that many material flows in systems of high integrity can be disrupted when the system is stressed by over-harvesting, pollution, or biotic impoverishment. This happens when nutrient cycles are changed by human activity, as in certain types of cultivation or by allowing nutrients to become too abundant (toxic) or too limiting (as in some of the responses to acid rain). When surface water flows are changed, and when plant or animal production is mostly harvested and removed (for example, with overgrazing or clearcutting), much of the normal cycling is removed, thereby depressing productivity and efficiencies in the system with a resultant loss of resilience and integrity (see Table 2.1). The scope of intrinsic value in this definition will be discussed in Chapter 5.

Natural Science Foundation of Sustainability: Health and Integrity of Resources

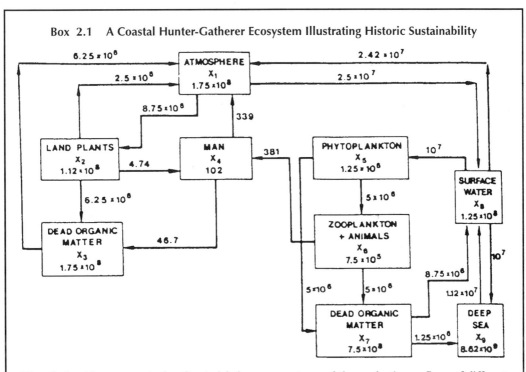

Box 2.1 A Coastal Hunter-Gatherer Ecosystem Illustrating Historic Sustainability

The relationships among stocks of materials in an ecosystem and the production or flows of different kinds of materials are shown in the above schematic. This is a summary of the carbon exchanges surrounding the Aleuts of southwestern Alaska some 40 years ago, a simple hunting and gathering society at the time, living sustainably on coastal shellfish, fish, seals and plant products.[8] Ecological systems, like economic systems, are concerned with material stocks (shown in the boxes as a stock of carbon, in metric tons), and flows (shown here along the arrows as carbon fixed, used, or decomposed, in metric tons per annum). When the rates of almost any of the flows is changed substantially, major changes occur in the subsequent (or downstream) parts of the system, higher in the food chain. The feedbacks in this system operate through nutrient cycling and, therefore, production of plants and the food chain can be affected. Under conditions of severe over-use or exploitation, ecological systems become so seriously degraded that changes can be irreversible.

But the significance of these results is in showing how a combination of data on stocks and flows over a few decades (a generation for humans) could be used to determine whether the system is being sustained. The Aleuts used only a small portion of the annual growth of seaside fish, shellfish and seals (the "zooplankton plus animals" box). What if this annual removal was increased tenfold, from 381 to 3810 tons per year, as happens with commercial exploitation of fish and shellfish. Obviously, when there are fewer animals there to grow, there is less annual growth for this compartment. The result is a gradual drawing down of the existing accumulated stock. Probably only 20 percent of the stock of 750,000 metric tons (i.e., 150,000 tons) of zooplankton plus animals are fish, seals, or mussels that are potentially usable by humans. The data show that a harvest of 3810 tons per year would not be sustainable any longer than about 30 years, the time it takes to deplete most of the useful stock of 150,000 tons (30 × 3800 = 114,000 tons). Further, once the stock is dramatically reduced it tends to be replaced by a few inedible species, and the degraded system may be incapable of recovering.

In an article titled "Global sustainability: toward measurement," Liverman et al.[7] cite a variety of indicators suitable for measuring large-scale sustainability. Among these they suggest per capita gross world product, per capita fossil fuel consumption, population density per unit area, and pollutant discharges per unit area. Notice that some of these are stocks and some are flows (compare Box

2.1), but none directly addresses the long-term (inter-generational) sustaining of natural system productivity or integrity. Criteria for selecting measures of large system integrity or sustainability will have to be similar to those that can be derived from data such as are outlined in Table 2.1 and Box 2.1. The hunting and gathering system shown in Box 2.1 is an open system, in that carbon density can be increased or decreased somewhat, but *it is a system characterized by essentially closed cycles. In a natural system with high integrity, everything is recycled,* and although the plants and animals grow, *there is no net increase or decrease in stocks for the system as a whole.*

Still, much work is needed to obtain agreement on what, specifically, we should measure and how to use the data. Liverman et al. recommend several essential characteristics of sustainability indicators: sensitivity to change over time; sensitivity to change across spatial aggregates or across demographic groups; predictive capability; availability of historical reference data; evidence of utility for threshold values; ability to measure reversibility (or its converse, irreversibility); and relative ease of data collection and use. All of these are useful criteria, but we still do not have an ideal set of measures for sustainability. As with our present economic indicators, some of the best measures are rates, others are surrogates for a combination of activities, but we readily accept the continuing need for more functional measures. In the following pages we will illustrate how and why the three models for resource use and their socio-economic contexts have operated at various times in our history.

Ultimately, the older models have had to be discarded. The lessons learned have been difficult, and decades or centuries have been required for change. Irreplaceable resources have been lost, and the resilience of resources and peoples over large areas has been impaired permanently. However, considering the changes in resource use that have already taken place and the further opportunities for innovation in measurement and efficient use of resources, human society and economic systems can continue to evolve while repairing the damaged systems.

2.2 NATURAL CAPITAL AND ITS CONVERSION TO HUMAN USE

The early vegetation, fisheries, wildlife, and fertile soils of the earth provided a natural wealth (also definable as natural assets or capital) for each of the world's early civilizations, as well as for settlers of the western hemisphere.[9] Natural assets, such as the forests of the midwest and northwest of the United States, and the fertile soils of the Great Plains, were the result of the accumulated productivity of ecological systems over hundreds to thousands of years. *These resources were viewed as rich, not just because goods required by early settlers could be produced from them, but because the goods could be produced with much less labor than in areas poor in resources.* Although the principle characteristics of resource use may have remained generally the same for hundreds of years, there have been important changes in access to the resources and the pattern of their exploitation. We illustrate this in the pages that follow through elaboration of three distinct sets of local conventions adopted by different societies to guide their use and allocation of natural assets.

2.2.1 Naive Uses of Resources

Appropriation of local resources for personal or communal gain has been a human instinct, apparently, from the earliest hunter-gatherer societies to the modern industrial age. When human communities used their resources sustainably to provide for local needs, as in the case of the Aleuts (Box 2.1), they simply converted the fisheries or plant foods they found into the goods they required including shelter and fuel, and except for the local accumulation of refuse, the community and the resources were sustainable over hundreds of years. With small populations, natural systems could tolerate the human extraction of resources without detectable loss of the systems' capacity to regenerate and repair.

Naive exploitation of resources also has led to the concepts of abundance and scarcity. Humans used the resource that was both the best quality and the easiest or least costly to obtain.[10] Until recent history, resources were found in abundance and the supply appeared limitless (if often distant). As human populations grew, supplies of the resource were drawn down and the next most available resource was appropriated if one was available, often with a new technology or development strategy. This pattern of resource use, depletion (beyond the capacity of the system to be resilient) and then substitution, is seen throughout history. Three examples will illustrate how resilience came to be exceeded, leading to impoverishment.

Although early hunter-gatherer societies had a minimal effect on the environment as long as human numbers were small and in a long-term equilibrium, the arrival of a new population of humans may have been responsible for the remarkable depletion and extinction of large mammals in prehistoric North America.[11] These nomadic peoples, newly arrived from Asia across the land bridge at the Bering Strait 12,000 years ago, were organized around an abundant resource, the many large mammals including mammoths and mastodons. They moved from area to area, following game that had never been exposed to humans as predators. Usually with this life-style, natural resource use can be dispersed, and, where animals have evolved to avoid humans, as in Africa, systems of wild game harvest can be maintained for thousands of years. However, where a new human population moves in with an advanced hunting capability (as with the arrival of these Asian peoples in North America at that time), the large mammals were taken easily and disappeared in little more than 2000 years. These early peoples, in effect, took advantage of the new wealth of wild game and drove about a dozen species to extinction.[11] Only the buffalo, elk, and grizzly bear remained in the western high plains.

Naive over-exploitation of resources has not been limited to the hunter-gatherer societies. Early agrarian societies in the Middle East lived sustainably for long periods, but eventually also over-exploited the natural resources surrounding their communities. As the populations grew, more land was needed for cultivation. Land that was only marginally suitable was put into production through technological improvements such as local irrigation. However, some of these innovations caused salt accumulation and erosion of the soils, exceeding their resilience, and leaving the original land unproductive and the now-larger communities unsustainable. The Sumarian society, located in the valley of the Tigris and Euphrates Rivers in Mesopotamia 4000 years ago, is a well-known example of over-exploitation of resources by an early agrarian society. The ecological system became degraded, no longer able to sustain the human and economic system[10] or to defend itself from surrounding tribes.

Like the soils of Mesopotamia, the forests of early Europe were depleted over hundreds of years to provide timber, fuel, and grazing lands for a growing population. Approximately 95 percent of western and eastern Europe was covered with forests prior to the formation of early settlements just over 2000 years ago.[10] Almost none of the original forest was left in Germany and other middle European countries by the 16th century with catastrophic hardships for the people and the economy. Virtually all of the forests of the region now are manual recreations after a long period of hardship.

Naive exploitation of resources, within limits, allowed use of natural resources to provide for the basic needs of local peoples. Impoverishment of resources took only a few hundred to a few thousands of years, until recently, when some resources have been impoverished in just decades. New or restored resources sometimes have been developed to provide for societal needs, although civilizations have collapsed in many cases when the social and resource system proved unsustainable.

2.2.2 LIQUIDATION OF NATURAL ASSETS FOR FINANCIAL GAIN

When societies became more specialized, and increasingly able to use a complex trading and economic system, natural assets were not used simply for local communities. Rather, the apparent local surpluses were sold or traded for other goods and services. In this way, *the apparent surpluses were converted to human system capital, roads, merchant ships, buildings and systems of financial services.*

Financial and infrastructure wealth was generated by the conversion of these assets for both individuals and nations. The resulting capital also grew to include the infrastructure of government and industry including advanced transportation systems, knowledge of the world, and higher education. Very often capital in the form of natural assets no longer existed, but through its conversion, society had created other relatively permanent capital assets.[9]

In the two examples described in Box 2.2, the production and trade in natural resources could have been sustained for the local communities had the *productivity* of the resources been managed in a sustainable manner. However, this would have involved a slower rate of conversion (development) to wealth and capital in infrastructure. Asset conversions are necessary in the development of a new country, but there are limits determined by the natural resources themselves.[12, 13] These limits have been recognized in the past, but were either poorly defined or inadequately documented for public consensus on management approaches, or to prevent loss of the resource and economic hardships. Rapid conversion of natural assets to financial or infrastructure wealth can provide a new generation's financial capital needs, but limits need to be determined before an irreversible collapse of the resources and socio-economic systems occurs.

A fundamental question remains: *Did the world's system of creating and transferring wealth require the liquidation of the great natural assets described in Box 2.2, or could the same wealth have been generated by converting the sustainable productivity of the assets to wealth without impoverishing ecological systems and local communities?*

2.2.3 THE SUSTAINABLE DEVELOPMENT PARADIGM: OPEN SYSTEMS AND CLOSED LOOPS

The discussion of Principles 1 and 3 in Chapter 1 illustrates that modern society is recognizing that there are limits to the liquidation of natural assets, and recognizes the unsuccessful attempts to

---

Box 2.2 Examples of High Natural Capital Systems Exploited for Financial Capital

Here we consider two examples of early businesses that converted natural capital to financial wealth: the Hudson Bay Company of England and Canada and, more generally, the Great Lakes timber industries. Both not only utilized natural assets to produce other forms of wealth, but the liquidation of one asset created new wealth in other parts of the country and the world.

The exploitation of fur resources in Canada by the Hudson Bay Company initially involved only trade with native people to obtain the desired fur species for markets in Europe. Harvesting fur-bearing animals in western Europe and parts of Russia had reached the point during the 17th century where many species were practically extinct.[10] Thus, when the prospect of a fur resource was recognized in North America, the Hudson Bay Company was incorporated in 1670 to exploit the resource, creating the prospect of wealth in England.[14] Area after area in North America was depleted as fur trading followed exploration into the interior regions. By the end of the 18th century the fur trade in North America was declining sharply owing to the reduced stock of fur animals.[10] The company was wealthy, however, and continues to this day as a leading department store in the major cities of Canada.

Harvest of the Great Lakes region's white pine forests began in Michigan in the early 1800s through operations of many relatively small companies. The timber harvest reached a peak in northern Wisconsin and in Minnesota in the 1890s. The peak in exploitation was followed by a rapid decline and abandonment of many settlements.[9] The valuable white pine (natural capital) of these three states was liquidated, but it was converted to various forms of infrastructure and financial wealth critical to the growth of cities such as Chicago, Detroit, Milwaukee, and Minneapolis. The consequences of the conversion of the Great Lakes forests must be assessed, however, at both the local and regional scale. The regional scale saw conversion into a midwest urban infrastructure, but the northern logging communities saw a short-term gain transformed into unemployment, extreme family hardships, and, ultimately, ghost towns. *These contrasts in the scale at which sustainability functions illustrate not only the complexity of the systems, but also the need to have measurements of sustainability at appropriate scales.*

achieve sustainability after major asset conversions. A new paradigm is emerging: *The sustainable use of natural systems differs from simple appropriation of resources in that it seeks to convert the long-term productivity of land and ocean systems and the renewability of air and water to human capital incrementally, without impoverishing the utility and productivity of those systems.*

The question of how judiciously to appropriate only part of the stock or productivity of resources relates to the concerns of Hardin and others in challenging over-use of resources, as described in a paper "The Tragedy of the Commons."[15] The paper notes the historic over-use of communal resources (such as a village pasture) that are common to the families living in local communities, land that is often known as the commons. In this old European practice, individuals stood to increase personal gain by adding a cow (or two, or more) in the commons without regard for the desire of the next individual to do the same. The practice gradually pushed grazing of the commons beyond its limits. In Hardin's words:

> "The individual benefits as an individual from his ability to deny the truth even though society as a whole, of which he is a part, suffers. Education can counteract the natural tendency to do the wrong thing, but the inexorable succession of generations requires that the basis for this knowledge be constantly refreshed."[15]

Hardin wrote "The Tragedy of the Commons" at a time when society was just beginning to recognize the need to move toward more sustainable use of all common property goods. The right to limited use of such common property goods as water, air, land, and biodiversity can become a political issue because these resources are still viewed as a part of our historic commons to be used by everyone. Without social contracts for sustainable use and consensus at local, national, and international levels, solutions to the "tragedy of the commons" (individual gain vs. collective loss) are thought by many people to be impossible. *Sustainable development with appropriate measurement, auditing, and public verification has potential to be a solution, however.*

But the problems of appropriate measurement and public verification are formidable (see the accounting and reporting principle, Chapter 6). The collapse of major timber and fishery harvests in parts of the United States since 1900 provide object lessons, particularly in relation to why past practices, even when regulated, have been so unsuccessful. Three major issues were identified by Gale and Cordray[12] as stumbling blocks for scientific management of sustainable yields. The first is in defining what truly is to be sustained: Is it the standing stock itself, or is it the *means of producing* that stock? The second issue has to do with measuring and forecasting the rate of asset replenishment, a particularly serious problem in ocean fisheries where one must estimate the resource dispersed over a very large area. The third is the consistent failure of managers and operators to take into account changing human elements such as the use of new technologies to achieve more thorough harvesting.[13]

Even the definition of what is a sustainable harvest can become controversial. The forest industry has worked quantitatively with this question for a century, seeking to establish a verifiable sustained yield in all regions of the United States. One problem has been an inability to find consensus on the main measure, or even the averaging area over which to define the sustainable yield.[12] For example, there are important differences between human-centered and ecosystem-centered management plans, differences owing to the geographic scale over which the sustained harvest (a farm, a county, or a state) is to be averaged. The scale and technology principle (Chapter 3) expands on this issue. In addition, there are related runoff and pollution effects problems from the forest harvest affecting communities and resources at great distances. Each definition appears to threaten the interest of some group and its perception of rights to the commons.

The human factors to be considered during over-exploitation result from the following: When harvests are high and economic conditions are good, businesses invest in capital improvements enlarging their operations. As an industry like marine fishing expands, so does the community's economic dependence on the higher harvest. Thus, when the available harvest goes down, as occurs naturally from time to time in ecological systems, the local economy suffers. To maintain jobs and economic stability, subsidies are often granted by the government in the form of low interest loans to buy *more* boats and equipment, encouraging more fishing while resource productivity is already low, further reducing the potential of the system to recover. *Sustainability requires that political solutions to natural variability be avoided by regulating the harvest within the capacity of the low productivity period.*

The modern goal of living within a sustainable level of resource use will mean not only a gradual shift away from the old approach of asset liquidation, but a commitment to using new technologies and more sophisticated information on how natural and social systems function and produce goods. This requires the balancing of scientific understanding of natural processes and better public understanding of economic systems and political processes as described in Chapters 3 and 7. A stepwise progression toward a new paradigm will foster sustainable life styles in developed countries as well as assuring long-term productivity in developing countries.

2.3 POLLUTION THREATS TO SUSTAINABILITY

The second major class of effects by humans on the natural environment, after natural capital conversion, is through disposal of wastes (pollution). These effects derive primarily from our use of land, air, and water resources to assimilate by-products of human activity. When the quantity of wastes to be disposed of exceeds the assimilative capacity of the receiving system, by-products accumulate in the water, soil, and air, degrading the natural system, and threatening its functions and productivity. As the functions of natural systems are weakened, the capacity of the resource either to assimilate the wastes or to produce useful assets is impoverished in much the same way as in overharvesting, thereby imposing a cost on society which economists refer to as an externality (see also Chapter 3).

This leads us to define a fifth principle (and metric) of sustainable socio-economic systems, the obligation to *quantify externalities:*

2.3.1 PRINCIPLE 5

> Externalities occur when the action of one person or company impacts the welfare of another person, company, species, or other systems component in a way that is not reflected in market prices. In a sustainable system, the impacts of all externalities are fully valued.

How humans have responded to externalities can be seen as changing over time, just as we saw change in our view of integrity in health and nature. The earliest view of externalities, historically, was the naive adding of by-products of human activity into natural systems. Initially, these externalities were unimportant because the settlement might move for other reasons. But the intensity of waste-accumulation effects increased as populations and settlement size increased and became permanent. A second view emerged during the 19th and 20th centuries, and by the decade of the 1970s, society, and later industrial managers, recognized the need to minimize externalities and to manage all waste disposal so as to sustain healthy ecological systems. Here we explore the core ideas and consider two views of externalities in relation to sustainability. (Aspects relating to government regulation are developed in Chapter 7.)

2.3.2 NAIVE DISPOSAL OF WASTE

As hunter-gatherer clans evolved into agrarian communities, the refuse from the resources they used came increasingly to influence the land and water into which the wastes were disposed. Even present-day hunter-gatherer groups usually relocate to avoid the problem, but large communities cannot do so. Nevertheless, these naive disposal practices continued in parts of the developed world well past the mid-20th century.

Early human societies produced three major types of residuals that overloaded natural systems: human waste; residuals from traditional agricultural production, including soil erosion; and residuals from the production of goods. Disposal of human wastes and agricultural residuals have plagued human society for centuries. On the other hand, disposal of wastes from the production of goods became an issue mainly during the past 200 years.

Wastes became a problem when agrarian societies replaced the hunter-gatherers. As settlements grew into small cities during the Greek and Roman eras, unsanitary waste disposal practices caused health problems such as intestinal diseases and cholera. Untreated waste was disposed of directly into rivers and streams, and drinking water was taken from the same sources by downstream communities. Providing clean drinking water and disposing of wastes became manageable after the centralization of these services in some Mediterranean cities more than 2000 years ago. As early as 312 B.C., when the Tiber River was too polluted to drink, the Romans built aqueducts to carry water into Rome. Later, in northern Europe, the first lead pipe water system was built in 1236 to supply London with water because the Thames River was polluted. Approximately 600 years after that, the first water filtration plants were built to provide clean intake water locally. However, it was not until the latter part of the 19th century that sewage treatment was undertaken on a broad scale to achieve rapid, local treatment of waste before discharging into streams.[10] Each of these developments illustrates how the size of the local population (scale) interacts with technology, as discussed further in the next chapter.

The second type of wastes, originating with agricultural activities, overloaded adjacent streams and rivers with residuals from soil erosion and high levels of livestock manure. Human cultivation, grazing, and logging of land all create disturbances leading to erosion. In dry regions, the disturbed soils are eroded also by wind, as evidenced by the profound effects of wind erosion centuries ago in the Middle East, and later on the Great Plains of the United States. Large amounts of topsoil were lost and soil particles have been found in rain washout a thousand miles or more from the sources.

The third type of residual comes from industrial production of goods and services. Although disposal of these wastes seemed at one time to be sustainable at the community level, and was improved upon by urban government's investment in centralized sewer and water systems in the late 19th century, public disposal services were rarely undertaken for large manufacturers. Residuals from industry were considered the inevitable by-product of creating goods for public use, and the by-products of their production were disposed into rivers, lakes, coastal water, and landfills. Early manufacturing and mining facilities were located downstream on rivers so that wastes could be carried away easily and cheaply by the water flow. The inexpensiveness of this practice made it difficult for industry to consider internalizing the costs of achieving a low-impact disposal.

As with the conversion of natural assets to infrastructure capital previously discussed, naive pollution was not an issue that society considered important. This was partly because resources such as air and water were viewed as so widely available that any amount of waste could be disposed. The fallacy of this view became evident only much later.

2.3.3 Sustainable Waste Disposal and Development

Hardin[16] discusses the long history of reluctance by business and industry to incorporate outside costs (externalities) into the price of their goods. At various times over the past 200 years, public debates have focused on industry's apparent need to externalize certain costs of raw materials (for example, mining injuries and related diseases), of labor (often favoring cheap child labor), of raising and educating a labor force, and now the cost of cleaning up and preventing pollution of the environment. With every new advance in social priorities (for example, clean water), the issue of how the costs should be born has had to be addressed. Industry once believed that these external costs were not their responsibility, and should not be incorporated into the price of their products. But, if clean water is a priority public good and goal, the costs ultimately must be born by society as a whole, whether through taxation and a government service, or through industry in the pricing of externalities from their product before reaching consumers. There were also dispersed individual costs; for example, the personal costs of black lung disease borne for many years by coal miners and their families, but recently incorporated into the market price of coal through corporate-wide prevention measures, taxes, and workmen's compensation fees. Similarly, *the costs of pollution formerly externalized to the "commons" are becoming internalized through abatement implemented and paid for by the industries themselves. The costs (if not eliminated by new technology) are incorporated into the price of products to the consumer.* Recovery of some natural resources is being achieved in this way.

Industries reacted to their responsibility for the environmental commons in two major ways described in the following sections: The first was in identification and removal from use of environmentally persistent chemicals, and the second has been in the initiation of industrial and public/environmental partnerships for community and public protection.

2.3.4 Ecologically Persistent Chemicals

The first chemical fertilizer firm was organized in England[17] in 1842 to assist in soil replenishment. Although the modern fertilizers and pesticides that comprise the agrichemical industry represented technological advances for improving food production, their overuse can create new residuals and new risks to the productivity of ecological systems, including agriculture itself. Chemical applications in excess of what can be taken up by the crop or decomposed by the soil become a non-point source of pollution through runoff, effecting groundwater, rural streams, lakes, and coastal areas. Today, many water resources, both surface and groundwater, are impoverished owing to the long-term applications of chemicals for agriculture.

It was not until the late 1960s and early 1970s that the harmful effects of chemical residuals in the environment began to be well documented and understood, and agri-chemical companies were asked to change their present product line. The problem chemicals such as DDT (see Box 2.3) and dieldrin did not biodegrade fully, and in some cases, the harmful effects observed in remote ecosystems were not apparent until several generations of organisms were exposed. Governmental regulations were used, initially, to limit certain uses of chemicals such as DDT, dieldrin, chlordane, and others, but overall use increased during the 1960s in spite of these regulations. At the same time, unintended releases of related, non-agricultural compounds such as dioxin, polychlorinated biphenols (PCBs), and chlorofluorocarbons (CFCs) began to affect remote ecosystems as well. The way in which society, government, and industry responded is illustrated by the challenge to the DDT registration for agriculture in 1970 (Box 2.3), by termination of PCBs in printer's ink, electrical transformers, and other uses during the 1970s, and by ending the production of CFCs (or freon) for refrigeration and other uses during the 1990s.

Commercial use of the gaseous CFCs began in the 1920s. These chemicals were used by manufacturers because they were viewed as non-poisonous, did not burn, did not react with other substances, and also were cheap to produce. With their low cost and apparent safety, there was little incentive to minimize waste or make any special arrangements for their disposal.[10] In the late 1970s the hazardous effects of CFCs for the stratospheric ozone layer were being discussed by a few scientists, but were unproven. Leaders in industry and government thought the linkage with the ozone layer was unlikely. A public protest took place, however, in the form of a successful boycott of spray aerosol products containing CFCs,[10] and the business community eventually responded by removing that form of CFC emission from the market.[20]

Undeniable evidence of the catastrophic effects of CFCs came in the 1980s when the ozone hole over the Antarctic was discovered and linked to chlorine by-products of CFCs in the stratosphere. A broad control strategy was undertaken through an international treaty, called the Montreal Protocol, agreed on in 1987. With leadership from the DuPont Company, a timetable was agreed to for terminating the use of CFCs and adopting benign substitutes, a milestone in the use of environmental science to achieve a measure of self-regulation for harmful chemicals.

Natural Science Foundation of Sustainability: Health and Integrity of Resources

> **Box 2.3 Bioaccumulation of Persistent Toxic Substances as a Factor in Sustainability**
>
> Although DDT was used extensively on agricultural fields, its persistence and capacity for long-distance transport led to its accumulation in streams, lakes, and estuaries at great distances from its application. By 1970 the governments of both Canada and the United States were forced to ban public consumption of fish harvests in many areas owing to the level of contamination by DDT. Shown here is an ecological food chain by which DDT is biomagnified and builds up high concentrations among the top carnivores around the Great Lakes.[18] Wildlife populations, particularly carnivorous birds, declined over a large geographic area. Deformities in other species increase owing to the combined effects of several toxic chemicals. Changes began to take place after a DDT trial in Wisconsin from 1969 to 70.[19] By the mid-1970s, major efforts were underway to control these toxic substances, seeking to restore the productivity and utility of the natural resources. The DDT trial represents one of the first attempts by society to require that the full cost of the use of a chemical (in terms of the damage to non-target resources) be taken into account.*
>
>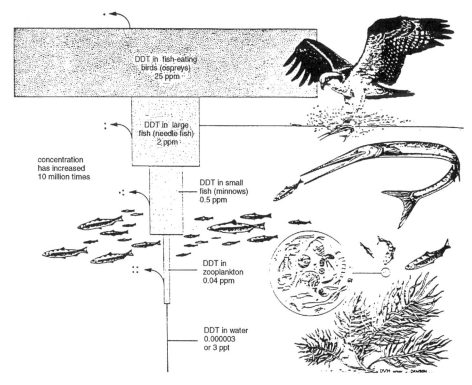
>
> *Source:* Reference 19
>
> *Bioaccumulation of DDT in the food chain of individual Great Lakes. Dots represent DDT and arrows show small releases of DDT through excretion.

Knowledge of the negative effects from overloading ecological systems with industrial chemicals was accompanied by many proposals for reforms. Some of the chemicals were banned from use, but to limit other sources many new environmental laws were passed throughout the 1970s and 1980s. Included here were the Clean Air Act (CAA) of 1970, amended in 1977 and 1990; the Resource Conservation and Recovery Act (RCRA) of 1976; amendments to the Clean Water Act (CWA) of 1977; the Comprehensive Environmental Response, Compensation, and Liability Act

(CERCLA) of 1980 (the beginning of what came to be known as "Superfund"); and the Superfund Amendments and Reauthorization Act (SARA) in 1986.[21]

Society was beginning to understand the serious consequences of externalities, that is, exceeding natural assimilative capacities for ecological systems, but it had yet to understand fully the effects and limits of command and control regulation of externalities. It was thought, for a time, that the costs of protecting the resources would come in the form of businesses closed, lost jobs, and slower economic growth.[22] Although the subject is still being debated, industry has recognized there is an alternative to the large cost of reducing pollutants at the end of the pipe, and that is through design and production of benign products, which we will discuss later.

2.3.5 Air, Water, and Biodiversity as Threatened Common-Property Assets

The earlier discussion of Hardin's[15] views on the risks to common property assets exploited by diverse public ownerships applies equally to the desire by many to dispose of wastes into air and water, our two most important public commons. We will discuss the solution to these problems later in Chapter 4, as part of a discussion managing natural resources for sustainability. Here, we want only to note the nature of the transition taking place in regard to recognizing the value of air, water, biodiversity (see Box 2.4), and land resources.

Box 2.4 Biological Diversity as a Resource to Be Used and Sustained

Conservation of biological diversity as a resource focuses on retaining, sustainably, the native plant and animal species and their habitats. The goal is to achieve sustainable reproduction of these species in nature. Understanding each species' needs, adaptations, genetic variability, and the components of whole ecosystems that afford the conditions needed for each life cycle, are all part of achieving successful outcomes. As with forests and fisheries, and as noted by Perrings,[24] a sophisticated predictive capability is needed to project what habitat amenities are required, and in what quality and duration, to sustain populations of uncommon species. Extensive work has been done to estimate how much old-growth forest is needed to maintain a breeding population of spotted owls in the Pacific northwest. Populations are sometimes at risk owing to natural fluctuations in the environment (weather, fires, etc.), which, when the population becomes small, could lead to extinction.

In the case of the American alligator, once a threatened species owing to intense harvesting for its valuable skins, control of the annual harvest of alligators was projected to bring about recovery, and it has. The population is now large, and a managed harvest by professional hunters is essential for limiting the population density. Society, the local hunters, and the alligator population have all benefited.

Hunting has never threatened the spotted owl, and human population density is not directly a threat. However, extensive clear-cutting of the forest habitat is a problem. The owl population has been declining since the 1950s, but the specific ways in which forest clearing affected this species are only now becoming reasonably well understood. Not only does the species live primarily by hunting for prey in the open spaces between tall trees, but the young, after they have matured and must seek a new nesting site, are unable to go long distances across large-scale clear-cuts to search for suitable new nesting habitat. The population seems able to tolerate some clear-cutting if it is not continuous, and can forage in forests that have had light cutting or have recovered from logging many years ago. The sustainability question, then, is to assure a *pattern* of partial cutting, intermixed with old-growth reserves that can maintain a population of spotted owls large enough to avoid extinction owing simply to bad weather, fire, or a disease event. *Sustainability, then, needs to provide for a defined margin of safety linked to the uncertainty about how unexpected events may affect a small population.*

Air quality, generally, has been thought of as protected by air primary standards, principally set at levels to protect human health. The predilection of individuals, and small or large businesses, to discharge pollutants cheaply into the air is constrained, in part, by a public consensus that emissions leading to violation of the air standards must be controlled. However, problems arise when it is recognized that many forest and crops species are more sensitive to the pollutants than are humans. Protection of these resources has been sought through definition of secondary standards, designed to protect crops and forests.[23] There is no consensus, however, for the enforcement of emissions controls designed to achieve these secondary standards, and, thus, measurable damage is imposed on individuals who own the forest and agriculture commons, rather than incorporate the costs of preventing the damage (through further pollution reductions) into the price of consumer products everyone buys.

For these risks to resources, sustainability involves not just the avoidance of long-term impoverishment from pollution. It also involves fairness among all stakeholders as to who should bear the economic burdens of damage, individual owners of resources, or consumers of the products who reap a benefit from externalizing the costs of air emissions disposal. (See also the ethics and equity principle, Chapter 5.) Business and industry are involved only as the brokers who do or do not pass along the costs. The costs ultimately are borne by some public sector: either by property owners indiscriminately or consumers who pay the costs in the purchase price of the products they use. Business also ensures that the entire global marketplace moves toward redistribution of costs in a fair and systematic way over time.

The overuse of the Great Lakes for water-borne disposal of production wastes, some intentional and some accidental, is a useful example. The Great Lakes is the largest freshwater resource in the world with a watershed covering 201,000 square miles and supporting 50 million people in two countries. Because of the size of the water system and the richness and abundance of aquatic life, its use for disposal was thought to be limitless. But this resource also is used to provide drinking water to hundreds of human settlements, to generate hydroelectric power, to provide for cheap, water-based transportation of goods, to provide water-based recreation including fishing and tourism, and as a coolant or process water for many industries.[21] Concern over restrictions designed to assure safe consumption of fish from these waters led to concern about the economic losses that may be incurred if the low-cost waste disposal opportunity of these waters was lost.

2.3.6 PUBLIC-PRIVATE PARTNERSHIPS AS A PARADIGM FOR SUSTAINABILITY

As businesses and society responded during the 1970s to concerns about the discharges of wastes into natural systems, there was increased acceptance of the need to protect the environment and pass along some minimal embedded costs to consumers. This support by business has emerged now in three major ways: *First, on the international front, industry leaders and environmentalists began to meet to discuss mutual goals. (See Box 2.5.) Second, senior business managers began implementing policies and operations so as to create products or production processes that reduce wastes and emissions. And third, documentation of the external costs borne by the public began to show benefits from internalizing these costs in products consumers buy, thereby minimizing command and control regulation.*

The World Commission on Environment and Development, chaired by Brundtland, the Prime Minister of Norway, presented *Our Common Future* to the U.N. General Assembly in September 1987. It represented four years of work by the commission, established to discuss "a global agenda for change." The commission, initiated by the U.N. General Assembly, led to a milestone in understanding the linkages between global economic and environmental issues. *Our Common Future* called for new economic growth and environmental stability. It recognized "humanity's ability to make development sustainable," and defined sustainable development as "meeting the needs of the present without compromising the ability of future generations to meet their own needs".[25] It also recognized for the first time in world history, that humanity has the power, in numbers and

> **Box 2.5   Steps Taken by the International Business Community to Embrace Sustainable Development**
>
> Significant changes in business' approach to sustainable development came at the international level in the 1980s. Multinational corporations, with a long-term view of their business and markets, were accepting the principle of a positive relationship between economic development and environmental protection. Two major international events marked the emergence of this new thinking. The first was the World Industrial Conference on Environmental Management (WICEM) in 1984,[22] sponsored by the International Chamber of Commerce and the United Nations Environment Program (UNEP). The second, a process, was begun in 1983 with establishment of the World Commission on Environment and Development (WCED), culminating in 1987 with a report entitled *Our Common Future*.[25]
>
> The first World Industrial Conference on Environmental Management was held in France in November 1984. This meeting was a milestone in the relationship between leading environmentalists and industrialists. For the first time, international environmental organizations, major industries, and governments were brought together to discuss global business and environmental concerns. It recognized the important role that industry would have in future approaches to sustainable development and the economic health of all countries. *The basic message of the conference was that "economic development and environmental protection are mutually supportive," and it called for industry to play an even greater role in sustainable development.*[22]
>
> The conference also recognized that industry had the knowledge, expertise, and infrastructure positioning to make a great positive impact on the resolution of environmental concerns. The conference also acknowledged, that while many industries were moving to improve environmental conditions, business and industry as a whole was still part of the problem. A major step for WICEM would be to facilitate transfer of information, not only among members of the environmental community, industry, and government, but also, between industry and the public, and between Northern and Southern countries around the world.

technology, to alter global systems in fundamental and lasting ways. Although the report has been criticized by some as wishful thinking, and supported by others as a remarkable breakthrough,[25] *Our Common Future* brought together the thoughts and concerns of leaders in economic development and environmental protection from throughout the world, which in itself was no small achievement.

Another way in which the business community initiated environmental protection was in modification of product design and production processes toward the end of the 1980s. The industrial environmental movement that had its beginnings in the mid-1980s became accepted by business leadership generally. Executives made clear that they expected their companies, as well as customers and their environment, would benefit from more sophisticated production processes with waste minimization. According to Smart,[20] industrial management was recognizing five important advantages to environmental protection:

1. Pollution is waste, and preventing it at the source can save money in materials and in end-of-the-pipe remediation.
2. Acting voluntarily now can minimize future risks and liabilities, make costly retrofits unnecessary, and aid in the design of more efficient regulations.
3. A company moving "ahead of the curve" on environmental issues will find competitive advantage over those struggling to keep up.
4. New "green" products and processes can increase consumer appeal and open up new business opportunities.
5. A reputation for being environmentally progressive improves recruitment, employee morale, investor support, host community acceptance, and management's self-respect.

Industry was finding that environmentally sound management could be profitable. The 3M Company and Proctor & Gamble are cited in Smart[20] as two examples of reducing production residuals and costs (which had been increased by a public demand for regulation) through innovative changes in products and production. 3M's Pollution Prevention Pays program initiated reductions in waste at the source through more environmentally benign processes and materials as early as the mid-1970s. Over 15 years, the savings from these programs total more than $530 million.[20] Other examples are being documented now in the case studies reported in this volume and in many other sources.

The shift to the modern paradigm of sustainable approaches to waste disposal also can be seen through new approaches in economics[4] (and see the next chapter). To achieve these changes, Daly and Cobb suggest several broad steps: *recognition that something is wrong with the widespread acceptance, indeed dependence, on production externalities, and that present policies do not work; recognition that most of the problems faced by humanity today are interconnected and indeed have a common source; and recognition that human beings still have the possibility of choosing a livable future for themselves and their descendants.* However, specific options to foster economic and environmental sustainability still have to be investigated and documented.

Rubenstein[27] indicates that new social contracts are necessary, and, in fact, are beginning to occur between society and business. In his view, the new contract, with new accounting and audit practices for business, will focus on the rights of invisible stakeholders in regard to common property assets that the business may be dependent on. (See Chapter 6 for an in-depth discussion.) Rubenstein also recognized the need to include natural capital in measurement of national productivity and gross domestic product (GDP) because many business and industrial operations are dependent on natural resources (including air and water) in important ways. Included in these new views of economics and accounting is the full internalizing of previously externalized costs of pollution, as suggested by Hardin[16] in 1969 and many others since that time.

Now, in the 1990s, there is even stronger evidence of the shift to sustainable practices in waste disposal, while fostering economic development.[28] Environmental organizations and business leaders are working together frequently to discuss issues of mutual interest at international as well as national and local levels. Economic goals, as well as environmental and ecological goals, are beginning to be met jointly through common measurement approaches and fundamental change, not compromise. *Industry has begun to take the lead in implementing changes in manufacturing processes and product design, forestalling the need for some of the command and control regulation while minimizing impacts on the environment.* The idea that environmental protection does not have to come at the expense of jobs or slower economic growth increasingly is being accepted.

2.4 REFERENCES

1. Daly, H. E., Operationalizing sustainable development by investing in natural capital, in *Investing in Natural Capital,* Jansson, A., Hammer, M., Folke, C., and Costanza, R., Eds., Island Press, Washington, D.C., 1994, 22.
2. Rapport, D. J. Regier, H. A., and Thorpe, C., Diagnosis, prognosis, and treatment of ecosystems under stress, in *Stress Effects on Natural Ecosystems*, Barrett, G. W., and Rosenberg, R., Eds., John Wiley & Sons, New York, 1981, 269.
3. Rapport, D. J. and Turner, J. E., Economic models in ecology, *Science,* 195, 367, 1977.
4. Daly, H. E. and Cobb, J. B., Jr., *For the Common Good: Redirecting the Economy Toward Community, the Environment, and a Sustainable Future,* Beacon Press, Boston, 1989.
5. Solórzano, R., de Camino, R., Woodward, R., Tosi, J., Watson, V., Vásquez, A., Villalobos, C., Jiménez, J., Repetto, R., and Cruz, W., *Accounts Overdue: Natural Resource Depreciation in Costa Rica,* World Resource Insitute, Washington, D.C., 1991.
6. Jansson, A. and Jansson, B.-O., Ecosystem properties as a basis for sustainability, in *Investing in Natural Capital,* Jansson, A., Hammer, M., Folke, C., and Costanza, R., Eds., Island Press, Washington, D.C., 1994, 74.

7. Liverman, D. M., Hanson, M. E., Brown, J. B., and Merideth, R. V., Jr., Global sustainability: toward measurement, *Environ. Manage.,* 12(2), 133, 1988.
8. Hett, J. M. and O'Neill, R. V., Systems analysis of the Aleut ecosystem, *Arctic Anthro.,* 11, 31, 1974.
9. Cronon, W., *Nature's Metropolis: Chicago and the Great West,* W. W. Norton, New York, 1991.
10. Ponting, C., *A Green History of the World: The Environment and the Collapse of Great Civilizations,* Penguin Books, New York, 1991.
11. Martin, P. S., Prehistoric overkill, in *Pleistocene Extinctions: The Search for a Cause,* Martin, P. S. and Wright, H. E., Jr., Eds., Yale University Press, New Haven, 1967.
12. Gale, R. P. and Cordray, S. M., What should forests sustain? Eight answers, *J. For.,* 89(5), 12, 1991.
13. Ludwig, D., Hilborn, R., and Walters, C., Uncertainty, resource exploitation, and conservation: lessons from history, *Science,* 260, 17 and 36, 1993.
14. Robert, R., *Chartered Companies and Their Role in the Development of Overseas Trade,* G. Bell and Sons, London, 1969.
15. Hardin, G., The tragedy of the commons. *Science,* 162, 1243, 1968.
16. Hardin, G., Not peace, but ecology, in *Diversity and Stability in Ecological Systems,* Brookhaven Symp. Bio. No. 22, Brookhaven National Laboratory, Associated Universities, 1969, 151.
17. Hillel, D., *Out of the Earth: Civilization and the Life of the Soil,* University of California Press, Berkeley, 1992.
18. Woodwell, G. M., Toxic substances and ecological cycles, *Sci. Am.,* 216, March 24, 1967.
19. Loucks, O. L., The trial of DDT in Wisconsin, in *Patient Earth,* Harte, J., and Socolow, R. H., Eds., Holt, Rinehart and Winston, New York, 1971, 88.
20. Smart, B., *Beyond Compliance: A New Industry View of the Environment,* World Resources Institute, 1992.
21. Council on Environmental Quality, *Environmental Quality, Twentieth Annual Report,* 1990.
22. World Industrial Conference on Environmental Management (WICEM), Outcomes and reactions, *The United Nations Environment Programme,* No. 5, 1984.
23. MacKenzie, J. J. and El-Ashry, M. T., Eds., *Air Pollution's Toll on Forests & Crops,* World Resources Institute, Yale University Press, New Haven, 1989.
24. Perrings, C., Biotic diversity, sustainable development, and natural capital, in *Investing in Natural Capital,* Jansson, A., Hammer, M., Folke, C., and Costanza, R., Eds., Island Press, Washington, D.C., 1994, 92.
25. World Commission on Environment and Development (WCED), *Our Common Future,* Oxford University Press, New York, 1987.
26. Environmental Liaison Centre (ELC), Reviewing *Our Common Future, Ecoforum,* 12(2), 1, 1987.
27. Rubenstein, D. B., Lessons of love, *CA Mag.,* 124(3), 34, 1991.
28. Erekson, O. H., Loucks,. O, L., and Aldag, C., The dimensions of sustainability for business, *Mid-Am. J. Bus.,* 9(2), 3, 1994.

3 Sustainability and Economic Well-Being

O. Homer Erekson

"Instruments register only through things they're designed to register. Space still contains infinite unknowns." Mr. Spock, The Naked Time, *Stardate 1704.2*

CONTENTS

3.1 Introduction 41
3.2 The Economic Approach to Sustainability 42
    3.2.1 The Economic Way of Thinking 42
        3.2.1.1 Allocation Function 42
        3.2.1.2 Distribution Function 45
        3.2.1.3 Discounting 46
        3.2.1.4 Stabilization Function, Economic Growth, and Issues of Scale 49
    3.2.2 Principle 6 50
3.3 Measuring Aggregate Economic Well-Being 50
    3.3.1 Principle 7 51
3.4 Toward Improved Measurement of Economic Well-Being 54
    3.4.1 System of National Accounts 56
    3.4.2 Genuine Progress Indicator 57
3.5 Conclusion 58
3.6 References 60

3.1 INTRODUCTION

Economics is defined often as the study of the use of scarce resources to satisfy humankind's insatiable wants. It focuses on the choices of individuals, businesses, governments, and other organizations within society, and how those choices determine the way resources are used by society.*

In this chapter, we consider how economists traditionally have approached environmental issues and the suitability of this approach for examining the idea of sustainability. In addition, we will incorporate the contributions of ecological economics, an emerging field within economics that seeks to meld ecology and economics. To a great extent, ecological economics may be conceived best as the science and management of sustainability. As we move through the chapter, we will see several of our principles of sustainability at work. Of particular emphasis will be our principles of *scaling and technology,* and *measuring and adapting.*

---

*In this book, we assume familiarity with the basic principles of economics. For a review of these principles, consult any standard textbook such as Mankiw.[1]

3.2 THE ECONOMIC APPROACH TO SUSTAINABILITY

Mainstream (or neoclassical) economists have typically taken a very anthropocentric approach to environmental concerns. The focus of ecologists has been on viewing natural systems as assets that serve as reservoirs of energy and materials. Their concern has been about the resilience of ecosystems, their ability to recover and return to a characteristic level or pattern following a significant stress or intervention. The economists' focus has been quite different. "Natural systems are simply the vessel within which agents interact and the state of repair of the vessel is set aside."[2] Thus, economists have concentrated on describing the workings of markets. In this context, the root cause of environmental problems lies in the failure of markets to work efficiently. Because many environmental services do not involve market transactions, environmental quality tends to be under-valued and under-provided. The solution is, then, simply to seek better methods of evaluation for environmental amenities.[3]

To understand the motivation for this approach to environmental problems, it is helpful to consider the basic systems approach followed by traditional economic analysis. A fundamental construct introduced in any basic course is the *circular flow diagram*. In this diagram, as shown in Figure 3.1-A, the two participants in the simple economy are households and businesses. Households provide inputs to businesses in the form of land, labor, and capital (e.g., machinery and buildings). Firms then transform these inputs into outputs, consumer goods, and services that flow back to households and provide the income for the next round of production. Although land is typically one of the inputs included in this simple conception, the emphasis is on market transactions and the use of manufactured capital and human capital, not on the quantity or quality of natural resources and the ecosystems.

This approach to understanding the allocation of resources is fundamentally an *open system approach*. As shown in Figure 3.1-B, the economic subsystem operates within the context of a larger physical system. Energy flows into the economic subsystem, primarily from the sun. In addition, natural capital provides the basic asset base in which consumption and production occur. Moreover, to the extent that inputs are not recycled within the economic subsystem, waste is generated as a byproduct of economic processes.

A natural question is the extent to which this simple open system approach is useful. In the recent past, natural capital, at least in developed countries, was relatively abundant. The scale of humankind's activities did not provide significant depletion of natural capital nor excessive stress to most ecosystems. From this view, economic growth and development was limited only by the provision of sufficient manufactured and human capital, and development of appropriate technologies. Now, however, there is increasing concern among scientists that the scale of human activity has risen to the point that natural capital has begun to become a limiting factor in economic development.[4]

This debate about the scale of human activity and its importance for sustained economic development is a very important topic. In fact, later in this chapter we will return to this discussion because it is central to how one thinks about sustainability. However, at this point, we need to devote some further energies to understanding the way economists think about the organization of production, consumption, and economic development.

3.2.1 THE ECONOMIC WAY OF THINKING

The economic way of thinking typically distinguishes three distinct functions of a socioeconomic system: the allocation function, the distribution function, and the stabilization function. In this section, we explain each of these functions and how they are important to thinking about sustainability.

3.2.1.1 Allocation Function

The allocation function focuses on the process by which scarce resources are divided among competing uses within the economy. On one hand, allocation includes the familiar microeconomic

Sustainability and Economic Well-Being

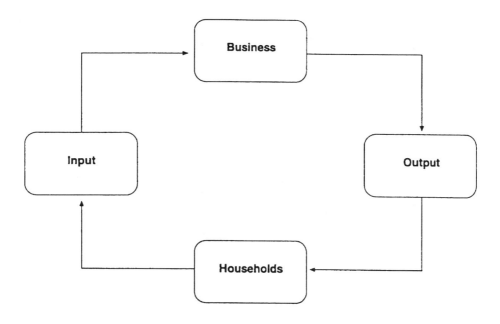

FIGURE 3.1-A Simple circular flow diagram.

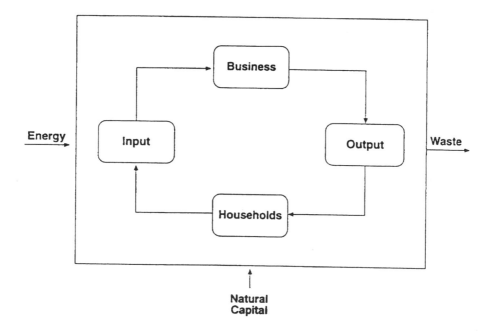

FIGURE 3.1-B Open system circular flow diagram.

processes behind supply and demand. In this framework, on the supply side, producers combine various inputs (labor, human-made capital, and natural capital) to produce goods and services. Economists often assume that producers have an objective of maximizing profit. On the demand side, consumers choose between different goods and services, maximizing utility subject to a budget constraint.

What brings producers and consumers together is the institution of markets. If markets work perfectly, then the economy achieves an *efficient* allocation of resources. This very important concept presumes three conditions[1]:

- *Exchange Efficiency*—Goods and services must be divided among individuals so that there are no remaining gains from trade. That is, consumers have all equated the marginal benefit from buying units of a good with its price, such that goods have flown to the consumers who most highly value them.
- *Production Efficiency*—The economy must be at its maximum production possibilities. This implies that all producers choose among inputs to maximize production so that the only way to increase the production of one good is to decrease the production of another.
- *Product-Mix Efficiency*—The economy must be producing the combination of goods and services that reflects the preferences of consumers.

The key to making this system work is a well-functioning price system. If markets work, prices guide inputs into their most highly valued uses.* In particular, the price of one good *relative to others* provides an important signaling function revealing consumer demands and thereby guiding production. At least at first blush, private markets, unregulated by government, effectively allocate many goods and services, such as basic food items. In fact, if all goes well, a perfect market system results in a condition that economists call *efficiency*, *allocative efficiency*, or *Pareto Optimality*. Efficiency occurs when all three of the above sub-conditions hold and may be best thought of as *a state of the world where no one can be made better off, without making someone else worse off*.

Unfortunately, markets do not always work perfectly. Market failure may arise for a variety of reasons. Markets fail and result in idle industrial capacity, or excessively high prices, in situations where a few firms dominate a market, or where consumers have imperfect information. But of most interest to environmental economists are uses made of public goods and the presence of externalities.

Public goods are goods or services quite different from private goods (those that would best fit the above discussion). Public goods are *nonrival* in consumption—that is, the consumption of the good by one person does not reduce the benefits that can be derived by another person. For example, the existence of grizzly bears may provide consumption benefits to many people jointly. Second, public goods are *nonexcludable*—that is, the cost of excluding a person from consuming the benefits of a good are extremely high. It is not possible to prevent someone from enjoying the existence of grizzly bears, although it may be cost-effective to charge a user fee that allows the right to see a grizzly bear. The market failure problem arising with public goods is that each consumer has an incentive to be a free rider, since they have access to the benefits of a good without paying for it, and the public good would be under-provided. Efficient provision of a public good requires government or some other collective organization to control the provision of the good and to appropriately charge beneficiaries of the good or service.

---

*In fact, the formal equation for an efficient allocation of resources for an economy with two goods, x and y, and two consumers, **a** and **b**, is as follows:

$$MU^a_x/MU^a_y = MU^b_x/MU^b_y = P_x/P_y = MC_x/MC_y,$$

where **MU** is the marginal utility each consumer receives from the goods x and y, **P** is the price, and **MC** is the marginal cost of the goods x and y.

Another related source of market failure are externalities. Externalities arise whenever one person or organization takes an action that significantly affects another person or organization. Thus when a firm emits effluents into a stream, it can affect those persons who use the stream for recreation, as a source of drinking water, or to produce other goods. The problem arises if the person creating the externality is not confronted with the cost (or benefit) to society of their action. If there is a negative externality that imposes social costs above private costs, and these costs are not internalized, the good in question would be overproduced.* As mentioned earlier, many economists have approached environmental issues from the perspective of "getting the prices right," of identifying externalities, determining the value of their effect on society, and implementing taxes or other regulatory means to internalize the externality (i.e., to confront the creator of the externality with the full cost of his/her actions). Clearly, this is an important part of the puzzle in moving toward a sustainable system. We will consider the problem of externalities and public goods more completely in Chapter 7.

3.2.1.2 Distribution Function

In Chapter 1 we referred to the notion of sustainable development in the Brundtland Report: "Sustainable development is development that meets the needs of the present without compromising the ability of future generations to meet their own needs." This raises a crucial question about the responsibility of current generations to future generations. Some would cast the problem as one of allocative efficiency. They would say, "Large-scale damages to ecosystems such as degradation of environmental quality, loss of species diversity, or destabilization from global warming are not intrinsically unacceptable. The question is whether compensatory investments for future generations in other forms of capital are feasible and undertaken[6]." In the next section, we will consider the validity of this approach. But in this section, we want to view the "future generations" problem from a different perspective—an equity perspective.

The second function of the economic system is the distribution function. Talking about the distribution of income and wealth (or resources across generations) is an equity issue. Howarth and Norgaard[3] note that "Intergenerational equity is manifested in how we acknowledge rights of future generations, assume responsibilities for our descendants, and otherwise demonstrate our care for those who follow us. If development is not sustainable, it is because the institutions through which the present provides for the future have not evolved in consonance with changes in social and economic structures, technology, and population pressure."

Although it is often difficult in practice to do so, economists attempt conceptually to separate allocative efficiency from equity concerns. Consider Figure 3.2. The graph is called a utility possibilities frontier and shows the tradeoff in utility between current and future generations as resources are transferred among generations. Each point on the frontier represents an efficient outcome. Resources would have been allocated within each generation in such a way as to maximize the net value to that generation's consumers and producers (including fully internalizing externalities). The efficiency problem is to move from a point like A to a point like B, or to move from a point like A^* to a point like B^*. Once point B (or B^* or any other point on the frontier) is achieved, the maxim of Pareto optimality clearly holds: future (current) generations cannot be made better off without making current (future) generations worse off.

Once efficiency is achieved then, the sustainability concern becomes an equity issue. Solow[7] has defined "a sustainable path" as one that "allows every future generation the option of being as well off as its predecessors" and argues that resources are valuable for their "generalized capacity to produce economic well-being . . . for what they do, not what they are." Similarly, Norgaard[8] has argued that "emphasis must be placed on the continuity of flow, not some measure of aggregate value . . . each generation is obligated to pass on to the next a mix of assets that provides equal or greater flows

---

*For a thorough discussion of the issues surrounding public goods and externalities, consult any public finance textbook such as Rosen.[5]

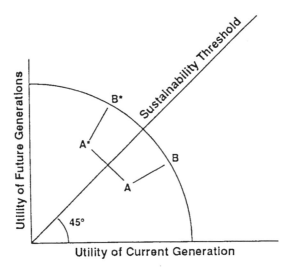

FIGURE 3.2 The relation between intertemporal allocative efficiency and the intergenerational asset distribution.[3]

to the next generation without greater effort on that generation's part to provide the same for the next." If this definition of sustainable is accepted, then a sustainable distribution of resources across generations would require being to the left of the sustainability threshold in Figure 3.2. If we started at inefficient point A, we could reallocate resources and move to point B. However, this point would not be sustainable since it would be impossible for future generations to enjoy a level of utility at least as large as that of current generations. However, we could redistribute resources among the generations such that we moved to a point like A^*. If society then moved from A^* to a point like B^*, we have achieved both efficiency and sustainability.

How do we provide for intergenerational equity? After all, equity poses the real problem of making someone worse off in the course of making someone else better off. Who or what social processes decide what is a fair distribution of resources within generations or across generations? Answers to this question are largely beyond the realm of economics and will be a topic of discussion in Chapter 5. However, economists have entered the discussion in an important way in thinking about how much the future should be discounted in living today.

3.2.1.3 Discounting

The choice of an optimal discount rate, or even whether to discount future benefits at all, is a highly complex issue and one filled with highly passionate debate. The extent to which we undertake actions to preserve old-growth forests or to protect an endangered species is dependent upon the degree to which we discount the future, or alternatively are willing to sacrifice present consumption so that benefits will accrue to future generations.

Some environmentalists argue that we should not discount the future at all. They argue that it is misguided at best, if not in fact unethical to discount the future. They would resist the benefit-cost type of calculus that leads economists to cast decisions about environmental issues in terms such as considering whether a dollar's worth of future benefits is worth less than a dollar's worth of present costs.[9] The ethical position argues that the present generation has no right to make evaluations about the optimal welfare of future generations. In fact, some would argue that the present generation has an obligation to sacrifice today for the sake of benefiting the future generations. They would argue that future generations should have *equal standing* with present generations and that the future should not be discounted at all.[10]

As further ammunition to their arguments, some environmentalists say that discounting can lead to absurd social policies, especially for long-lived projects. Lesser and Zerbe[11] provide an example

> **TAKE THE MONEY AND RUN!**
>
> Suppose you win the state lottery and have a choice of receiving $500,000 today or $1,000,000 in twenty years. Which should you choose? It depends upon how much you discount the future. The present value of a one-time benefit received t years from now is:
>
> $$PV_t = B_t/(1 + r)^t,$$
>
> where r is the discount rate. In this case, if your discount rate is 5 percent, the present value of that $1,000,000 twenty years from now is $376,889. If your discount rate is 10 percent, the present value is just $148,644. Better take the money and run!

of undertaking a project that has a toxic waste time bomb that will cause catastrophic environmental damage far into the future (as much as 10,000 years ahead). Their example involves waste disposal technology that can contain the waste for 5000 years. Then sometime after this point, the waste will escape its container and cause estimated damages of $6,000 trillion, about 10^3 times the current value of Gross Domestic Product in the United States. Discounting the value of this damage at any positive discount rate, at a rate such as 3 percent, would result in a present value of damage of far less than a penny, and would lead to the damages being essentially ignored in a present value analysis.

While the arithmetic of this example is essentially sound, it, along with some other arguments of environmentalists, jumbles several issues related to discounting. It will be useful to think more carefully about discounting and the choice of discount rates, and the implications for sustainability.

Let us consider first the choice of discount rates. If we were to invest $100 in a savings account that earned 3 percent interest per year, it would take 24 years for those funds to grow to $200. Similarly, if we were to discount a project that would have $200 of benefits in 24 years by a 3 percent discount rate, we would have a present value of $100. So there seems to be a relationship between interest rates and discount rates. Let us look at this relationship more closely.

Interest rates in financial markets are determined principally in one of two ways.* The first approach, the opportunity cost of capital, involves three fundamental principles. Interest rates will be higher, first, if projects are riskier; second, if inflation is higher (such that creditors will be paid back in dollars worth less than what they lend); and, third, if funds are borrowed under tight economic conditions (as other debtors bid for the funds). All of these factors have to do with the opportunity cost of capital, based on the marginal productivity of capital. The lower the discount rate, the greater the demand for borrowing funds and the more likely a project will be undertaken. If this opportunity cost basis for discounting is used, it is not clear that lower or even zero discount rates serve sustainable objectives best.

On one hand, if we were considering the benefits to be derived from an old-growth forest, we may be considering a horizon of 200 or more years. In this time frame, it is clear that lower discount rates would encourage development and preservation of old-growth forests. However, using lower discount rates also can encourage the over depletion of natural capital. Creation of buildings and other manufactured capital requires the use of natural capital. If we over invest in the creation of human-made capital by using an inefficiently low price (i.e., lower discount rates than the opportunity cost base would suggest), we also may be drawing down natural capital stocks faster than is desirable. Norgaard[8] notes that the same low discount rates that encourage a larger stock of standing forests can also be used to make transformation of diverse tropical rainforests into single-species plantations relatively less expensive.

---

*For a more complete discussion of issues surrounding discounting and implications for environmental analysis, see Markandya and Pearce.[12]

Interest rates also are determined within a different framework on the consumption side of the market. This framework often is called the pure time preference framework, and refers to the "apparent fact that people require more than $1 in promised future benefits in order to be willing to give up $1 in goods today."[10, 13] Most economists agree that private interest rates resulting from consumers borrowing for present needs are likely to be too high. The argument here has at least two dimensions. First, people are fundamentally myopic, caring most about direct, immediate benefits. Cropper and Portney[14] conducted a survey of 1000 Maryland households asking about their willingness to trade-off lives saved in the future for a life saved today. They found, for example, that one life saved today was equivalent to six lives saved 25 years from now and 44 lives saved 100 years from now. This implied a discount rate of 7.4 percent with a 25-year time horizon and 3.8 percent with a 100-year time horizon.

A second reason that private market discount rates are likely to be too high is that transfers for future generations have an important public good quality.[8] To the extent that we provide resources for our children and grandchildren, other parent's offsprings and the economy as a whole benefit without a mechanism for compensation. This creates a classic free-rider problem and would lead to under-provision of resources even for one's own offspring. As a result of this and the argument in the preceding paragraph, it is generally accepted that the *social discount rate* used to make decisions about projects with implications for future generations should be lower than the private market discount rate. Martinez-Alier[15] makes an interesting argument about the relationship between consumption and investment. She pointed to the sacrifice of consumption by present generations in the Andes in building terraces and irrigation systems during the Inca empire. These projects increased the capacity to use solar energy for photosynthesis and increased crop production. However, she warned that a zero discount rate would lead to too much sacrifice of present consumption and undermine the sustainability of the people themselves. Indeed, some opponents of discounting have argued for negative discount rates. Because of the uncertainty of future consequences of our present actions and our obligation to the future, the more distant we look into the future, the greater would be the value today. This is an absurd argument leading to the impoverishment of the present generation to protect the consumption possibilities of future generations.[11] Although we believe that a valid case can be made for a social discount rate lower than the private market discount rate, we hope it is clear that some positive rate of discount is appropriate.

Before leaving this section, however, we want to provide a cautionary note. Some neoclassical economists argue against the public goods argument made above, and note that people regularly leave estates for their children and support environmental projects and others with long term implications. Thus, they completely dismiss the zero discounting argument and support their argument with the observation that each generation has lived better than the one before.[16] We believe this argument can be just as limiting as the zero discounting argument. The discount rate used in evaluating a project should be viewed in a comprehensive systems (general equilibrium) view. Humans leave many forms of capital for future generations besides natural capital.[9] These include manufactured capital, human capital, and basic knowledge. Economists would argue that maximization of the value of all of these forms of capital for future generations should be the goal, not just natural capital. Environmentalists often argue that natural capital and ecosystems deserve special attention. On one hand, they argue that the use values that humans place on natural assets is undervalued. Second, as discussed in Chapter 1 and below, there are limits to the substitutability of human-made capital for natural capital.

This brings us to an important point. The fact that we do have time preferences does not tell us very much about how we should allocate resources among present and future generations. Sustainability requires both efficient use of resources and an equitable distribution of resources within and among generations. The claim that future generations are better off than the one before is not an inviolable truism. Brennan[10] summarizes many of the issues discussed above when he says, "to say that present and future generations have equal standing in an ethical sense does not necessarily imply that they have the same claim on present resources, because the general level of wealth or well-being may

Sustainability and Economic Well-Being

be changing over time." In fact, as we move to macroeconomic considerations, we will raise the very real possibility that human behavior at its current scale is not sustainable.

3.2.1.4 Stabilization Function, Economic Growth, and Issues of Scale

The first two functions of the socioeconomic system described above, the allocation and distribution functions, are normally described within the context of microeconomics, and include the decisions of individual consumers, firms, and other agents within an economy. The third function of a socioeconomic system that economists often utilize is the stabilization function. This function traditionally is approached from a macroeconomic framework and focuses on the promotion of full employment and maintenance of price stability (little or no inflation). There is considerable debate among economists as to the effectiveness of monetary and fiscal policy by governments in achieving these objectives. The macroeconomic goals of full employment and minimal inflation are significant to sustainability to the extent that they impact economic or social resiliency. For instance, economies with high levels of unemployment or underemployment do not maximize the productivity possible from the labor force, and thus reduce economic resiliency.* Economies with high levels of employment or high levels of inflation typically face significant inequities in the distribution of income and wealth, thus compromising social resilience.

But of even greater significance is economic growth, most often defined as the rate of increase in gross domestic product per capita, also considered to be a macroeconomic concern by economists. But as was discussed earlier in Chapter 1, economic growth may not be consistent with sustainable development. Simply because a country increases its gross domestic product, it does not necessarily improvement its quality of life. Daly[17] has made a clear distinction between growth and development. To him, growth emphasizes physical expansion resulting from depletion of natural capital. An economy that is growing is one with an increase in the scale of the physical dimension of the economy, the rate of flow of matter and energy through the economy (the throughput), and an increase in the population and human artifacts. On the other hand, development emphasizes qualitative improvement in life, with improvement in the structure, design, and composition of physical stocks and flows resulting from greater knowledge, both of technique and purpose. Thus, sustainable development focuses on increasing the capacity to satisfy human wants, not by increasing the scale of human activity (that is, by increasing resource throughput), but rather by increasing the efficiency of resource use. In fact, Daly goes on to argue for strong sustainability, which requires that natural capital stocks be held constant, and supports a steady-state, no-growth economy. Most economists opt instead for weak sustainability, which involves substituting human-made capital for natural capital.[18] As discussed in Chapter 1, this substitutability is limited. However, appropriate technological advance and more efficient management of resources (human-made and natural) makes weak sustainability increasingly feasible. However, substitution becomes increasingly costly as the process of substitution proceeds, as natural capital becomes more of a limiting factor.[7]

Implicit in this discussion of sustainability is a notion of optimal scale for an economy. To the extent that human activity grows to a level that makes it impossible to preserve natural capital or to replace it sustainably with human-made capital, it is not possible to achieve sustainability. Achieving a sustainable scale of human activity involves both microeconomic and macroeconomic dimensions. The microeconomic dimension relates to the allocation function discussed above and involves using all resources within the economy in the most efficient way possible. The macroeconomic dimension relates to the physical size of the economy per capita. Daly has provided a useful metaphor using a boat's Plimsoll line:

> The micro allocation problem is analogous to allocating optimally a given amount of weight in a boat. But once the best relative location of weight has been determined, there is still the question of the absolute amount of weight the boat should carry. This absolute optimal scale is . . . the Plimsoll line.

---

*Underemployment involves people working in jobs below their skill levels.

When the watermark hits the Plimsoll line the boat is full, it has reached its safe carrying capacity . . . if the weight is badly allocated, the water line will touch the Plimsoll mark sooner. But eventually as the absolute load is increased, the watermark will reach the Plimsoll line even for a boat whose load is optimally allocated. Optimally loaded boats will still sink under too much weight—even though they may sink optimally! . . . The major task of environmental macroeconomics is to design an economic institution analogous to the Plimsoll mark—to keep the weight, the absolute scale, of the economy from sinking our biospheric ark.[19]

As we move to a sustainable system whereby economic growth and improvements in the quality of life occur in a coherent, unified system that is complementary with, rather than antagonistic to the maintenance of renewable and non-renewable resources, we must consider the optimal scale of human activity. This leads us to our sixth Principle of Sustainability:

3.2.2 Principle 6

Sustainable systems require minimizing the degradation of natural capital at scales from local to global, immediate to long term. Changes in technology can put natural capital at risk or facilitate its efficient use at all scales.

Our understanding of sustainability recognizes an important role for technology when it is introduced in a benign manner that facilitates the use of natural capital. Moreover, we recognize that scale has dimensions not only defined in terms of geographical limitations, but also with respect to time. The impact of human activity may be understood and measured differently depending upon the environmental challenge at hand (e.g., maintaining an isolated natural habitat versus managing global warming). Some of these challenges are defined as local concerns, others on a global basis; similarly some environmental problems may only be manifested as a result of chronic accumulation of negative impacts over a considerable period of time. Another example of the importance of time in defining the scale reference for sustainability involves the sustainable growing season for crops. For instance, agronomists have argued that a corn field is sustainable over a period of two years to approximately half a century, but not for a single growing season or for periods longer than a few centuries, because of immediate problems like frost or drought or long term problems such as climatic change.[20]

3.3 MEASURING AGGREGATE ECONOMIC WELL-BEING

When approaching the measurement of aggregate economic well-being from the perspective of sustainability, emphasis must be on the continuity of productive capacity for the economy. That is, the measure of aggregate welfare must be cast in terms of flows, as opposed to stocks. In accounting, stocks are measured on a balance sheet, reflecting the accumulated wealth of an individual, business, or nation. The balance sheet shows the assets that reveal the wealth of the entity in question. The assets also produce a flow of income which is measured periodically (most often annually). The Lindahl-Hicks definition of aggregate income includes the premise that any measure of welfare should measure the value of goods and services that can be consumed during a given period without reducing future consumption opportunities.[21] This conception of income is quite consistent with the idea of sustainability presented in this book. It implies that income or welfare may be considered as a return to capital. In this sense, each generation is obliged to pass on to the next generation a stock of natural and human-made capital that will provide equal or greater welfare to the next generation without greater effort on their part.[8]

Approaching the measurement of well-being from this standpoint implies a broad measure that incorporates full accounting of the effects of human activity. In Chapter 6, we will consider valuation issues confronting businesses and governments as they make microeconomic decisions about

TABLE 3.1
Gross National Product and Gross Domestic Product, United States, 1960–1994 (Selected Years, Billions of Dollars)[22]

Year	Gross National Product	Gross Domestic Product
1960	529.8	526.6
1970	1042.0	1035.6
1980	2819.5	2784.2
1990	5764.9	5743.8
1994	6922.4	6931.4

the allocation of resources to produce goods and services within the economy. In this chapter, we focus on measures of welfare that characterize comparatively how well nations succeed in promoting aggregate economic well-being.

However, income measures do not fully reflect economic well-being. A business or nation may be producing a high level of income, but its wealth may be declining. If the entity is not reinvesting its assets or preserving its natural capital, the high level of income cannot be sustained into the indefinite future. Thus, the measure of well-being must include the degradation of the stock of assets, as well as the flow of income. This leads us to the next principle of sustainability on valuation. For either the microeconomic or macroeconomic perspectives, sustainable valuation implies the following:

3.3.1 Principle 7

Equitable and efficient allocation of resources across space and time requires that the values of goods and services reflect all of the costs and benefits from their production, consumption, and disposal. Inherent in this valuation process is the internalizing of externalities, appropriate discounting of future costs and benefits, and consideration of other issues related to intergenerational and intragenerational equity.

The most common measures of aggregate economic activity used for a cross-country comparison of economic well-being, as well as to track economic growth for one nation, are the national income accounts. In the same way that businesses have a set of ledgers that measure the output and sales of the firm, Gross National Product or Gross Domestic Product are used as comprehensive measures of economic activity.* Gross National Product (GNP) is defined as the value of final goods and services produced by the citizens of a country no matter where the citizens happen to live. Gross Domestic Product (GDP) is defined as the value of final goods and services, produced within a country, whether the production is by the country's own citizens or that of another country. Although you encounter both measures, we will use GDP in the discussion that follows. As shown in Table 3.1 for the United States, the measures are very similar.

Before considering the usefulness of GDP as a measure of economic well-being, it is useful to consider further how it measured. It is calculated as follows:

$$GDP = C + I + G + (X - M)$$

where

C = consumption by households of goods and services; approximately 65 percent of GDP in the United States,

---

National Income is another measure of aggregate economic activity that measures the sum of earnings paid to labor, natural capital, human-made capital, and entrepreneurship, the factors of production used to create output.

I = investment, that is, business purchase of human-made capital, residential construction and change in business inventories; approximately 15 percent of U.S. GDP,

G = purchases of goods and services by federal, state, and local governments; approximately 20 percent of U.S. GDP

$(X - M)$ = exports minus imports, approximately 1 percent of U.S. GDP.

It is important to realize that these components are measured in expenditure terms. That is, consumption is valued as the sum of all goods and services purchased in a period of time multiplied by the price of those goods and services. Thus, the extent to which prices are accurate and reflect all costs of human activity is a significant factor in assessing the adequacy of GDP as a measure of aggregate welfare. For instance, if a nation's over use of a resource causes a scarcity that increases the price of that resource, the resulting increase in value of the resource may provide flawed indicators as to the true contribution of that resource to economic well-being.

In fact, most economists acknowledge that GDP is not a complete measure of economic well-being. It omits many factors that affect quality of life. For instance, GDP only measures market activities. Thus, the value of leisure and non-market production (e.g., unpaid activities in the home) are not included. Of direct concern here is that depreciation of assets is not included as part of GDP.* This is especially significant because degradation or reduction of assets, whether human-made or natural, reduce the income potential of the assets. As shown in Figure 3.3, the treatment of natural capital resources and environmental assets is very incomplete in the national income accounts. Related to the treatment of assets are three significant limitations to current national income accounting practices:

1. Changes in the value of natural capital are not included in many cases, especially where the natural capital is publicly owned.
2. Depreciation of environmental resources and natural capital are not included in the flow accounts; therefore, GDP overstates the true income gains related to use of these resources.
3. Defensive expenditures to mitigate the negative effects of environmental degradation (e.g., pollution abatement activities) are included as positive income flows in GDP without concurrent accounting for the negative effects on human quality of life or on the quality of natural capital, thus again overstating true income flows.[23]

In thinking about measuring economic well-being, it is important to consider the context in which the indicators are developed. Milon and Shogren[24] argue that "economic and environmental indicators are instrumental concepts that must be linked to social objectives." The national income accounts were developed in the middle of the twentieth century when unemployment problems of the Great Depression in the United States and the need to measure the capacity of the economy to wage World War II were primary social concerns.[25] At the time, with the exception of localized concern about shortages, environmental concerns were not important, and the scale of human activity was not seriously threatening the resiliency of the system. However, when viewing economic well-being within the context of sustainability where public concerns about stress to the natural environment are prominent, one needs to rethink the approach to measuring economic well-being.

A preferred approach, from the perspective of sustainability, is rooted in the notions of resilience and integrity introduced in Chapters 1 and 2, respectively. In measuring the change in

---

*Net National Product is an alternative national income measure that includes a standard depreciation expense for human-made capital. However, it does not includes adjustment for elimination in the stock of natural capital or for degradation of existing natural capital (e.g., the reduced value of the water supply caused by pollution from the production of goods and services).

Resource Type	Asset Values Included	Asset Values Not Included
Geologic Resources		
Minerals	Privately owned or leased proven deposits	Publicly owned proved deposits; possible or undiscovered deposits; user cost/net accumulation
Land/soil	Privately owned or leased agricultural/forest land	Publicly owned tracts; user cost/net cost/net accumulation
Freshwater aquifer	Partially capitalized in value of privately owned or leased land with defined user rights to resource	No capitalization in privately owned land with uncertain user rights; publicly owned tracts; user cost/ net accumulation
Biologic Resources		
Forests	Privately owned timber stock	Public owned forest stocks; user cost/net accumulation
Fisheries	Private owned aquaculture	Publicly owned fishery stocks; user cost/net accumulation
Wildlife	None	Publicly owned stocks
Marine systems	None	Publicly owned systems
Biospheric cycling resources		
Atmosphere	Air pollution abatement equipment	Atmospheric gases, chemical balancing processes
Hydrologic	Water pollution abatement equipment	Publicly-owned water systems (rivers, lakes, etc.); precipitation

FIGURE 3.3 Treatment of environmental and natural resources in national income accounts by resource type.[23]

economic well-being with the basic objectives of producing goods and services sustainably, three elements must be in place and thus included in the quantification. First, the integrity and resilience of ecological systems must be maintained. This requires full accounting for the changes in the stock and productive capacity of natural capital and ecological systems. Second, the integrity and resilience of economic systems must be maintained. The productivity of the economic subsystem is dependent, not only upon the stock of natural capital, but also the stock and productive capacity of human-made capital and labor, the state of technology and the management of these factors of production. Implicit in this measurement is full accounting of the externalities associated with human activity. Finally, the integrity and resilience of social systems must be maintained. The social, legal, and organizational infrastructure for economic activities and conflict resolution must be healthy, at local, national, and international levels. The social aspect of resilience is most troublesome in even conceptualizing how to measure effects on economic well-being. However, there are indicators such as the effects of high inflation or expenditures on national security that may provide partial indications of the impacts on economic well-being of the social infrastructure.

Bartelmus[26] has stated the following operational definition for *sustainable economic growth*: "increase in (real) domestic product, allowing for the consumption of produced capital and the depletion and degradation of natural capital, taking into account that past trends of depletion and degradation can be offset or mitigated by technological progress, discoveries of natural resources and changes in consumption patterns." This definition provides some structure as we think further about developing measures of economic well-being. However, it ignores the social dimension mentioned above. In the following section we will review some of the efforts that have been made to improve the measurement of economic well-being. The basis for our assessment of their desirability will be to think about sustainability and development, as opposed to economic growth, and will require consideration of both quantitative and qualitative aspects of aggregate well-being.

3.4 TOWARD IMPROVED MEASUREMENT OF ECONOMIC WELL-BEING

Developing a reliable aggregate measure of economic well-being is indeed a tall order, as suggested by the brief overview above. In fact, some scholars suggest that "analysts abandon the notion of a society-wide sustainability index. The search for such a 'Holy Grail' already has consumed the careers of too many talented scientists."[27] Although there will be no attempt here to present a single perfect index of human welfare, this section will present selected attempts at quantifying different aspects of economic well-being and related sustainability components that will provide additional information about economic welfare than GDP alone. The primary focus will be on increased attention given to natural capital and environmental resources, and to social infrastructure aspects that contribute to understanding sustainability. Since "economic indicators are the main feedback loop to national policy," it is important that they identify and measure variables likely to be significant to the political process.[28]

As mentioned above, one of the fundamental problems of GDP is that it does not take account of the depletion or degradation of natural capital. In fact, to the extent that natural capital is used to produce goods or services valued in the market, a *positive* value is assigned to the depletion of these natural resources. Moreover, GDP does not account for the social costs of decreasing environmental quality, and again, in fact, pollution and other environmental effects are given a positive value to the extent that abatement activities require purchase of equipment and personnel valued in the market.

The World Bank has sponsored significant research that has begun to provide indicators of these effects on natural capital and environmental quality. In Table 3.2, the results from two studies are consolidated for comparative purposes.* The first part of this table provides information for a select number of countries as to the wealth per capita and the composition of wealth divided into four categories: *land* (crops, etc. . . .); *other* (subsoils, minerals, and water); *human-produced* (plants and machinery manufactured); and *human* and *social* (education, skills, and other human capital dimensions).[29] It is apparent in looking at these numbers that there is significant variation in the composition of wealth among countries. These, in turn, give rise to different potential sustainable paths for each of these countries to follow.

The right-hand side of Table 3.2 shows two environmental indicators that are intended to adjust the basic GNP data to be more sensitive to the types of concerns mentioned here. GNP per capita is given for each country. The first adjustment recommended is to account for the impacts of pollution on economic well-being. This adjustment involves first developing a Pollution-Adjusted Economy Indicator (PAI). In developing this indicator, Rodenburg et al.[30] used per capita carbon dioxide emissions as a proxy indicator for the whole range of pollutants that degrade the commons, multiplying GNP by the ratio of world median per capita emissions (2.1984 tons) to the actual per capita emissions for the country in question. This calculation results in the Pollution-Adjusted GNP per capita shown. In making the adjustment for natural capital, the authors focused only on depletion, not degradation of the natural resources. The Natural Capital Indicator (NCI) was based on the country's remaining natural areas (forests, coastal zones, natural wetlands, relatively unmanaged rangelands,

---

*Wealth figures are 1990 values; the global environmental indicators are GNP in 1991, per 1990 population.

TABLE 3.2
International Wealth Comparisons and Environmental Indices

		Composition of Wealth				Global Environmental Indicators				
Country	Wealth per Capita ($1,000)	Land	Other	Human-produced	Human and Social	GNP per Capita	Pollution-Adjusted GNP per Capita	Natural-Capital Adjusted GNP per Capita	Share of Gross World Product	Adjusted Share of Gross World Product
Australia	835.0	64%	8%	7%	21%	17480.7	2543.0	138366.9	1.38	2.96
Canada	704.0	64	5	9	22	21040.7	3043.0	12763.2	2.61	1.47
Japan	565.0	1	—	18	81	27029.0	6756.8	4576.9	15.62	8.44
United States	421.0	22	3	16	59	22552.0	2538.7	10479.8	26.31	13.44
Germany	399.0	3	1	17	79	24430.8	4428.3	143.5	8.87	4.07
Singapore	306.0	—	—	15	85	14411.8	2095.6	33.5	0.18	0.08
Saudi Arabia	184.0	26	28	18	28	8513.8	1337.6	2153.4	0.56	0.28
Mexico	74.0	11	5	11	73	2851.0	1599.3	16290.5	1.18	2.31
China	6.6	3	5	15	77	372.2	372.2	1532.7	1.98	3.78
India	4.3	2	9	25	64	331.4	903.7	929.5	1.32	3.71
Ethiopia	1.4	12	27	21	40	127.9	3811.9	2571.3	0.03	0.74
World	86.0	15	4	16	64					

Source: References 29 and 30

protected areas, protected watersheds, and other areas that are left in or managed in a natural state), multiplied by the ratio of actual biodiversity to average biodiversity. Obviously these adjustments result in significant changes in the ranking of the countries shown. Countries (often developing countries) rich in biodiversity with low per capita energy consumption gain most from these adjustments. In addition, it is interesting to note the resulting changes in the share of Gross World Product, and the adjusted (both with the PAI and the NCI) share of Gross World Product for these countries. In fact, the industrialized OECD countries share of Gross World Product falls from 80.15 percent to 45.61 percent, while that of the non-OECD countries increases from 19.85 percent to 54.39 percent when all 148 countries in the study are aggregated into these categories.

Further evidence on the importance of accounting for depreciation of natural capital in aggregate national income measures comes from the work of Repetto et al.[31] Although it is important to apply depreciation measures consistently in resource-dependent countries, evaluations of economic well-being may be distorted seriously by failure to account for resource depletion or degradation. Repetto and his associates performed an extensive case study for Indonesia examining the change in the stock of natural capital for three natural resource sectors (petroleum, forestry, and soil), and the resulting impact on aggregate national income. One of the advantages of the approach, in comparison to the World Bank efforts, is that they focused on the net change in natural capital. Natural resources must be valued on their long term ability to generate income, not on the overall value of the resource stocks to society. We will discuss existence value in Chapter 6, where natural capital may have a value simply because people derive pleasure from knowing that a resource exists. But in the context of measuring the flow of economic welfare, we must value resources for what they do, and not for what they are.[7,32] As shown in Table 3.3, the growth rate of Gross Domestic Product increased by 7.1 percent annually from 1971 to 1984. However, after adjusting for changes (positive and negative) in the stock of natural capital for just three natural resource sectors, the average growth rate of Net Domestic Product (NDP) was only 4.0 percent. This significant change reflects the degree to which natural resources have been depleted to finance current consumption expenditures. It is unsustainable for a country to continue to "eat its seed corn" in this way *ad infinitum*.

TABLE 3.3
Natural Capital Reserves in Indonesia and Gross Domestic Product[31]

		Net Change in Natural Capital				
Year	GDP	Petroleum	Forestry	Soil	Net Change	NDP
1971	5545	1527	−312	−89	1126	6671
1972	6067	337	−354	−83	−100	5967
1973	6753	407	−591	−95	−279	6474
1974	7296	3228	−533	−90	2605	9901
1975	7631	−787	−249	−85	−1121	6510
1976	8156	−187	−423	−74	−684	7472
1977	8882	−1225	−405	−81	−1171	7171
1978	9567	−1117	−401	−89	−1607	7960
1979	10165	−1200	−946	−73	−2219	7946
1980	11169	−1633	−965	−65	−2663	8506
1981	12055	−1552	−595	−68	−2215	9840
1982	12325	−1158	−551	−55	−1764	10561
1983	12842	−1825	−974	−71	−2870	9972
1984	13520	−1765	−493	−76	−2334	11186
Average Annual Growth	7.1%					4.0%

All data are in constant 1973 rupiah (Billions).

While these data provide useful perspectives on the depletion of natural capital and the resulting impact on Gross Domestic Product, they are only partial indicators of aggregate economic well-being. As noted above, economists have long been critical of the value of GDP or GNP in measuring economic welfare. One early attempt to offer a substitute measure was Nordhaus and Tobin's[33] Measure of Economic Welfare (MEW). This measure addressed several of the weaknesses of GNP, including estimating the value of leisure and housework, reclassifying several categories of consumption, investment, and government spending to recognize certain "regrettable" expenditures such as police services that are "overhead costs of a complex industrial nation-state," and including an estimate of the value of urban disamenities.

While the MEW was a more precise measure of economic well-being than GNP or GDP, Nordhaus and Tobin found that it correlated rather positively with them and thus were not convinced as to the value of changing our basic approach to national income accounting. However, during the almost 30 years since their initial research, environmental externalities and depletion and degradation of natural capital have become increasing concerns. As a result, numerous attempts are being made to develop alternative aggregate measures more attuned to environmental concerns. While we cannot cover all of these alternative approaches, it will be useful to consider two of them.

3.4.1 System of National Accounts

The System of National Accounts (SNA) is a comprehensive framework for recording estimates of the aggregate flows and stocks of the economy. The SNA includes the national income accounts as used in the United States, but it also includes input–output accounts, and other measures intended to record the stocks and flows for each major economic sector. The SNA is not intended to produce one comprehensive index (like GDP). Rather it is intended to provide guidance to national statistical agencies as they design their own macroeconomic measures and to provide a basis for international reporting of data. It builds in flexibility for satellite accounts developed for areas such as health, education, environment, research and development, tourism, and agriculture. The satellite account on

Sustainability and Economic Well-Being

the environment includes estimates of pollution abatement and control. However, the means for incorporating natural resource accounting into the SNA is still in an evolutionary period.[34]

The satellite system for integrated environmental and economic accounting (SEEA) was developed at the request of the Statistical Commission of the United Nations and served as the basis for the United Nations *Handbook on Integrated Environmental and Economic Accounting*. The SEEA has three primary objectives.[35]

1. To segregate all flows and stocks of assets in national accounts related to environmental issues and to estimate the total expenditures required for the protection and restoration of the environment.
2. To link physical resource accounting with monetary environmental accounting and balance sheets.
3. To assess the depletion of natural capital in production and final demand, and the changes in environmental quality resulting from pollution and other adverse effects of production, consumption, and natural events, and from environmental protection and restoration.

The SEEA is intended to be flexible so that it can be linked to other methods of integrated environmental-economic accounting. Although it is a broad-based measure, it does not include phenomena that occur outside the economic system. The designers of the SEEA believe that such phenomena are better dealt with by biophysical resource accounts and systems of environmental statistics and monitoring. Although the actual system behind SEEA is fairly complex, Table 3.4 provides summary data on the basic accounting identities that emerge for a hypothetical country. The use of natural capital and adjustments for other environmental activities are reflected in these figures. One of the major goals of the SEEA is to calculate the Environmentally Adjusted Net Domestic Product (EDP). The EDP represents the net value added to aggregate economic well-being after adjusting for changes in natural capital and environmental activities. In this example, the EDP is 37 percent smaller than the GDP.

3.4.2 Genuine Progress Indicator

One of the more comprehensive attempts to develop a measure of socio-economic well-being and sustainability is the Genuine Progress Indicator (GPI).* It is intended to measure a wide range of

TABLE 3.4
Accounting Identities under the System for Integrated Environmental and Economic Accounting (SEEA)[35]

Gross Domestic Product (GDP)	293.4	Domestic Final Demand (SNA)	294.2
−Use of fixed capital	26.3	−Consumption of fixed capital	26.3
−Use of natural capital	59.8	−Government restoration costs	5.0
−Environmental adjustment of final demand	22.2	+Net capital accumulation of natural capital	−73.9
Environmental Adjusted Net Domestic Product (EDP)	185.1	Environmentally Adjusted Domestic Final Demand	189.0
+Import of Products	74.5	+Export of Products	73.7
+Import of Residuals	−1.6	+Export of Residuals	−4.7
Environmentally adjusted primary inputs	258.0	Environmentally adjusted final demand	258.0

---

*The Index of Sustainable Economic Welfare (ISEW) was a precursor to the GPI. The GPI was a revision extended to address several criticisms by scholars and policymakers. For a detailed description of the ISEW, along with a critique by various analysts, see Cobb and Cobb.[36]

social and ecological effects to provide a more precise indicator of economic well-being. Starting with consumption as a base, it includes adjustments for more than 20 aspects of economic well-being that GDP ignores. These include, but are not limited to the following[28]:

- depletion and degradation of natural capital
- adjustments for environmental abatement activities
- value of household and volunteer work
- costs from crime (e.g., hospital expenses and property losses)
- adjustments for changes in the distribution of income
- changes in hours of leisure (valued at an average wage rate)

In considering how the GPI provides a measure that helps evaluate movement toward sustainability, it clearly goes well beyond simply accounting for changes in natural capital and environmental activities. However, sustainability involves consideration of many broad-based social indicators. To the extent that these require variables important in quantifying social resiliency, the GPI provides useful information about the social infrastructure and how it affects aggregate economic well-being. Pointing to social trends where families and communities show responsibility, for instance, in the care and nurture of children, the authors note that a monetized service sector has emerged, falsely resulting in an increase in measured GDP. "Both the government and the private market grow by cannibalizing the family and community realms that ultimately sustain us," they say.[28]

Table 3.5 shows the calculation of the GPI for 1992. Although we will not describe in detail the calculation for each category within the table, the additions and deletions should be fairly straightforward. Table 3.6 shows comparative figures on annual growth rates for GDP per capita and GPI per capita, while Figure 3.4 plots GDP and GPI over time. When the social and ecological costs of economic activity are accounted for, the GPI shows a steady decline in economic well-being since the 1970s, as opposed to the increase shown for GDP. Depletion of non-renewable resources and long-term environmental damage have a significant effect on GPI. When both of these factors are included, per capita GPI increases by 29 percent from 1950 to 1973, and then decreases by 73 percent from 1973 to 1992. Omitting resource depletion, per capita GPI increases by 44 percent from 1950 to 1973 and then decreases by 28 percent during the later period. If the effects of long-term environmental damage are excluded, per capita GPI rises by 34 percent during the early period and then declines by 36 percent.[37]

The GPI also is not a perfect measure of economic well-being. It omits many factors that affect aggregate economic welfare. These include such variables as the value of goods and services that are exchanged as part of the underground economy and the depletion of genetic diversity. Moreover, Gottfried[38] has noted that measures of aggregate welfare such as the GPI address the persistence or stability of the economy, that is, the ability of the system to remain *constant* over time, but do not truly address issues of resiliency fundamental to sustainability. Nonetheless, the GPI does move us closer to a more precise, and comprehensive measure of economic well-being.

3.5 CONCLUSION

In this chapter, we have considered the approach that economists have taken to thinking about sustainability. Discussion has centered around three basic areas: allocative efficiency, distributional equity, and stabilization, economic growth and their questions of scale. With respect to efficiency, we noted that markets and a well-functioning price system provide the fundamental organizing forces that lead to optimal allocation of goods and services. However, to the extent that markets fail by the existence of public goods, externalities, or failure for prices to reflect true costs of human activity, alternative means or organizations must be found to achieve efficiency.

TABLE 3.5
Calculation of the Genuine Progress Indicator for 1992[37]

Consumption (in constant 1982 billions of dollars)	2719.2
Income distribution index (100 base) (change in inequality	113.6
Weighted personal consumption Consumption income distribution index)	2392.9
Value of housework (+)	1202.6
Sevices of household capital (+)	403.3
Services of government capital (+)	40.5
Loss of leisure time (−)	141.5
Cost of underemployment (−)	136.4
Cost of consumer durables (−)	442.7
Cost of commuting (−)	115.8
Cost of personal pollution control (−)	11.0
Cost of auto accidents (−)	70.0
Cost of water pollution (−)	35.3
Cost of air pollution (−)	42.0
Cost of noise pollution (−)	11.3
Loss of wetlands (−)	142.4
Loss of farmland (−)	77.2
Depletion of non-renewable resources (−)	760.0
Long-term environmental damage (−)	616.0
Cost of ozone depletion (−)	208.3
Loss of forests (−)	57.6
Net capital investment (+)	24.2
Net foreign lending or borrowing (+)	−116.4
Genuine Progress Indicator	1079.5
Gross Domestic Product	4192.5

TABLE 3.6
Annual Growth Rates—Gross Domestic Product and Genuine Progress Indicator[37]

Years	GDP/Person	GPI/Person
1950–60	1.8%	0.7%
1960–70	2.2%	1.5%
1970–80	1.7%	−1.0%
1980–92	1.4%	−3.3%

With respect to distributional concerns, we noted that concern about inter-generational equity is at the very heart of discussions about sustainability. Each generation has an obligation to pass on to future generations a mix of assets such that future generations can be at least as well off as present generations without exerting more effort. This is an ideal. Hartwick's rule regarding resource use was that inter-generational equity can be achieved by investing all profits or rents arising from the depletion of natural capital in the acquisition of human-made capital assets.[39] In reality, it is increasingly costly to substitute human-made inputs for natural capital. In some cases, where a

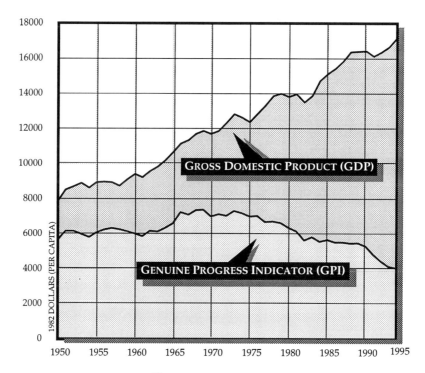

FIGURE 3.4 GDP and GPI over time.[37]

resource is complementary to human-made capital or where any degradation or depletion of the resource results in irreparable damage, any use of that natural capital may not be sustainable. To understand fully the potential for sustainable systems that promote efficiency and inter-generational equity, it is necessary to consider both questions of scale (geographic and time), and the role that technology may play in hindering or supporting optimal use of inputs.

This concern for the asset base available to future generations led us to consider the valuation of natural capital within the context of national income accounting. Although no perfect measure of aggregate economic well-being exists, we introduced several attempts to improve the measurement of social and economic welfare. One specific measure focused on natural capital such as the World Bank Global Environmental Indicator, and more comprehensive attempts included the SEEA and Genuine Progress Indicator. The fundamental notion imbedded in these approaches is that the value of an incremental unit of natural capital should increase as its scarcity increases. Most existing measures of economic well-being, such as Gross Domestic Product, fail to account for the degradation of natural capital and the environmental costs imposed by human activity. Although we applaud these attempts to more fully measure economic well-being, much work still remains to be done both conceptually in developing alternative measures and in collecting data for cross-country comparisons.

3.6 REFERENCES

1. Mankiw, N. G., *Principles of Economics,* Dryden Press, Fort Worth, TX, 1998.
2. Crocker, T. D., Ecosystem functions, economics, and the ability to function, in *Integrating Economic and Ecological Indicators,* Milon, J. W. and Shogren, J. F., Eds., Praeger Press, Westport, CT, 1995.
3. Howarth, R. B. and Norgaard, R. B., Environmental valuation under sustainable development, *Am. Econ. Rev.,* 82(2), 473, 1992.

4. Costanza, R. and Daly, H. E. Natural capital and sustainable development, *Conserv. Biol.*, 6(1), 37, 1992.
5. Rosen, H. S., *Public Finance,* Irwin, Chicago, 1995.
6. Toman, M. A., Economics and "sustainability": balancing trade-offs and imperatives, *Land Econ.*, 70(4), 399, 1994.
7. Solow, R., An almost practical step toward sustainability, Lecture, fortieth anniversary of Resources for the Future, Resources for the Future, Washington, D.C., 1992.
8. Norgaard, R. B., Issues in asset aggregation and intergenerational asset transfer, in *Toward Improved Accounting for the Environment,* Lutz, E., Ed., The World Bank, Washington, D.C., 1993.
9. Jaeger, W. K., Is sustainability optimal? Examining the differences between economists and environmentalists, *Ecol. Econ.*, 15(1), 43, 1995.
10. Brennan, T. J., Discounting the future: economics and ethics, *Resources,* 120, 3, 1995.
11. Lesser, J. A. and Zerbe, R. O., Jr., What can economic analysis contribute to the 'sustainability' debate?, *Contemp. Econ. Pol.*, 13(3), 88, 1995.
12. Markandya, A. and Pearce, D. W., Development, the environment, and the social rate of discount, *World Bank Res. Obs.*, 6(2), 137, 1991.
13. Lind, R. C., Reassessing the government's discount rate policy in light of new theory and data in a world economy with a high degree of capital mobility, *J. Environ. Econ. Manage.*, 18, S8, 1990.
14. Cropper, M. L. and Portney, P. R., Discounting human lives, *Resources,* 108, 1, 1992.
15. Martinez-Alier, J., Distributional issues in ecological economics, *Rev. Soc. Econ.*, 52(4), 511, 1995.
16. Norgaard, R. B. and Howarth, R. B., Sustainability and discounting the future, in *Ecological Economics: The Science and Management of Sustainability,* Constanza, R., Ed., Columbia University Press, New York, 1991.
17. Daly, H. E., Steady-state economics: concepts, questions, policies, *Gaia: Econ. Perspect. Sci., Hum., Econ.*, 6, 333, 1992.
18. Wackernagel, M. and Rees, W., *Our Ecological Footprint,* New Society Publishers, Gabriola Island, BC, 1996.
19. Daly, H. E., Elements of environmental macroeconomics, in *Ecological Economics: The Science and Management of Sustainability,* Costanza, R., Ed., Columbia University Press, New York, 1991.
20. Allen, T. F. H. and Hoekstra, T. W., Toward a definition of sustainability, in *Sustainable Ecological Systems: Implementing an Ecological Approach to Land Management,* Covington, W. W. and DeBano, L. F., Eds., U.S. Department of Agriculture, Forest Service, Rocky Mountain Forest and Experiment Station, Fort Collins, CO, 1994, 98.
21. Hicks, J., *Value and Capital*, 2nd ed., Oxford University Press, Oxford, 1946.
22. Economic Report of the President, Table B-1, 1996.
23. Milon, J. W., Environmental and natural resources in national economic accounts, in *Integrating Economic and Ecological Indicators,* Milon, J. W. and Shogren, J. F., Eds., Praeger Press, Westport, CT, 1995, 131.
24. Milon, J. W. and Shogren, J. F., Economics, ecology, and the art of integration, in *Integrating Economic and Ecological Indicators,* Milon, J. W. and Shogren, J. F., Eds., Praeger Press, Westport, CT, 1995, 3.
25. Cobb, C. W. and Cobb, J. B., Jr., *The Green National Product: A Proposed Index of Sustainable Economic Welfare*, University Press of America, Lanham, MD, 1994.
26. Bartelmus, P., Accounting for sustainable growth and development, *Struct. Change Econ. Dyn.*, 3(2), 241, 1992.
27. Kaufmann, R. K. and Cleveland, C. J., Measuring sustainability: needed—an interdisciplinary approach to an interdisciplinary concept, *Ecol. Econ.*, 15(2), 109, 1995.
28. Cobb, C., Halstead, T., and Rowe, J., If the GDP is up, why is America down?, *Atlantic Mon.*, 276(4), 59, 1995.
29. Passell, P., The wealth of nations: a "greener" approach turns list upside down, *New York Times,* September 19, 1995.
30. Rodenburg, E., Tunstall, D., and van Bolhuis, F., Environmental indicators for global cooperation, The World Bank, Washington, D.C., 1995, Working paper no. 11.
31. Repetto, R., Magrath, W., Wells, M., Beer, C., and Rossini, F., *Wasting Assets: Natural Resources in the National Income Accounts,* World Resources Institute, Washington, D.C., 1989.
32. No accounting for tastes, *Economist*, September 23, 1995, 64.

33. Nordhaus, W. and Tobin, J., Is growth obsolete?, in *Economic Growth,* National Bureau of Economic Research, New York, 1972.
34. Carson, C. S. and Young, A. H., The ISEW from a national accounting perspective, in *The Green National Product: A Proposed Index of Sustainable Economic Welfare,* Cobb, C. W. and Cobb, J. B., Jr., Eds., University Press of America, Lanham, MD, 1994, 111.
35. Bartelmus, P., Stahmer, C. and van Tongeren, J., Integrated environmental and economic accounting—a framework for an SNA satellite system, in *Toward Improved Accounting for the Environment,* Lutz, E., Ed., The World Bank, Washington, D.C., 1993, 45.
36. Cobb, C. W. and Cobb, J. B., Jr., *The Green National Product: A Proposed Index of Sustainable Economic Welfare,* University Press of America, Lanham, MD, 1994.
37. Cobb, C. and Halstead, T., *The Genuine Progress Indicator: Summary of Data and Methodology,* Redefining Progress, San Francisco, 1994.
38. Gottfried, R. R., Some reflections on the ISEW, in *The Green Product: A Proposed Index of Sustainable Welfare,* Cobb, C. W. and Cobb, J. B., Jr., Eds., University Press of America, Lanham, MD, 1994, 135.
39. Hartwick, J. M., Intergenerational equity and the investing of rents from exhaustible resources, *Am. Econ. Rev.*, 67(5), 972, 1997.

4 Natural Resource Conflicts And Sustainability

Orie L. Loucks and Esi Wurapa

CONTENTS

4.1 Introduction 63
    4.1.1 Principle 8 64
    4.1.2 Principle 9 65
4.2 Forest Resources 65
    4.2.1 Forest Status: The Problem 65
    4.2.2 Local and National Outcomes 66
    4.2.3 Solutions: Steps toward Sustainability 68
4.3 Fisheries 70
    4.3.1 The Problem 70
    4.3.2 Social Consequences 71
    4.3.3 How to Achieve Sustainable Development 72
4.4 Sustaining Fresh Water 73
    4.4.1 Use and Abuse of Water Resources 74
    4.4.2 Implementing Sustainable Water Management 76
4.5 Air Resources 76
    4.5.1 The Air Problem 77
    4.5.2 Approaches for Solving Air Problems 78
4.6 Discussion 79
4.7 References 80

4.1 INTRODUCTION

Previous chapters have shown that sustainable development, expressed either as a sustainable economy or as sustainable access to the resources our society needs, requires understanding of both social systems and natural systems. We will explore sustainability of natural systems in this section, including forests and fisheries, air, and water, all of which are renewable systems. The case studies in the later sections of the book consider minerals and energy, which are thought of as non-renewable, and biological diversity, which in different ways is both renewable and non-renewable.

Forests and fisheries are called renewable because they grow, and in principle, harvests for human use can be continued in perpetuity, while leaving the resource as healthy in the end as it was in the beginning. Such sustainable use has not been a common experience, however (as seen in Chapter 2). Nearly all of the so-called renewable resources are seen now as having not been used sustainably. Rather, they have experienced substantial impoverishment, some of which probably is irreversible. Conversely, however, supplies of some non-renewable resources, such as iron and aluminum, are being met by recovering discarded material and reusing it as input for new products.

Thus, in sustainable development, these mineral resources can be seen as "renewable" resources, while fisheries and forests must be restored to achieve a renewable state.

This chapter explores how sustainable levels of use can be accomplished for both renewable and non-renewable resources. Only energy from fossil fuels is truly a non-renewable resource now, as opposed to energy from wind, solar, and hydropower, which clearly are renewable. None of these resource systems stand alone, however, independent of other resource uses. All are linked through human values and preferences that have led to industrial linkages among the renewable and recyclable resources that comprise inputs to our economic system. Seen in these terms, *sustainable development ensures long-term access to resources, undiminished in availability over multiple generations, and facilitated through new cost-effective technologies.*

Thus, sustaining resource systems such as fisheries, forests, air, water, metals, energy, and biological diversity involves more than understanding the social and economic systems that diminished their quality. Also required is technical understanding of why impoverishment is an outcome, and how renewal or regeneration can be accomplished through appropriate investment and industrial development. A further goal of this chapter, therefore, is to consider, from a scientific perspective, how the loss of "productive capacity" in natural resource systems has taken place, and how the loss can be measured and understood for its economic impact on society. We will also consider the extent to which regenerative capacity of resources can be engineered, and the cost to society of such regeneration. Since impoverishment of resources took place over many decades, providing a financial stimulus initially, the possibility of recovery also will take time, and could be an economic burden. Sustainable development, of course, does not imply a commitment to rehabilitation of resources damaged in the past. The interesting technical dilemma is, instead, that *pursuit of a policy of no further degradation of natural resources requires just as deliberate a decision to minimize the economic burden of rehabilitation as to minimize the economic advantage from liquidation of natural capital.*

To have business and public consensus on distinctions among environmental costs requires precision in measurement over generations, and understanding of social and environmental processes. These needs are the basis of an eighth principle of sustainability, *measuring and adapting,* as developed in the following sections:

4.1.1 PRINCIPLE 8

> Sustainable processes require that system characteristics and outcomes be measured in quantitatively precise ways and that mechanisms be included to allow the resulting information to modify human behavior so as to maintain system integrity.

To explore these principles, we need not review all the natural resource systems used by our economy. Instead, we will discuss only a few, two (forests and fisheries) that illustrate biological resources used directly in commerce, and two (air and water) that illustrate renewable resources used by society for low cost disposal. All four have experienced over-exploitation, to the advantage of our economy at one time and disadvantage at another time. Our goal is to understand in what ways these resources are sustainable, over what time periods, at what risk to present or future generations, and whether their sustainability is being measured (or not).

The material in the following sections gives us an appreciation of what is at risk in the environment, as to human health, and for businesses, when the productivity or assimilative capacities of resources are overestimated. The field of risk assessment is an interdisciplinary approach to estimation of the likelihood of damages (explosions, disease, or financial loss) from improbable events in air, water, forest, and fishery systems. These "improbable" events need to be considered when projecting the sustainability of resource systems, or society's use and ultimate disposal of resources. These considerations lead to a ninth principle of sustainability, *risk assessment,* developed in this chapter and in others, and defined here as follows:

4.1.2 PRINCIPLE 9

Human impacts on the environment involve uncertain consequences owing to the imprecision of measurements and scientific models. Sustainable systems must include identification and quantification of risks and determination of the level of acceptable risk to system components.

4.2 FOREST RESOURCES

Forests are a long-lived biological system, susceptible to one-time or periodic harvests, but also, over time, capable of renewal. Over periods as short as one year, a growing, healthy forest can have a net gain of one to two metric tons per hectare of carbon (and other elements) in its above-ground mass through the process of growth, but at the same time annual losses of carbon and nutrients occur naturally through the shedding of foliage and small branches, and the death of some trees. Mass balance accounting of changes in "forest stocks," therefore, have to consider annual credits as well as debits, as it were. These accounts need to be maintained for each large land owner, as well as for political units such as counties and states, and for the United States (or the world) as a whole. Such accounts, when compared over a generation or two (20 to 40 years) document whether the forest stock has gained or lost over the period, and therein lies a measure of sustainability.

4.2.1 FOREST STATUS: THE PROBLEM

In *Trees: The Yearbook of Agriculture,*[1] C. E. Behre reported that the area of commercial forest in the U.S. at that time (excluding high-elevation or rough land) was 461 million acres. However, he notes that, even then, 75 million acres, one-sixth of the forest land, was "denuded or so poorly stocked with seedlings as to be unproductive for decades." In addition, the timber on 30 million acres was too small for sawlogs, and 58 million acres of second-growth timber had less than 40 percent of the number of trees needed for full stocking. Therefore, this area was growing new forest at less than half the expected rate. Most of the denuded and poorly stocked land was in the East then, with southern forests the most deficient. Idle land, he noted, "contributes little to the maintenance of schools, roads, or other community services, and no jobs."

What must be understood here is that when timber is grown as a crop, *the amount that can be harvested year after year depends upon the volume of stock or standing timber on which growth can be added.* Up to a point, the more growing stock or forest capital there is, the greater the crop available for harvest each year. Removing that capital base beyond a certain threshold is like eating the seed corn, or spending a corporation's capitalization. Also, to maintain an annual harvest of merchantable timber, there must be a succession of age classes from seedlings up to full-grown timber, so that as mature trees are cut and removed, new trees take their place. Thus, to sustain a high output of timber products, a substantial volume of standing timber must be maintained as "working forest capital." When we liquidate forest capital, we reduce the capacity of the system to accrue growth.

As of 1945, Behre[1] reported that saw-timber in the United States totaled 1,601 billion board feet (470 billion cubic feet), about half of which was in virgin stands. He noted that despite the progress in forestry at the time, U.S. timber supplies were decreasing. Saw-timber volume in the Pacific Northwest and Lake States regions had declined 14 percent over the previous 11 years, but he viewed the crux of the country's forest problem to be the continued shrinkage and deterioration of forest growing stock. Annual growth of timber was then estimated at 13.4 billion cubic feet.

Behre's data show that the saw-timber drain *exceeded annual growth by 50 percent* in the 1940s. He describes this drain as a measure of the rate at which harvests were "overdrawing the forest bank account." If the 1944 trends were to continue with no changes in forest practices for 20 years, he said, unlikely because scarcity of accessible timber would make it increasingly difficult to sustain output, the saw-timber stock would drop 27 percent. However, he added that balance in itself, nationally, is not an adequate goal. There was a near-balance then between growth and drain across all the types of timber for the nation as a whole, but this derived from a small accumulation of forest growth in the northeast, and deficits in the west.

In any event, Behre saw the need for timber products in the United States to be considerably greater in 1949 than the timber cut at that time. Studies of long-term growth potentials and demand suggested a demand approaching 65 to 72 billion board feet (about 20 billion cubic feet, compared with the 13.4 billion cubic feet of growth noted above), and that this was a reasonable goal for *annual growth* on U.S. forest land. This would have been a 40 percent increase over the actual growth at the time.

4.2.2 Local and National Outcomes

Reports in 1994[2] provide recent data on the forest harvests of the United States. These results are for 1989–1991, and can be compared with similar data in the same reports for 1979–1981. For the recent data, the total U.S. annual harvest is 18 billion cubic feet, but this includes fuelwood as well as export logs, lumber, and paper products, a significantly more comprehensive summary of harvests than that of Behre[1]. The reports show the 1989–1991 harvest to be a 22 percent increase over the annual harvest of 1979–1981, a cut of 14 billion cubic feet at that time. This is probably a smaller total than in the 1940s after considering the differences in wood uses included.

The imbalance, or non-sustainable patterns of use, regionally, can be seen in Box 4.1 and Table 4.1, based on 1970 surveys. The problem is evident by comparing the net growth, less removals, for the North (central and northeastern United States), with the growth less removals in the Pacific Coast

TABLE 4.1
Net Annual Growth and Removals of Growing Stock in the United States, by Species Group and Section[1] (in *Billion Cubic Feet*)

Section	All Species 1970	Softwoods 1970	Hardwoods 1970
North			
Net growth	5.5	1.4	4.2
Removals	2.4	0.6	1.8
Ratio of growth to removals	2.3	2.2	2.3
South			
Net growth	8.6	5.4	3.2
Removals	6.5	4.0	2.5
Ratio of growth to removals	1.3	1.4	1.3
Rocky Mountains			
Net growth	1.4	1.3	0.1
Removals	0.9	0.9	([2])
Ratio of growth to removals	1.5	1.4	26.2
Pacific Coast			
Net growth	3.1	2.6	0.5
Removals	4.2	4.1	0.1
Ratio of growth to removals	0.7	0.6	4.1
Total, United States			
Net growth	18.6	10.7	7.9
Removals	14.0	9.6	4.4
Ratio of growth to removals	1.3	1.1	1.8

[1]Columns may not add to totals because of rounding. Ratios calculated from growth-removal data before rounding.

[2]Less than 0.05 billion.

Source: Reference 3

> **Box 4.1 70-Year Projections of Harvests**
>
> A long-term view of changes from the 1950s to 2020 can be seen in Table 4.2. The data on hardwood and softwood "wood supplies" (equivalent to growth) are reported for 1952, 1962, and 1970, and projected for 1980, 1990, 2000, and 2020.[3] The interval from 1952 to 1970 saw just over a 10 percent increase in total supply, about half of which came from increased harvests of the Pacific Northwest forests at that time. By 2020, however, total projected supplies have increased by nearly 100 percent owing largely to projected increases in supplies from the North Central and Northeast, the South, and the Rocky Mountains regions. All of these projections, however, assume that current clear-cutting and forest regeneration practices, and air pollution effects on growth, do not limit the new stand densities or their growth rates. However, numerous studies have shown that neither of these assumptions hold for large parts of the U.S. timberland. Thus, truly sustainable harvesting of forests requires consideration of unexpected outcomes among internal forest regeneration processes, as well as externalities (pollution) from the industrial system itself.

TABLE 4.2
Roundwood Supplies[1] from U.S. Forests, by Section and Species Group, 1952, 1962, and 1970, with Projections to 2020 (in *Million Cubic Feet*)

Section	1952	1962	1970	1980	Projections 1990	2000	2020
North	1,981	1,812	1,988	3,231	4,107	4,954	4,912
South	4,983	4,283	5,413	7,273	8,226	9,095	9,204
Rocky Mountains	506	698	863	1,090	1,204	1,364	1,320
Pacific Coast	3,274	3,386	3,890	3,724	3,472	3,437	3,605
United States	10,745	10,179	12,154	15,318	17,009	18,849	19,040

[1]Includes supplies from growing stock and other sources such as rough and rotten trees, dead trees, and trees from noncommercial forest land; excludes logging residues and nonproduct removals.

Source: The Outlook for Timber in the United States, USDA Forest Service, 1977.

States. The former regions shows a ratio of 2.2 times as much growth as removals in 1970, while the Pacific shows growth of only 60 percent as much growth as the harvest. This unsustainable rate of harvest in the Pacific region was justified at the time because so much of the Northwest was dominated by one age-class, old growth, in which very little of the annual growth accumulates, owing to balancing of growth and mortality within the forest itself. It is also clear, however, that after some decades of aggressive liquidation of old growth, the region must return to an annual harvest that is *not* larger than the growth. This would mean a reduction in harvests there by at least half, steps that have been introduced in 1994 and 1995. Thus, the harvest of 18 billion cubic feet cited for 1989–1991 above, must be seen as abnormally high owing to the high rate of liquidating old growth forest in the Northwest. The next question we should ask is whether future growth rates have been impaired by the aggressive harvest pattern, and if so, for how long?

The outcomes from rapid local liquidation of mature forests, such as took place in northern Michigan and Wisconsin in the late 19th century[4] have been summarized by many authors. The expectation of a transition to sustainable local farm communities in the forest areas after the liquidation logging did not materialize, for reasons of poor (or unstable) soils and short growing seasons. Ghost towns and family hardships prevailed in the logged-out areas. Even small port cities on Lake Superior, such as Ashland, Wisconsin, declined from a peak population of nearly 15,000 people to 9,000 after the forests had been removed.[5] Twining cites a source as follows:[5]

The sack of the largest and wealthiest of medieval cities could have been but a bagatelle compared with the sack of the North American forest and no medieval ravisher could have been more fierce and unscrupulous than the lumberman. His lust of power and wealth have changed the face of the country, built cities and railroads and created a sort of civilization—as impermanent as the material with which it was concerned, wood, but out of destruction perhaps bringing forth certain fruits of abiding worth.

Today, the risks of non-sustainable development are similar. We still seek to meet deficits owing to inadequate timber supplies in one area with temporary excesses in another area (e.g., compare the growth to removals ratio of the Pacific Coast vs. the North, Table 4.1). Any regional deficit, however, as was induced in the North from logging 100 years ago, leads later to a loss of jobs and deficiency in local timber output and related secondary manufacturing that cannot be made up by regional imports. Sustainable development nationally and internationally, therefore, requires policies that do not lead to future risks of impaired annual growth, but which, instead, encourage the gradual recovery of all degraded forest systems.

4.2.3 Solutions: Steps toward Sustainability

Scientific forestry has operated for decades in Europe to assure sustained yields in forest harvests, but without comparable success in most of North America. Many problems arise: Sometimes fires or insect outbreaks occur, reducing the growing stock below expectations; at other times, community or other political decisions have led to reducing the growing stock below its most productive baseline; and in still other areas industrial pollutants have reduced productivity below expectations (as in the mountains of the U.S. northeast and in Central Europe during the 1980s). This short list of problems illustrates, however, the difficulty in considering all the major influences, while seeking to project sustainable forest harvests 50, 100 or 150 years into the future. Many authorities in the field are now using "adaptive ecosystem management," as outlined in Box 4.2, to describe the comprehensive approach needed for modern industrial forestry on either publicly owned or private timberlands.[6] According to Maser,[7] adaptive ecosystem management is, at a minimum, a continuing cycle of action, based on studying, planning, monitoring, evaluating, and adjusting. The adaptive part of ecosystem management focuses on using current scientific and societal information to plan, act, and accumulate new information as an integral part of management, using it to improve future decisions. The adaptive concept underscores that knowledge and societal values are changing ever more rapidly, and that management must keep abreast of these changes.

Adaptive ecosystem management, therefore, is designed as a two-path process: one part explicitly promotes rapid, comprehensive learning by scientists, managers, society, and policy makers; the other path promotes rapid change in management in response to the learning. These parallel processes are central to managing for the sustainability of timber supplies. Sustainability in the face of changing demands and values is the ultimate goal of adaptive ecosystem management, as opposed to traditional scientific forestry. What people and industry want over the long run, and what they can actually have depends on what is ecologically possible long term, within the limits of productivity of the system itself.

These goals must be set by society because successful implementation of management—adaptive, sustainable, or otherwise—depends on societal approval.[6] Broad management goals and concise objectives can be seen in the directives given by President Clinton at the 1992 Forest Conference in Portland, Oregon:

1. Protect the long-term health of forests, wildlife, and waterways.
2. Be scientifically sound, ecologically credible, and legally defensible.
3. Plans must seek sustainability and predictability.
4. Always remember the human and economic dimensions.
5. Make the federal government work together for you.

Box 4.2 Ecosystem Management Example

In 1994, the General Accounting Office (GAO) of the U.S. Congress[6] released a report analyzing the approaches available to achieve "ecosystem management" as articulated by the U.S. Forest Service and other land management agencies. The report focused on several areas of intense conflict over how to balance necessary use and judicious protection of forests and related resources. The report concludes:

> Neither the administration's fiscal year 1995 budget document nor the task force's draft "Ecosystem Management Initiative Overview" clearly identifies the priority to be given to the health of ecosystems relative to human activities when the two conflict. Definitions developed by BLM, FWS, and others leave no doubt that greater priority will have to be given to maintaining or restoring a minimum level of ecosystem integrity and functioning over non-sustainable commodity production and other uses. The practical starting point for ecosystem management will have to be to maintain or to restore the minimum level of ecosystem health necessary to meet existing legal requirements.

The report goes on to say that implementing ecosystem management will require taking practical steps that clearly identify what must be done and which parties must be involved. They suggest these steps:

1. Delineating, on the basis of reasonable ecological and management criteria, the boundaries of the geographic areas to be managed as ecosystems.
2. Understanding their ecologies (including their current conditions and trends, the minimum level of integrity and functioning needed to maintain or restore their health, and the effects of human activities on them).
3. Making management choices about desired future ecological conditions, about the types, levels, and mixes of activities that can be sustained, and about the distribution of activities over time among the various land units within the ecosystems.

Despite enactment of numerous laws to protect individual natural resources, the report concludes that ecological conditions have declined on many federal lands. For example, a federal interagency team found "many forests in the Pacific Northwest have become so damaged by timber harvesting that species are disappearing and many streams no longer provide adequate habitat for fish." Similarly, the GAO report cites a BLM study showing that sedimentation in streams has increased; rangelands have become less productive; plant, animal, and fish habitats have been damaged; the health of forests has declined; and the range and numbers of many native flora and fauna have decreased. Numerous other reports indicate that such problems are neither isolated nor diminishing.

The diminishment of resources, the opposite of sustainability, requires that federal land managers reconsider their levels of timber harvests, livestock grazing, and other uses on some land units or risk complete collapse of the resource, as seen in other areas and in other countries. The reductions, in turn, have had some adverse economic and social effects on some nearby communities whose economies are dependent on uses associated with federal lands. Other communities also have been adversely affected because they depend on commodities—such as the Pacific salmon—whose stocks have been reduced by declining ecological conditions on federal lands.

The GAO report notes that concern over declining conditions and reduced commodity production on federal lands has led to an increasing number of judicial challenges to federal land managers' decisions by environmental as well as industrial organizations. As a result, a growing number of agency officials, scientists, and resource policy analysts believe that the new, broader approach—ecosystem management—will balance timber needs and long-term productivity interests. They believe that maintaining or restoring ecosystems, rather than managing by legislative directives, would better address declining conditions and ensure the sustainable, long-term availability of forest resources.

Thus, ecosystem management does not necessarily alter the federal land management agencies' basic legislative mandates—sustaining multiple uses of federal lands and protecting natural resources. The extent to which ecosystems receive protection above the minimum levels necessary to maintain or restore their integrity will depend on public policy decisions involving trade-offs among ecological and socio-economic considerations. In reaching these decisions, however, scientists, corporations and policy makers all need to understand the technical and socio-economic relationships within the system (as we discussed in Chapter 1), and their variability from place to place and over time. Sustainability can be measured, and winners and losers identified, but the outcomes are quite different depending on scale (see Scale and Technology Principle, Chapter 3), and whether the balances are local and short-term, or national and long-term.

4.3 FISHERIES

4.3.1 THE PROBLEM

Much of the world has benefited from locally rich fisheries, and from early times humans have taken advantage of ocean resources. Like other common-property goods, these fisheries do not belong to any one person or company, but to everyone, and governments generally have not been able to regulate the level of harvest of fish. There are millions of tons of fish in the oceans, but it is difficult to enforce a limit to the millions of tons that can be removed annually and still maintain the growth and reproductive capacity of desirable species.

Measurement and prediction of the stocks available is difficult because of the large areas and natural variation in ocean systems. Methods for measuring the current inventory of fish in regions of the ocean are expensive and hard to analyze. As summarized in Box 4.3, each regional resource requires data every year, and when industry is allowed to "experiment" on each system, the "optimum levels of exploitation can only be determined by trial and error".[8]

Stakeholders in a common resource such as individual local fisheries must look at the systems in terms of their annual production, seeking an annual fish harvest that corresponds with the rate at

---

Box 4.3   Responses to Claims of Sustainability Drawn from Experience with Fisheries and Forests

Although past efforts to provide sustainable yields in both forestry and fisheries have been unsuccessful, important lessons have been learned and are being incorporated into current management approaches. Ludwig, et al.[8] suggest the adoption of five principles as an aid in determining resource use practices that promote sustainability and avoid the *tragedy of common property assets:*

1. *"Include human motivation and responses as part of the system to be studied and managed";* humans generate the need for natural resource use, and without understanding their motivations, establishing sustained yields is virtually impossible.
2. *"Act before scientific consensus is achieved";* complete scientific consensus in the face of diverse measurements and variable data is difficult to achieve, but waiting for consensus before acting endangers the resource. Risks are minimized by early, prudent action.
3. *"Rely on scientists to recognize problems, but not to remedy them";* resolving sustainability issues requires the collaboration of many disciplines, including business, economics, and political leaders, and, therefore, technological answers alone do not meet the need for consensus on long-term measures.
4. *"Distrust claims of sustainability";* sustained yields have not been accomplished in the past apparently because the systems are complex and human factors are ignored, so new approaches need to address fully the underlying demographic, social, and political issues.
5. *"Confront uncertainty";* plans to harvest only the productivity of an ecological system must incorporate the scientific uncertainties associated with variable natural processes, as well as the uncertainty of political and social systems.

which the species can replenish themselves. Overfishing means taking more of a particular fish out of the ocean than its population can replace naturally, thereby reducing the stocks and their capacity to grow, and affecting the food base of other species in the food chain. Overfishing breaks an otherwise closed loop in a self-sustaining system. Although fisheries also can be affected by degradation of oxygen in the water, sediment deposition, excessive nutrients in runoff, and pesticides or other persistent chemicals in the food chain, overfishing is recognized as the most serious problem facing this resource and its dependent industries.

To state the problem simply, there are too many vessels chasing too few fish. Fishing stocks have collapsed world-wide, and it is unclear now whether the complete bans imposed on fishing in several areas can bring about regrowth. The Food and Agriculture Organization[9] (FAO) reported in 1989 that world-wide catches had grown to 86 million tons per year. They attributed this growth to increases in human population, an increase in demand for protein from the ocean, expanded markets, and the development of more sophisticated technology to detect and catch fish (e.g., underwater radar). There are no data, however, to indicate what is the world-wide annual growth of fish, and as seen in the forestry example just described, the most important determinant of annual growth is the existing stock on which growth will take place.

The process of fish stock depletion is largely as follows: Each country begins by overfishing their own coastal waters and exhausting their own stocks of the most desirable species. Then fishermen begin a movement outward into the open ocean with new technologies to catch smaller fish scattered over larger geographic areas. To make up for short-falls in their catch most countries have extended their territorial waters out 200 miles into the ocean, so as to control more of the coastal fisheries. In the open ocean beyond 200 miles conflicts arise between countries because, although the ocean fisheries are a "common asset," each country seeks to appropriate as much as possible of the asset, exploiting to the maximum. Fishermen with larger boats, or countries with fleets and advanced water radar technology, can do this quite successfully. With radar, even small schools of fish at a considerable distance can be detected and harvested, and the fleet moves on. Some of the ships are essentially floating factories, processing, refrigerating, or freezing the catch on board.

It is interesting to note, as well, that the fishing industry maintains considerable political weight, world-wide. As return-on-investment from fishing has declined, many countries have subsidized their fishing industries heavily. Because many fishing communities have no other form of employment, their governments are inclined to subsidize the industry, sponsoring loans for more boats to help the fishers catch the fewer and fewer fish that remain in the commons. All over the world, national fishing fleets have increased to a size that is unsustainable. The only reasons to keep the industry at this level are social, but the open ocean resource is being wiped out.

The depletion of fish stocks also causes a change in the survival patterns among other species of fish and their predators. The commercial species on Georges Bank off Maine and Massachusetts no longer function in their normal role in the ocean ecosystem. Instead, unusable species such as skate and dogfish are coming to dominate the system, potentially precluding recovery of the desirable species even with little or no ongoing fishing pressure. These changes in the ecological systems owing to scarcity of the primary resource have led to a situation where it may be too late to save the fishing community or induce sustainability in the resource.

4.3.2 Social Consequences

One important consequence of the above pattern of exploitation was the closing of the Canadian cod fishery on the Grand Banks off Newfoundland in 1993, costing the industry 42,000 jobs, and the Canadian government $1.8 billion in unemployment support. This fishery had been highly-valued internationally, economically, and ecologically. Unfortunately, now, instead of a sustainable harvest from these fisheries contributing to GDP in Canada, they have become a drain on the economy and society.

Then followed the closing of a major portion of New England's fisheries in 1994, costing 14,000 jobs, $350 million/year in business incomes lost, and millions in unemployment payments granted by the government. Some in-shore fishing continues, but the goal is to reduce the open-ocean catch of the most sought-after species to as close to zero as possible. Even as recent regulations reduced the allowable catch to 50 percent of prior totals, the problem worsened.[10] In March and April, 1995, the Grand Banks fisheries beyond the 200-mile territorial limit of Canada became the site of "gunboat diplomacy" between Canada and Spain with armed boarding of Spanish fishing vessels that were found to be in violation of agreed-upon international equipment limitations. Over several weeks, leaders of the European Union negotiated a new sharing of the resource and a strict on-ship procedure for enforcing the new regulations.

4.3.3 How to Achieve Sustainable Development

The dramatic decrease in cod, haddock, and herring stocks of the Northeast U.S. coast during the 1980s indicated then that there was an urgent need for conservation, despite the risk of job loss and

---

Box 4.4 North Pacific Fish Management Example

The North Pacific Fishery Management Council attempted to manage the northwest fishing industry with an "open to all at no fee" policy for many years. This policy led to the introduction of 65 factory trawlers, catching 35,000 lbs each, up from 12,000 lbs in earlier ships. New equipment was introduced with the expectation that fishing at this level could be continued. Unfortunately, it could not, and many of the individuals involved are now on the verge of bankruptcy, some fishing only 5 out of 10 fishing months, and others fishing for only a few days. By the time there were 5000 boats, the fishing season was reduced to an incredible 2 days per year. The type of fishing practice also changed as the fishing season shortened; it became a faster process, more wasteful of other species in the catch, and resulted in more spoiled fish. The conclusion now being reached is that, *for this industry to survive, 95 percent of the boats will have to leave the industry*. As a result of the patterns summarized here, U.S. fishing in the northwest has declined 60 percent from its previous total.

Figure 4.1 shows data on trends in value of harvest by the fishing industry off western Canada from 1935 to 1990. During the middle part of this period, the effects of harvesting beyond sustainable levels began to be expressed (in the late 1960s). However, new management was introduced, focused on sustainable yields, and Figure 4.1 shows the beneficial effect for this fish species. The results could be extrapolated to fisheries in other parts of the world.

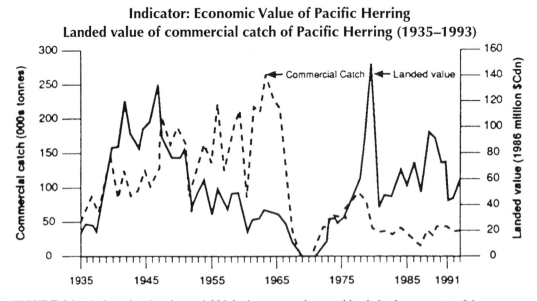

FIGURE 4.1 A chart showing the partial biologic recovery, but good landed value recovery of the Canadian Pacific herring following 20 years of restoration.[12]

damage to communities that relied on fishing. A sustainable harvest strategy had to be found. Scientists needed to estimate fish stocks accurately in the face of fishing pressure, and managers and governments needed to adopt long-term regulations that reflected these forecasts.

The Magnuson Fisheries Management and Conservation Act of 1976 had given authority to the U.S. government to claim the area out to 200 miles from shore and manage its fisheries. The objectives of the act were to eliminate foreign fishermen from U.S. waters, and to restore and conserve fish. The first of these objectives has been achieved, but the second has not, because of badly conceived and poorly implemented programs and policies.[11] One conclusion is that the Act failed in its intended goals (of conservation) owing to the requirement of having the people being regulated (the fishermen) make up a majority of the governing "councils" created under the act. The situation has been worsened by government *guarantees of fish harvests and investment incentives, overcapitalizing, and, consequently, overfishing the resource.* Members of the various stakeholder-based agencies and councils that govern the fishing industry have been heavily criticized for poor use of science and for serious conflicts of interest.

The striped bass is a once abundant fishery of eastern in-shore coastal areas and rivers, from the southeast to the northeast United States. Its recovery from a catastrophic decline in the 1980s is an example of a fishery management success that gives hope to sustainable fishing. The management plan was quantitative and mandated by state as well as federal legislation. It ensured that local fish harvests either followed the plan or were shut down. The program was well-defined and stringent, and it worked. The striped bass population has returned over much of its range to levels that are better than were predicted, but still short of its former abundance.

Another step toward sustained fisheries is evident in the fishing industry's recognition and acceptance of the root problem: over-exploitation. It is probably too late for many of the individuals involved and others who depend on fisheries, but not for the remaining industry as a whole. To correct the mistakes, the fishing industry will have to exist under constraints for some years, not only to allow for the fish stock to recover, but for the ecosystem that supported the fish historically to recover a sustainable equilibrium.

4.4 SUSTAINING FRESH WATER

A world map shows, clearly, much more water available than land. However, 97 percent of the earth's water is in the oceans: salt water that cannot be used as drinking water or for irrigation. The remaining 3 percent is fresh water, but much of this water is polluted with high sediment loads, is too far underground to be extracted economically, or is locked up in polar ice caps. What is reassuring, however, is that fresh water is continually renewed, purified, and distributed over the continents by the natural hydrologic (water) cycle. The freshwater system has the potential to keep water supplies clean and replenished, provided that our economic activities do not withdraw or pollute the supplies faster than the cycle can replenish it.

Freshwater supplies can be separated into two major components, surface water (including lakes and streams) and groundwater. Surface water includes all rainfall and snow that does not immediately infiltrate the ground or evaporate back into the atmosphere. Thus, it includes temporary ponds, wetlands, rivers, and reservoirs. The water that infiltrates the soil and deep rock cavities is called groundwater. Nearly half of our drinking water supplies come from deep groundwater.

To use fresh water sustainably, human uses should be constrained to remove water only at the rate it is being replenished, and, depending on the type and quantity of pollutants, discharges of wastes should be limited to those that can be purified by the system itself over short time-spans. The key is in maintaining a system at a strong, positive rate of renewal, while limiting activities that degrade water quality for downstream uses. What we use water for is not as critical as ensuring that it be recycled or renewed. Thus, we should see both fresh water and its renewability as a public good that is not owned by any one individual, although individuals or water companies own treatment and distribution facilities. Therefore, restricting the use of water by individuals and industry is difficult, but restricting the privilege of discharging wastes and degrading water is more generally agreed upon.

4.4.1 Use and Abuse of Water Resources

Societal use of water can be measured in terms of withdrawals, human consumption, and volumes of wastes discharged into water. Withdrawal refers to taking water out of the surface or groundwater supply and transporting it to the place of use. Much of the withdrawal volume is used for irrigation or industrial cooling water, and is returned after some evaporation loss. Consumption refers to water that is withdrawn and then is not available for immediate reuse at the site, partly owing to evaporation, but also owing to the discharge of wastes as with urban sewage collection systems (Figure 4.2). Globally 70 percent of the water withdrawn each year is used for irrigation and lawn watering. It is important also to recognize the high concentrations of natural salts in the remaining water after irrigation. The remaining 30 percent of water withdrawn globally is used for industrial processes, residential use, and businesses.

For various reasons, however, our need to grow low cost food products has been allowed to override our capacity to sustain freshwater supplies in many low rainfall areas. Water development has been subsidized to the point where there is no incentive at all to find ways to reduce the volume of water used. The cost of water to such users in many parts of the world is much lower than its actual cost to society, a significant non-market outcome. With no incentive to reduce usage, the 70 percent of water going into irrigation could still increase over time, despite nearly two-thirds of this water

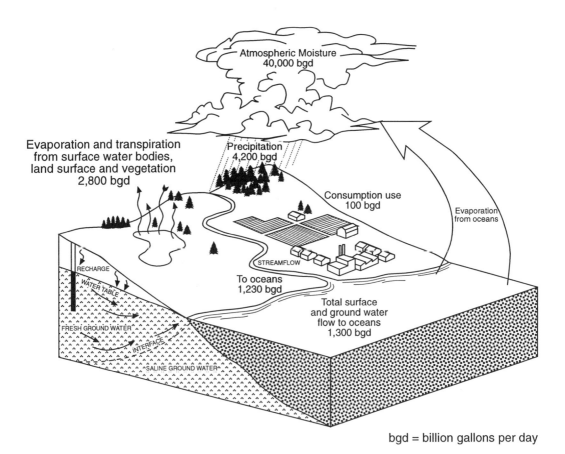

FIGURE 4.2 Hydrologic cycle showing the gross water budget of the conterminous United States.[13]

being wasted by unnecessary evaporation. Various methods have been developed to reduce the quantity of water needed to grow crops, and some corporate farms are adapting to these methods.

Other industries that use large quantities of water also could reduce usage, depending on the technologies available for their processes. Some processes already provide for recycling of the water they use. In these cases, the need for clean, pure water is not always paramount, as it is for food, beverages, and pharmaceuticals. Depending on the processes involved, industry needs for water as a solvent and as a conveyance of waste also can be changed. Thus, it is increasingly important that industries search for new technologies that reduce their water needs and ensure long-term economic viability for themselves, the resources, and other users.

Water pollution is becoming the most widespread problem in water resource management (see Box 4.5). According to the World Health Organization, 1.5 billion people globally do not have safe drinking water. This is the direct result of water being returned into the cycle in a contaminated state exceeding the purification capabilities of the system. Before the aquatic system has the chance to purify the water naturally, it often is withdrawn for use again. This process of over-use and contamination has been compounded to the degree that many public uses of the water cycle are no longer possible.

Various types of water pollution also need to be distinguished. One main type can be thought of as biological and therefore, health-related: bacteria, viruses, protozoa, parasites, organic (biological) oxygen demand; another is inorganic (phosphorus, nitrogen acids, salts, lead, and mercury); and still others include radioactive wastes (water soluble radioisotopes) and heat (water used in industrial cooling processes). The pollutants can come from specific discharges, point sources, or from area-wide non-point sources. Point source pollution is pollution produced at a location identifiable as a factory or sewage treatment plant. Non-point source pollution usually comes from large areas (agricultural or urban areas). Groundwater pollution is even more complex and difficult to detect or prevent. It is not visible from above ground, and once the groundwater has been contaminated, it is very hard to clean.

Thus, water pollution, and sustainable uses of water for waste disposal, are among the most serious problems being faced by business and industry now. Damage to resources from surface water being used as a medium for disposal is now well understood by the public, and there is pressure to change these practices. The problem is, to what degree? Most industries see the erosion of their option to discharge wastes into water (which the general public does every day) as an added cost that has no immediate commercial benefit. Water is a public good and using it for disposal is a major economic advantage. Also, those who pollute water are not always the ones who have to pay for its clean up. Business and industry are being asked to internalize the costs of modifying their production

---

Box 4.5 Global Water Supply Issues

Around the world, as in the United States, there also are large-scale problems of water shortage (drought) and floods. Both of these problems are linked to unsustainable land use practices such as deforestation, overgrazing, and many aspects of development. These practices lead to reduction in the natural capacity of the land to hold or absorb water, thereby contributing to flooding at some times during a decade, and inadequate continuity of water supplies at other times.

In several important coastal regions of the world such as the highly populated Mediterranean area, groundwater supplies also are degraded by saltwater intrusion after overdevelopment. This is a result of pumping groundwater from sources near the sea faster than these sources can be recharged with fresh water from land areas. Such excessive use represents a break in the hydrologic system that can also cause subsidence (lowering of the land surface) from the reduction of water volume in deep strata. These problems of irregular supply and unsustainable water use are expensive to remedy after they occur, but can be avoided by keeping the hydrologic system relatively intact.

processes to reduce water pollution, thereby conveying a different benefit to their consumers. Technological innovation, however, cannot only reduce pollution but also improve relations between industry and local communities, benefiting everyone's competitiveness over time.

4.4.2 Implementing Sustainable Water Management

In the United States, regulations such as those deriving from The Clean Water Act of 1977 seek to deal with both point and non-point pollution. There are two options:

1. Not releasing pollutants faster than natural systems can purify or bury them.
2. Removing the pollutants first to a level such that the remainder can be absorbed into the natural cycle.

The first is rarely feasible because it is nearly impossible to limit community or industrial pollutant discharges directly within levels nature can purify. A waste treatment industry has been created instead, installing purification systems that help natural processes keep up with the rate of waste generation.

Other approaches to water resource management have been developed, however, to increase available freshwater supplies and prevent pollution. Water is sometimes diverted from one watershed with surpluses to another where there may be shortages. For example, some of the flow from the Colorado River is diverted at Parker Dam to supply the Phoenix region, a distance of 248 miles horizontally. Other water from the Colorado is diverted to the Los Angeles basin and other parts of southern California. Reclamation of contaminated water, as well as desalinization, while expensive, can make a locally important contribution to increased water supplies. Water reclamation sometimes allows the use of dirty water over and over again instead of returning it to the source. Desalinization is the removal of salt from seawater or high-salt rivers so that it may be utilized as fresh water.

Sustainable water management, as with sustainable fisheries and forests, requires comprehensive measurement of the system and mathematical models to predict the total quantity and quality of water in the system.[14] These models are used to understand the dynamics of water quality and quantity (considering both discharges and renewal), and to provide other information needed for effective planning and management by an industrial society. Most importantly, computer models can find the most equitable solutions among the competing water uses in a very complex system.

The models are mathematical representations of both the components of the hydrologic system and its human interventions. For example, the building of a dam will increase the size of a reservoir above the dam, but it also increases the evaporation, *thereby decreasing the water available to flow beyond the dam.* The dam may increase the number of fish in the reservoir compared to the river, but it also reduces the level of water in the riverine wetlands downstream, thereby reducing habitat for other species including fish. A model of the hydrologic outcomes from an intervention such as the dam is a series of equations describing the status of the water system each month of the year, and expressing the relationships among all the factors involved. A good model also can be used to evaluate the conflicts among different objectives in water use, to ensure adequate, long-term water supplies, to provide a buffer against drought and floods, and to create options for pollution abatement. Thus, water management models are a key tool in achieving sustainable development of water resources.

4.5 AIR RESOURCES

Air as a resource often is taken for granted because it is not seen, smelled, tasted, or touched by any of us. Still, we are surrounded by it, and air is essential for nearly all forms of life. Air, like water, is used frequently as a vehicle for disposal of wastes. What is less frequently considered are the limitations (non-sustainability) of using air for waste disposal.

Natural Resource Conflicts And Sustainability

4.5.1 THE AIR PROBLEM

Like the hydrologic cycle, air also has its own cleansing and purifying processes, involving the scavenging of wastes by fog, rain, and snow, and the uptake of gases by vegetation. These natural cleansing processes can remove only certain amounts of pollutants within a specified time frame. When this "renewal capacity" is exceeded, the remainder of the pollutant load should be reduced in advance of discharge. Air pollutants remaining in the atmosphere for several days are a sign of some exceedance of the renewal capacity of the natural system. The goal for a sustainable society, therefore, should be to intercept the excess emissions, and "close the loop" to within the assimilative capacity of the air resource system.

Air pollution also can be defined as an atmospheric condition in which substances (gases, liquids, or solids) are present at concentrations high enough above normal levels to produce a measurable effect on visibility, vegetation, climate, materials, human health, or the condition of other animals. Among the most common pollutants are particles from automobile and truck exhaust, wind-blown soil, nitrogen oxides, carbon monoxide, carbon dioxide, and sulfur dioxide.

Air pollution, and the regeneration of clean air, therefore, can be expressed as a system composed of three basic parts:

The emission "sources" can be from trucks and automobiles, electric power generation, refuse burning, or industrial processes, as described in Box 4.6. Associated with these emission sources are

---

Box 4.6 Balancing Emissions and Renewal

The cleansing processes involved in air renewal (as noted previously) mostly occur during transport over long distances. Also occurring is chemical transformation into alternate pollutant forms. Cleansing responses by plant and animals, either of the original pollutant or its form in rainfall and soils, as well as wet washout, can result in many other modifications of natural resource systems (Figure 4.3).

The dispersal of pollutants released to the air often depends on the movement of the air both horizontally and vertically. The normal dispersal of air depends on temperature patterns, with cool air normal at higher elevations and warm air at lower elevations. Dispersal usually prevails when there is sufficient vertical mixing in the atmosphere, but it is limited when warmer air overlies cooler air. This inverse stratification is called a temperature inversion, a situation that keeps ground layer emissions trapped near plants, animals, and vegetation. It is during inversions that the most serious air pollution problems occur.

Sunlight and atmospheric humidity also are important factors in understanding air pollution risks. Sunlight, reacting with chemicals in the emissions, contributes to reactions that result in smog and ozone formation. Atmospheric humidity contributes to air pollution when water and sulfur oxides combine to form sulfur compounds such as sulfuric acid, which then can be is scavenged in precipitation and removed as acid rain. Small quantities of certain air pollutants, however, can originate from natural sources. Included here are smoke from forest fires, hydrocarbons from coniferous trees and shrubs, dust, volcanic eruptions, and pollen.

Societal sources of air pollution can be stationery or mobile. Stationary sources are site specific, such as the smoke stacks of large electric generating facilities or residential stack emissions. Mobile sources are not site specific, and include automobiles, trucks, trains, and airplanes. Acid in rain and loss of visibility are just two of the problems resulting from these pollutants. The consequences of acid washout in rain include effects on visibility, roofing, wires, instruments, soils, forests, and aquatic life.

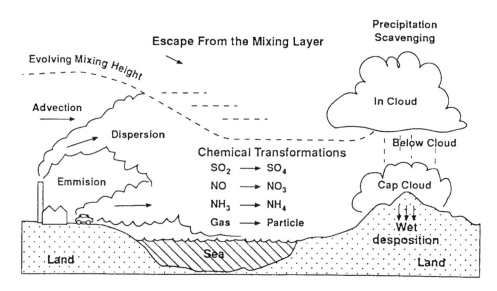

FIGURE 4.3 Processes involved in the transport and deposition of atmospheric pollutants.[15]

potential source controls, devices that prevent some of the pollutants from reaching the atmosphere at some expense, of course.

Strict regulation of industrial emissions seems unfair when one considers the large quantity of emissions released from personal autos and trucks, and the proximity of these emissions to residential areas. Every source represents a small or large departure from relatively "closed loop" relationships between uptake and release, the central idea of sustainability. Accordingly, some electrical power generators have committed to "biomass accumulation reserves," to store carbon dioxide at a remote site and balance the CO_2 released from their fossil-fuel based power stations with partial removal. Many industries do not agree with such proposals, and feel they should not be held directly responsible for maintaining such highly balanced air systems.

Thus, we see that gaseous waste emissions have many externalities associated with them, but at the same time everyone needs to use the air commons for some level of disposal. On the other hand, few can take responsibility and pay for reducing emissions. Also, some of those who do pay for damages often do not acquire any of the benefits. The general public, including plant and animal resources owned by private individuals, suffer from the externalities of air pollution. Eventually, public pressure on industry may lead them to accept responsibility and close the loop for air systems renewal.

4.5.2 Approaches for Solving Air Problems

There are three points in the pollution-air renewal system where control measures can be implemented: at the source of the emissions (interception and reduction), in the atmosphere (using wind flow patterns and/or washout to alter retention in the nearby airsheds), or at the receiving user site (receptor), where air filtration systems can be added to the air intake of buildings. Of these three, control at the emission source is the only feasible and practical option on a large scale. Basically, the best way to maintain air quality and sustain air renewal capacity is to prevent much of the pollution from entering the atmosphere in the first place.

Several approaches have been taken in the search for solutions to air emissions that exceed removal capacity, nearly all involving external regulation (command and control). In this approach, strict blanket regulations are set for all major emissions sources in order to main maintain overall competitiveness on a "level playing field." The U.S. Environmental Protection Agency (EPA) has agreed to accept a "bubble approach" for multiple stationary sources in large industrial complexes, however. This approach views the entire industrial area as a point source instead of regulating each smoke stack separately. In a bubble approach, each stack can emit more or less pollutants than a strict reading of regulations for that stack might allow, as long as the total emissions for the plant do not exceed a set regulated amount. This allows large plants to average their emissions from all stacks and make internal decisions about how to reduce overall emissions in a cost-effective way.

Another approach utilizes pollution permit trading. The purpose of trading in pollution allowances is to give business more options about when and where to reduce emissions, and to expand the economic incentives for reducing long-term pollution discharges. By giving companies pollution permits that are tradable on the market, some companies will make internal decisions to reduce now and sell their advantage (in reduced emissions) to other companies who need the pollution allowance for a few years until they can achieve their own emission reductions.

The Clean Air Act legislation of 1970 and its subsequent amendments has had many positive effects. Particulate levels in urban areas have decreased by 25 percent, sulfur dioxide emissions have been decreased by 40 percent in some areas, nitrogen oxides have decreased by 14 percent, ozone has decreased (in some cities) by 10 percent, and carbon monoxide has decreased by 30 percent.[15] However, these reductions have been the result of regulations that were costly to business and sometimes seemed inefficient.

A more sensible approach to pollution regulation may be achievable through industry commitments to pollution prevention. Businesses already are creating their own internal pollution prevention programs, allowing broad options on how to reduce pollution and reduce costs. Business can use new technology to modify processes internally, or to develop industry-specific approaches that reduce pollution at a modest cost while creating a competitive edge domestically and globally. With increasing public awareness, industry has found it difficult to ignore major stakeholder issues such as air pollution. In the long run, business seeks not only increased economic viability and efficiency, but also an overall advantage in the marketplace.

4.6 DISCUSSION

These four reviews of renewable resource systems, each used by industries in quite different ways, all illustrate resources as public goods or commons, and thus, the need for some public intervention to avoid over-exploitation of these public goods. In forests and fisheries, mechanisms leading to sustainable development are just now being put in place, but for water and air, substantial accomplishments already are in place.

For example, in 1985 alone, $73.8 billion was spent on air pollution abatement and control, despite the evidence that many people were still debating the value of air pollution abatement. While it is unrealistic to speak of having no air pollution at all, it is now reasonable to expect to see reductions in observable adverse effects from air pollutants. In light of the great financial expenditures required to attain these reductions, the political and social components of business systems must be recognized as having played a large role in meeting the goals.

Specific energy industries such as the electric power companies can be seen as emerging models. The New England Electric System is one company that has promoted eco-efficient residential behavior, using changes in its incentives framework and competitive pricing to make energy conservation more profitable than going out and building new power plants. They have plans in place to reduce emissions by 45 percent by the year 2000.[17] Other companies such as Exxon discovered that installing floating roofs on their fuel tanks could save 680,000 lbs per year of chemical emissions, worth about $200,000. (The lids cost only $5,000–$13,000 each). Riker labs, a pharmaceutical company, found that they could switch from a solvent-based tablet coating to a water-based coating and reduce air pollution by 24 tons per year, saving $18,000.[16] These are only brief illustrations of the very extensive changes now underway in the relationships between resources and businesses.

4.7 REFERENCES

1. Behre, C. E., Today and tomorrow: forest land and timber resources, in *Trees: A Yearbook of Agriculture,* U.S. Department of Agriculture, Washington, D.C., 1949.
2. World Resources Institute, *World Resources 1994–95: A Guide to the Global Environment,* Oxford University Press, New York, 1994.
3. USDA Forest Service, *The Nation's Renewable Resources—An Assessment, 1975,* U.S. Department of Agriculture, Washington, D.C., 1977.
4. Cronon, W., *Nature's Metropolis: Chicago and the Great West,* W. W. Norton, New York, 1991.
5. Twining, C. E., The lumbering frontier, in *The Great Lakes Forest: An Environmental and Social History,* Flader, S. L., Ed., University of Minnesota, Minneapolis, MN, 1983, 121.
6. General Accounting Office, *Ecosystem Management: Additional Actions Needed to Adequately Test a Promising Approach,* Report to U.S. Congress, Washington, D.C., 1994.
7. Maser, C., *Sustainable Forestry: Philosophy, Science, and Economics,* St. Lucie Press, Delray Beach, FL., 1994.
8. Ludwig, D., Hilborn, R., and Walters, C., Uncertainty, resource exploitation, and conservation: lessons from history, *Science,* 260, 17 and 36, 1993.
9. UN Food and Agriculture Organization, *Yearbook of Fishery Statistics: Catches and Landings,* United Nations, Rome, 1989.
10. Cushman, J., Fisheries: north east council recommends "virtual" shutdown, *New York Times,* October 27, 1994.
11. Safina, C., Collapse of fisheries in the northwest U.S.: an interview with Carl Safina, *Environ. Rev.,* 1(11), 8, 1994.
12. Environment Canada, Sustaining marine resources: Pacific herring fish stocks, State of the Environment Bulletin No. 94-5 (Canada), 1994.
13. United States Geological Survey, *National Water Summary 1993—Hydrologic Events and Issues,* Water-Supply Paper 2250, U.S. Government Printing Office, Washington, D.C. 1994.
14. Office of Technology Assessment, *Use of Models for Water Resource Management Planning and Policy,* Congress of the United States, Washington, D.C., 1982.
15. Barker, J., and Tingey, D. *Air Pollution Effects on Biodiversity,* Van Nostrand Reinhold, New York, 1992.
16. Council on Environmental Quality, *Environmental Quality, Twentieth Annual Report,* U.S. Government Printing Office, Washington, D.C., 1990.
17. Smart, B., *Beyond Compliance: A New Industry View of the Environment,* World Resources Institute, Washington, D.C., 1992.

5 Environmental Ethics and Corporate Decision Making for Sustainable Performance

Jan Willem Bol, Pamela C. Johnson, and James M. Childs, Jr.

CONTENTS

5.1 Introduction 81
5.2 Importance of Ethics in Business: An Example 82
    5.2.1 Principle 10 83
5.3 Steps in Corporate Responses to Environmental and
    Ethical Issues 84
5.4 Sustainable Development: The Nexus of Environment,
    Economy, and Equity 86
5.5 Comparing Values from Three Worldviews 89
    5.5.1 Principle 11 91
    5.5.2 Anthropocentrism, The Dominant Social Worldview 94
    5.5.3 Deep Ecology as a Worldview 94
    5.5.4 Sustainable Development, an Emerging Worldview 96
5.6 Integrity of Natural Systems: A Global Ethic for the
    Environment and Business 97
    5.6.1 Holism 99
    5.6.2 Scale 99
    5.6.3 Compatibility. 99
    5.6.4 Responsibility 100
    5.6.5 Commitment 100
    5.6.6 Stewardship 100
5.7 The Need for Dialogue 101
    5.7.1 Principle 12 101
5.8 Conclusion 103
5.9 References 103

5.1 INTRODUCTION

Up to this point we have been concerned with sustainability issues from the perspective, primarily, of natural scientists and economists without providing explicit consideration of how corporations behave. The purpose of this chapter is to look at environmental ethics as a framework for understanding corporate environmental responsibility. We will consider many aspects of sustainability as core values. We focus on a specific corporate issue, namely *ethical and value dimensions of working in and with the environment, and the implications of these values for corporate decision-making.*

Because the values of decision-makers within the corporation have major implications for business and the environment, we need to understand how different values and ethical orientations influence corporate strategy. We will find that a focus on the need for corporate survival (often at the expense of nature) can define the types of technology and products that a corporation develops. Similarly, an ethic that recognizes the intrinsic value of nature will influence an organization's perspective about its role in society and, hence, ways of interacting with its environment. This chapter argues that, unless we consider the ethical dimensions of corporate decision-making, we will not have defined or evaluated fully the sustainability of a corporation's performance. Then we explore the proposition that without the foundations evident in a global integrity ethic, which we explain, one does not fully frame the ethics of personal or corporate decisions.

The chapter is structured as follows: It begins with an overview of different ways businesses have responded to ethical and related environmental issues. The overview is followed by a discussion of ethical questions that are central to sustainable development, and the way in which economic, environmental, and equity considerations come together in it. That discussion also considers whether sustainable development is sufficient as an ethical premise for strategic decisions in corporations or in policy decisions by governments and we compare the value orientation of three worldviews: sustainable development, deep ecology, and anthropocentrism. Integrity then is proposed as a new "categorical imperative" to ground sustainable development morally and as an environmental ethic. The chapter concludes with a proposal that dialogue between leaders in business, science, philosophy, public policy, environmental advocacy, and religion can be the basis for establishing the authority for corporate responses to current and future environmental challenges.

Cross-cutting horizontally through this structure, we develop more fully the interconnections among the principles of sustainability introduced in Chapter 1. This chapter focuses more specifically on the principles of Ethics and Equity (Principle 10), Stakeholder Interests (Principle 11), and Information, Dialogue, and Expectations (Principle 12). Central to the discussion, however, are principles developed in other chapters: Principle 4, the Integrity of Individuals, Communities, and Nature; Principle 5, Quantifying Externalities; Principle 1, the Generality of a Systems View; and Principle 7, Valuation. All of these concern amenities that humans derive from nature, and all are themes to be woven into a unified whole.

5.2   IMPORTANCE OF ETHICS IN BUSINESS: AN EXAMPLE

In the early 1980s, Union Carbide (UC) manufactured patented pesticides in Bhopal, India. Having experienced a downturn in the market for pesticides, the plant in Bhopal had been allowed to run down; safety equipment was in disrepair, and the performance standards for some personnel had been eroded. Over the years, the Indian government had convinced a reluctant UC to make a number of technical and managerial changes at the Bhopal plant, with the result that by the time of the accident, the plant was run entirely by local people.[1]

When a leaking tank with gas was left alone by workers during a tea break (which was against stated corporate policies), the tank exploded and the gas escaped killing over 2,000 mostly poor local citizens and injured 10,000 more. In a statement issued shortly after the accident UC assumed moral responsibility for the disaster, even though it argued that the calamity was caused by a disgruntled worker. The Indian government, on the other hand, felt the disaster was owing to faulty operating procedures and designs. Although UC offered immediate assistance in addressing both human and environmental needs, the Indian government initially declined UC's help in order to control the investigation and clean up.

In spite of the obvious financial implications of the calamity, net profits for UC between 1986 to 1988 averaged $463 million per year, up from $237 million over the three years before the accident. Shortly after the accident, GAF Corp. attempted but failed to take over UC, and a number of directors left the company with "golden parachutes."

After four years of legal wrangling, the Indian Supreme Court ordered a settlement of $470 million. Litigation continued for years, but UC was never able to identify the alleged saboteur. In response to the disaster, in part, UC did enforce a company-wide policy that assumed total future responsibility for the safety of all its services, products, and supplies.

The Bhopal accident highlights many legal and moral dilemmas. With respect to the latter, the case identifies a number of important questions.[1] For example:

1. Should a company have cradle-to-grave and absolute responsibility for its technology?
2. Under what circumstances should a company be morally accountable for the acts and omissions of its subsidiaries?
3. Should or can a company accept moral responsibility beyond its legal responsibility in the event of a disaster?
4. What safety and environmental standards should apply to subsidiaries in foreign countries? Those of the parent country or the host country?
5. What is the company's moral obligation to respond to requests by foreign countries to make changes to its technology and/or procedures?
6. What are the moral responsibilities of the owners (shareholders) of the organization in case of a calamity?
7. Is it morally acceptable that management and/or the owners can benefit from a mass disaster?

Answering these questions is difficult for anyone, but the questions suggest ethics and moral values are an integral part of a corporation's responsibility for "closing the loop" (Principle 3) in information and ethics as well as in materials. When corporate managers perform an analysis of the value of natural resources or neighboring communities, they are incorporating their values about the appropriate role of human activity in nature and how the rightness or wrongness of their decision may be judged. The moral or philosophical foundations for such decisions are what we refer to as "ethics."

We can now consider an example of how business decisions can be integrated with the "ethical dimension." It is, in fact, related to the Bhopal tragedy, in that with active leadership by UC the entire U.S. chemical industry moved toward explicit means of taking responsibility and reducing risks. In 1988, the Chemical Manufacturers Association (CMA) adopted ten "Guiding Principles for Responsible Care®" (see chapter 7), and commitment to these principles has been made a condition of membership in the CMA.

The initiative of the CMA and other organizations described in the case studies later in this book suggest much progress has been made with respect to the level of awareness by corporate leadership about environmental issues. Notwithstanding this progress, there is arguably limited understanding of the changes in value orientations on the part of many consumers, which, in turn, underlie environmental issues. The result is that many corporations, having developed mostly reactive strategies to environmental dilemmas, continue to pursue economic opportunities in collision with the public perception of its environmental interests.

This example, then, introduces the significance of Principle 10, Ethics and Equity, a key element of sustainability being examined in this chapter.

5.2.1 PRINCIPLE 10

Sustainable systems include human appreciation of the intrinsic value of nature and the ethical positions assumed by business and its stakeholders. A sustainable system provides for consideration of equity among humans, locally and globally, as well as other species, and between present and future generations so that system integrity can be maintained.

5.3 STEPS IN CORPORATE RESPONSES TO ENVIRONMENTAL AND ETHICAL ISSUES

The Bhopal disaster describes only a few of the environmental pressures that bear upon a company, its employees, and its decisions. Fortunately, not all companies are faced with pressures of such magnitude. Other environmental pressures have been identified in earlier chapters. As we will show, the way these pressures are perceived and internalized shape much of a corporation's subsequent strategies and policies. We recognize the following four types of pressures on the corporation:

1. Regulatory pressures (federal, state, and local regulations pertaining to environmental safety and health).
2. Credibility pressures (including those coming from communities and the workforce; these generally concern failure to meet goals or relate to accidents).
3. Market pressures (such as consumer needs for green products; coming mostly from consumers and suppliers).
4. Financial pressures (such as those coming from the investor community).

The array of pressures from interest groups or stakeholders that influence the extent to which managers of an organization respond to environmental concerns are illustrated in Figure 5.1. In terms of the operations of the business, decisions to respond to environmental and ethical concerns influence the way in which resources are extracted and transported, the way in which technological processes transform raw resources into goods and services, and the way in which goods are marketed, packaged, and eventually put into the hands of consumers. At each stage, production and handling methods can be more or less environmentally friendly. While stakeholders play an important role in influencing corporate decisions related to the environment, not all companies have reached similar stages in their approaches and responsiveness to environmental pressures. An examination of the approaches employed by companies in response to these pressures reveals at least five stages that companies can move through.[3]

1. *The Beginner:* Company pursues no environmental strategies to reduce risk; little or no awareness of the need to do so; total lack of management support.
2. *The Firefighter:* Little funding and little protection against risk; work only on issues as they occur.
3. *The Concerned Corporate Citizen:* Moderate protection; the company expresses commitment, but it is not wholeheartedly implemented; no organizational change.

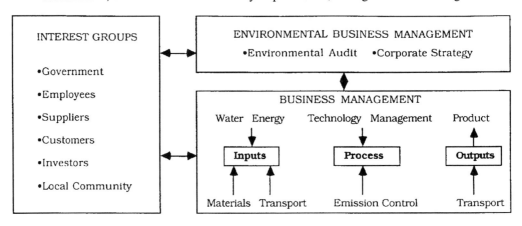

FIGURE 5.1 Environmental pressures on the firm, illustrating the interactions between interest groups (stakeholders) and corporate management.[2]

4. *The Pragmatist:* Comprehensive protection; active management of environmental protection functions; adequate funding, but policies still are not a top priority.
5. *The Proactivist:* Maximum risk reduction; environmental issues are top priority; company has moved beyond policing and prevention.

Analysis of how corporations have responded to environmental issues during the past 25 years reveals two distinct phases.[4] The first phase lasted roughly from 1970 to 1985, and was characterized by local compliance, as needed, without much involvement or commitment by top management. The absence of environmental measures of performance typically led companies merely to comply with regulations and, as illustrated by the Bhopal case, respond to calamities through crisis management (Beginner and Firefighter, above). In effect, to comply with the law was considered by companies during this time period as a cost of doing business. Few companies adopted a forward-looking, ethically responsive environmental management backed by strong corporate support, and directed toward more than compliance. The period can be characterized by an absence of willingness to internalize environmental issues.

During the second phase, which lasted from 1985 to the mid-1990s, many companies appeared to embrace environmental concerns, although they did not integrate them into their strategic business goals and policies (thus, Concerned Corporate Citizen). These companies developed more formal policy statements and clearly moved beyond mere compliance. Some even appointed a vice president for environmental affairs (Pragmatist). However, their primary motivation seems to be long-term survival of the company, with a focus on "reuse and recycle" rather than on developing new alternative products and technologies. During this second phase, however, three positive trends can be observed: First, environmental concerns are beginning to be institutionalized within firms. Secondly, many firms perceive environmental problems as their own problems to solve, and the search for environmentally friendly solutions has become part of the business operation at all levels. Finally, a significant number, but still a minority, of companies have begun to move beyond compliance-oriented approaches.

Considering the stages some companies have gone through in response to environmental pressures, we can observe that most corporate responses have been limited to technical, that is, mechanistic actions. For example, environmental audits, new water recycling operations, and product redesign are some of the means employed to effect necessary change. Still, technical problems remain with respect to environmental concerns, and those companies that have responded are primarily the larger companies. Clearly, if an industry structure is to be developed that meets the criteria of sustainability world-wide, all companies will need to adopt the broadest model of environmental responsiveness, the "Proactivist." These are companies that would base decisions on an in-depth understanding of their moral place in the world. Such companies would develop technologies and products in accordance, not only with the marketplace, but with the results of their own self-analysis. They would engage many stakeholders in defining the present state, scanning the future, and challenging comfortable assumptions, while emphasizing experimentation and feedback to examine how they define and solve problems.

How can industries be motivated to adopt an internal environment that ultimately leads to a more sustainable performance? One way is to adopt the "normative model," illustrated by the "proactivist" stage among the five listed above. A normative model is one that is generally accepted by people as the way business should be done—an implicit norm or standard for business conduct. The International Chamber of Commerce's program "Business Charter for Sustainable Development" cited in Chapter 1 is an example that corresponds closely to Stage 5.

Another approach is built around the notion of environmental standards and is implemented through the marketplace with economic instruments such as effluent charges and taxes. As discussed in Chapter 7, such taxes can force companies to deal with their environmental effects (externalities) by forcing them to pay for their share of the total damage from pollution. Assuming that these costs can be determined fairly, and that realistic environmental standards can be agreed upon, most companies prefer this market model because it forces a level playing field among all companies contributing to the damage. The market model credo is that the polluter pays and the incentives are sufficient for companies to address and resolve excesses in their environmental impact. As we can see from Box 5.1, an organization's ability to reduce its environmental impact through business management can be turned into a competitive advantage, and this is the essential reason for the market model's existence.

A number of large companies have adopted the normative model, but it is evident that the process is difficult and takes a long time. In addition, some evidence suggests that the market model may not work, in part because the costs of pollution taxes (or controls) levied on corporations are still small compared to other fixed costs. Also, there is a lack of data and information about the extent of pressures which, in turn, allows a lack of awareness to prevail, especially among small companies.

So, the question remains: How do we get to a rigorous internal environmental policy in combination with a market-oriented environmental policy? In this book we have dealt already with economic and market-oriented approaches, so the remainder of this chapter deals with internal elements of corporate policy: ethics, values, stakeholder interests, and dialogue. The need to go beyond state-of-the-art requires that attitudes in business shift through better understanding of the root causes of environmental problems. Therefore, we suggest that the relationship of humans to nature, the place they feel they have in nature, and beliefs and attitudes toward nature, all need to be examined if changes in the relationships between business and the environment are to be achieved. As illustrated by the Bhopal case, ignoring or discounting the ethical implications of decisions or policies can lead to acceptance of anthropocentric practices, with often disastrous results. Thus, what follows is a brief examination of underlying value relationships toward nature, with a view to considering corporate strategies for sustainable performance.

5.4 SUSTAINABLE DEVELOPMENT: THE NEXUS OF ENVIRONMENT, ECONOMY, AND EQUITY

As outlined earlier in this book, the term sustainable development was introduced as a new strategic vision in 1987 through a report by the World Commission on Environment and Development (WCED). The WCED, comprised of twenty-two members from industrialized and developing countries, conducted over four years of research and concluded, ultimately, that environmental issues were complex in space and over time, and interdependent with cultures and issues of economic development. Sustainable development is the term adopted to reflect the concurrent need for economic systems that allow people to meet basic material needs and, at the same time, allow some use of natural resources. The goal of both, in an ethical and responsible balance, was to assure that resources are sufficiently available to meet the needs of future generations.

Three main ethical or value-related elements are embedded in sustainable development. First, sustainable development is a response to an ecological imperative. Life in a semi-closed ecosystem, if it is to be viable for long periods, requires that all of its components be sustained for equally long periods. Limits to the effects of human activity on these system components are necessary to maintain the functions of ecological support systems on which humans depend.*

Inter-generational equity, i.e., the equal interest of a future generation with that of the present, is another central premise of sustainable development. Inter-generational equity imposes a moral obligation on present generations to refrain from using natural resources in ways that reduce access

---

*The following material is excerpted from Shrivastava.[5]

> **Box 5.1: Competitive Advantages of Greening the Corporation[6]**
>
> Many businesses are beginning to realize that "greening" the corporation is more than the "right" thing to do—it is also a potential way to gain competitive advantage by lowering costs, differentiating products and satisfying market–niche demands. For example, 3M's "Pollution Prevention Pays" program illustrates the potential for cutting costs through environmental programs. In its first 15 years—1975 to 1989—this program established 2,511 pollution prevention projects and cut pollution per unit of production by half. It prevented discharge of more than 500,000 tons of pollutants and saved the company $500 million. The 3P program involves many well-defined projects. Each project must meet four criteria to receive formal recognition and funding. It must:
>
> 1. Eliminate or reduce a pollutant.
> 2. Benefit the environment through reduced energy use or more efficient use of manufacturing materials and resources.
> 3. Demonstrate technological innovation.
> 4. Save money through avoidance or deferral of pollution control equipment costs, reduced operating and material expenses, or increased sales of an existing or new product.
>
> Examples of projects vary widely in objectives, scope, design, and benefits. In one case, a resin spray booth was producing 500,000 pounds of overspray annually. The wasted resin had to be collected, transported, and incinerated. A 3P project redesigned the spray booth and installed new spray equipment to eliminate excessive spray. The new design cost $45,000 in a one time investment, but saves the company $125,000 every year. At Riker, a pharmaceutical unit of 3M, medicine tablets were coated with a solvent-solution coating. Riker developed a water-based coating as a substitute for the solvent. This change eliminated the need for $180,000 in pollution control equipment, saved $15,000 per year in materials cost, and eliminated 24 tons of air pollution a year. The changeover cost the company $60,000.
>
> In addition to cost reductions, greening can be used as a product differentiation strategy. Loblaw International Merchants, Procter & Gamble, and The Limited are differentiating their products for environmentally conscious customers. Other companies such as The Body Shop and Seventh Generation sell *only* green products, focusing on niches of environmentally sensitive customers in their industries. Still other companies are producing pollution control equipment, such as gas scrubbers, air filters, waste incinerators, sewage treatment plants, and bioremediation systems.
>
> Private sector expenditures on environmental protection equipment in the United States now exceed $50 billion a year. The global demand for environmentally friendly products is more than $200 billion a year, and growing exponentially. Such demands and opportunities have encouraged companies to make fundamental product and packaging changes to remain competitive.
>
> For example, by replacing ozone-depleting chlorofluorocarbons (CFC's) with safer alternatives, Dupont has maintained its competitive leadership in this industry segment. The Body Shop has gained competitive edge over giant cosmetic companies such as Unilever, Procter & Gamble, and Revlon by pioneering an "all natural" line of cosmetics and body care products. In 1991, McDonald's abandoned Styrofoam clamshell hamburger packaging in response to consumer demands and to overcome Burger King's paper packaging.
>
> Five years ago, the green consumer segment was a tiny market niche. Today it is becoming a mainstream trend in many consumer goods industries. Customers are demanding green products and packaging more friendly to the environment. Some consumers are willing to pay higher prices for environmentally sound products, and they are seeking more information about contents, use, disposal, and recyclability. As a result, hundreds of new and reformulated green products are now on the market, along with dozens of green consumer guides to evaluate them and organizations like Green Seal and Blue Angel to certify them. Selling green consumer products clearly presents a strategic opportunity.

to resources or other elements of the quality of life for future generations, and the degrees of freedom they will have to discover new uses for the resources at their disposal.[6,7] Balancing between the needs of present generations and future generations requires a long-term perspective, extending the boundaries of a community through time into the future. Some ethicists have suggested that current generations are duty bound to minimize harm to real future persons.[8]

The third and final element of sustainable development as outlined here is that of socio-geographic, or intra-generational, equity. In their 1987 report, the United Nations Commission on Environment and Development[7] pointed out that sustainable development meant two things:

1. that " . . . those who are more affluent adopt life styles within the planet's ecological means . . . " (p. 9).
2. " . . . sustainable development requires meeting the basic needs of all and extending to all the opportunity to fulfill their aspirations for a better life. A world in which poverty is endemic will also be prone to ecological and other catastrophes" (p. 8).

This was an especially significant statement because it articulated, for the first time, that economic development and ecological protection were viewed as complementary rather than competing goals. Access to economic resources and opportunities for livelihood is a requirement for people throughout the world if sustainable development is to be possible. Still, there are many places in the world where uneven trading patterns and the international debt crisis create a condition of stable poverty, defined as that amount of income below which it is impossible to accumulate wealth.[9] Thus, the dilemma posed by the gap between economically over-developed and under-developed societies is an environmental and ethical issue as well.

Increasingly, sustainability is viewed as the maximizing of both socio-economic security and eco-resource security, as is shown in Figure 5.2. Socio-economic security means moving from conditions of acute poverty, population growth, mass migration, income disparity, unemployment, and rights-deprivation to increases in civic and social cohesion, participation of affected stakeholders, social justice, economic stability, and the obligation to assure basic need fulfillment before luxury need fulfillment. Eco-resource security includes reducing levels of toxic contamination, land degradation, biodiversity loss, global climate change, hazardous risks, and acute resource conflict, moving toward increased biological integrity, human health, economic productivity, material supply stability, growth accommodation, and option preservation.

Thus, we see that sustainable development views social, economic, and ecological goals as related and interdependent. Environmentalism is viewed as a global concern, not merely a local or regional one. Human activities cause ecological harm that affects people in areas far beyond national borders, as in the case of the pollution of the Rhine River or the release of radioactive contamination after the Chernobyl accident.[10]

We are seeing, therefore, that the ideas of equity, and concern for nature and resource security, lie at the heart of sustainable development. The shift in human attitudes needed to achieve sustainability is really an ethical shift, a transformation of values about the moral obligations that guide the relationship of human beings with nature and with one another, in an extended community of present and future generations. This is not primarily a technological shift or a matter of new financial considerations. Modern science and technology can offer solutions to enhance the likelihood of equity, but science alone cannot achieve equity. The question relates more to what kind of science and what kind of technology will be used. The choices depend on whether moral principles are applied that promote positive inter-generational and intra-generational relationships based on respect.

The presence of these choices leads to important concerns for business practice. Conventional strategies for corporate survival are necessary but not sufficient. A business strategy needs to be built upon a strong ethical foundation or it will shift continually in response to public opinion, market forces, and the contradictions and uncertainties of science and technology change. The goals and activities that drive corporate activities and strategic decision-making are likely to change, as illustrated by the Declaration of the Business Council for Sustainable Development (Box 5.2).[12] The

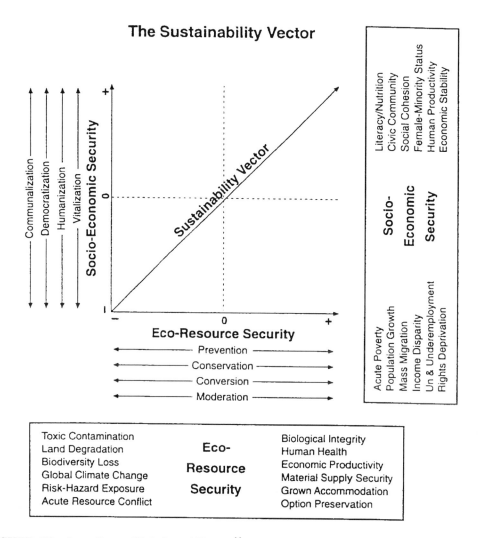

FIGURE. 5.2 According to Gladwin and Krause,[11] the need for balance between socio-economic security and eco-resource security suggests sustainability is a vector running diagonally between increasing needs.

principles set forth in this Declaration represents an important departure from business-as-usual, and the conventional wisdom that marginalizes environmental and human concerns. These practices from the past increasingly are in conflict with many public values, and leaders in business and public service organizations are creating a new collective ethic to guide action, one that is motivated by more than survivalism and expediency.[13]

5.5 COMPARING VALUES FROM THREE WORLDVIEWS

If ethical foundations are to be established for environmentally proactive firms, a coherent framework for values is needed. One central question to be addressed has to do with why we value nature. Is it simply an unlimited resource to be exploited for whatever human ends become dominant in a given society? Should we value it instrumentally as useful material for some human purpose? Should we care about nature only to the extent that it affects our personal self interest or our health and well-being as a species? What alternatives exist? How should we regard our moral obligation to others within our species (present and future generations) and to species other than our own? We will consider global integrity as one foundation for answering these questions in a later section. For now,

> **Box 5.2 Declaration of the Business Council for Sustainable Development[12]**
>
> In 1990, Stephan Schmidheiny was approached by the Secretary General of the 1992 U.N Conference on Environment and Development and asked to serve as a principal advisor for business and industry, to present a global business perspective on sustainable development, and to stimulate interest and involvement of the international business community. After accepting the invitation, Schmidheiny invited 50 business leaders from throughout the world to become members of the Business Council for Sustainable Development. Together, they participated in preparations for the U.N. Conference. After selecting a number of issues to study, conducting 50 conferences, symposia, and workshops in more than 20 countries, the Business Council for Sustainable Development (BCSD) completed a book, including this "Declaration." It was endorsed by each member as a blueprint for the role of business in helping to create a more sustainable world.
>
> *The declaration of the Business Council for Sustainable Development*
>
> *Business will play a vital role in the future health of this planet. As business leaders, we are committed to sustainable development, to meeting the needs of the present without compromising the welfare of future generations.*
>
> *This concept recognizes that economic growth and environmental protection are inextricably linked, and that the quality of present and future life rests on meeting basic human needs without destroying the environment on which all life depends.*
>
> *New forms of cooperation between government, business, and society are required to achieve this goal.*
>
> *Economic growth in all parts of the world is essential to improve the livelihoods of the poor, to sustain growing populations, and eventually to stabilize population levels. New technologies will be needed to permit growth while using energy and other resources more efficiently and producing less pollution.*
>
> *Open and competitive markets, both within and between nations, foster innovation and efficiency and provide opportunities for all to improve their living conditions. But such markets must give the right signals; the prices of goods and services must increasingly recognize and reflect the environmental costs of their production, use, recycling, and disposal. This is fundamental, and is best achieved by a synthesis of economic instruments designed to correct distortions and encourage innovation and continuous improvement, regulatory standards to direct performance, and voluntary initiatives by the private sector.*
>
> *The policy mixes adopted by individual nations will be tailored to local circumstances. But new regulations and economic instruments must be harmonized among trading partners, while recognizing that levels and conditions of development vary, resulting in different needs and abilities. Governments should phase in changes over a reasonable period of time to allow for realistic planning and investment cycles.*
>
> *Capital markets will advance sustainable development only if they recognize, value, and encourage long-term investments and savings, and if they are based on appropriate information to guide those investments.*
>
> *Trade policies and practices should be open, offering opportunities to all nations. Open trade leads to the most efficient use of resources and to more development of economies. International environmental concerns should be dealt with through international agreements, not by unilateral trade barriers.*
>
> *The world is moving toward deregulation, private initiatives, and global markets. This requires corporations to assume more social, economic, and environmental responsibility in defining their roles. We must expand our concept of those who have a stake in our operations to include not only employees and shareholders, but also suppliers, customers, neighbors, citizens' groups, and others. Appropriate communication with these stakeholders will help us to refine continually our visions, strategies, and actions.*
>
> *Progress toward sustainable development makes good business sense because it can create competitive advantages and new opportunities. But it requires far-reaching shifts in corporate attitudes and new*

ways of doing business. To move from vision to reality demands strong leadership from the top, sustained commitment throughout the organization, and an ability to translate challenge into opportunities. Firms must draw up clear plans of action and monitor progress closely.

Sustainability demands that we pay attention to the entire life cycles of our products and to the specific and changing needs of our customers.

Corporations that achieve ever more efficiency while preventing pollution through good housekeeping, materials substitution, cleaner technologies, and cleaner products, and that strive for more efficient use and recovery of resources can be called "eco-efficient."

Long term business-to-business partnerships and direct investment provide excellent opportunities to transfer the technology needed for sustainable development from those who have it to those who require it. This new concept of "technology cooperation" relies principally on private initiatives, but it can be greatly enhanced by support from governments and institutions engaged in overseas development work.

Farming and forestry, the businesses that sustain the livelihoods of almost half the world's population, are often influenced by market signals working against efficient resource use. Distorting farm subsidies should be removed to reflect the full costs of renewable resources. Farmers need access to clear property rights. Governments should improve the management of forests and water resources; this can often be achieved by providing the right market signals and regulations and by encouraging private ownership.

Many countries, both industrial and developing, could make much better use of the creative forces of local and international entrepreneurship by providing open and accessible markets, more streamlined regulatory systems with clear and equitably enforced rules, sound and transparent financial and legal systems, and efficient administration.

We cannot be absolutely sure of the extent of change needed in any area to meet the requirements of future generations. Human history is that of expanded supplies of renewable resources, substitution for limited ones, and ever greater efficiency in their use. We must move faster in these directions, assessing and adjusting as we learn more. This process will require substantial efforts in education and training, to increase awareness and encourage changes in lifestyles toward more sustainable forms of consumption.

A clear vision of a sustainable future mobilizes human energies to make the necessary changes, breaking out of familiar and established patterns. As leaders from all parts of society join forces in translating the vision into action, inertia is overcome and cooperation replaces confrontation.

We members of the BCSD commit ourselves to promoting this new partnership in changing course toward our common future.

Table 5.1 contrasts sustainable development with two other worldviews that influence the relationship of human beings and the environment, and explores the business implications of each.

These worldviews also influence how we judge the tradeoffs when the needs of a future generation must be considered, or when the risks to neighbors around a chemical manufacturing facility must be evaluated. How the tradeoffs are balanced varies widely from one nation to another, and varies as well between employees and shareholders of a company, and between customers and suppliers. All of the above can be seen as stakeholders with an interest in how the firm does business and how values are framed. In the following sections, we compare three ways of looking at the values shaping business, society and its range of stakeholders. Considering these values from the perspective of stakeholders leads us to another cross-cutting principle of sustainability, Stakeholder Interests:

5.5.1 Principle 11

All stakeholders, human and non-human, play an integral role in the functioning of a sustainable system. Therefore, these stakeholders must have an effective voice in decision-making processes affecting the system.

TABLE 5.1
Elements of Three Alternative Environmental Worldviews

Dimension	Dominant Social Worldview Anthropocentrism	Deep Ecology Biocentrism	Sustainable Development
Dominant societal imperative	Pursuit of unlimited economic growth, material consumption, scientific, and technological progress	Holistic balance with nature, recycling environmental and social justice, voluntary simplicity on behalf of greater self realization, doing with enough and recycling	Sustainable development, economic development that reduces societal inequities and assures long-term environmental viability
Dominant threat	Hunger, poverty, disease, war, natural disasters	Ecosystem collapse and unnatural disasters—extraordinary urgency	Resource degradation, poverty, population growth, ecological uncertainty and global change—some urgency
Human/nature relationship View of nature	Domination/control World is vast, offering unlimited opportunity for humans to fulfill goals	Harmony with nature Natural resources are limited, humans are subject to natural laws that constrain goal fulfillment beyond basic needs	Stewardship of nature World is finite with limits that constrain economic growth and social progress
Approach to natural environment	Instrumental/utilitarian use of nature for human purposes	Spiritual, aesthetic, intrinsic value for nature	Instrumental and intrinsic values for nature, use of science as a tool in resource conservation and development
Moral consideration/time horizon	Present generations (human); short-term	Equal moral consideration for all species (biospecies equality); long term	Present and future generations (human), species biodiversity, medium-long term
Social organization	Hierarchical, centralized authority, competitive, individualistic; nation-states	Egalitarian, decentralized, participatory, communalism, collectivistic; bioregions	Hierarchical, centralized with stake consultation, competitive/collaborative; holderindividual/collectivistic; nation-states
Economic assumptions	Frontier economics Open access/free markets; pursuit of undifferentiated growth; exploitation of infinite natural resources	Spaceship economics anti-growth, steady state economy	Green economy Resources management and eco-development; ecological economics

Environmental Ethics and Corporate Decision Making for Sustainable Performance

	Optimism	Pessimism	Skepticism
Technological assumptions	Faith in human intellect and ability to develop technological solutions for most problems	Technology must be appropriate to local ecology/culture, non-dominant role of science in solving problems	Technological solutions always have unintended consequences, approach technology critically before adopting
Primary stakeholder obligations	*Own society:* Stockholders, interest groups, customers, suppliers, regulators, unregulated free market	*Local bio-region: Sustaining* appropriately scaled economic activity and natural functioning in exact balance *Global ecosystem/biosphere.*	*Own Society:* Natural environment, stockholders, interest groups, customers, suppliers, regulators, semi-regulated market *Global society:* Other cultures/societies *Global environment:* Large scale environmental impacts (e.g., global warming)
Goal/mission	Efficient accumulation and allocation of wealth/resources	Simplicity, self-awareness, ecosystem health and vitality	Quality of life
Dominant organizing principles	Centralized authority/hierarchical Top-down decision making Large scale Efficiency/productivity Motivation through control (extrinsic)	Small is beautiful Consensual decision making Small-scale—local products/marketing Craftsmanship/quality/sufficiency Motivation of spirit (intrinsic) through service, simplicity, self-awareness, quality of living	Decentralized/flattened/organization Participative decision making Large/medium scale—chunking Eco-efficiency/productivity/quality Motivation through commitment (extrinsic and intrinsic)
Corporate moral consideration	Organizational survival/shareholders	Socio-ecological responsibility at local and global levels	Organizational survival, social responsibility, sustainability

Source: Derived from several sources. See References 6, 14, 15, 22, 24, and 25.

5.5.2 Anthropocentrism, The Dominant Social Worldview

Our society's present dominant worldview is anthropocentric (human-centered) because it values nature primarily as a resource to be used instrumentally for human purposes. Nature is regarded as a set of resources to be extracted, exploited, and transformed into goods (or services), discarding any residual. This instrumental view of nature has been termed "resourcism".[16] Human beings as stakeholders are assumed to be unique among all species, "masters of their destinies," and, therefore, having the right to dominate nature for selected goals.[17] Further, the world is considered vast, offering unlimited resources for fulfillment of human goals. The temporal orientation of this worldview is short term and oriented toward satisfying human goals in the present. When nature is viewed as inert, no moral consideration is owed to species of plants and animals sharing our space.

The economic system that underlies the dominant social worldview is that of "frontier economics."[18] Undifferentiated economic growth is viewed as necessary to achieve progress and societal prosperity, and businesses support open access to free markets. Prosperity is necessary to trigger patterns of material consumption that fuel economic activity in the business sector. Those numbers drive the stock market and investment opportunities, and there is a continuing sense that consumers need to purchase more, as advertisers attempt to translate wants/luxuries into needs. The standard of living is influenced considerably by how much one can possess or purchase.

The term frontier economics is used here because it captures the sense of no resource limitations. It assumes that the cycle of production, purchasing, and prosperity can continue indefinitely. In addition, frontier economics makes the assumption that American standards of living could be available to all stakeholders, in every country and for the majority of the world's population, if they would only engage in the type of economic activity and resource use (including the externalities discussed in previous chapters) that appear to have been successful in industrialized countries. As noted in Chapter 2, we have believed until late in this century that if society damaged its natural resources or faced a limiting supply, it could always develop a technological fix or find a new frontier by going somewhere else. Finally, and as described in Chapter 3, frontier economics utilizes a system of valuation that fails to incorporate the costs of environmental degradation into measures such as GDP or into the pricing of products and services. All increases in economic activity are valued equally, regardless of environmental impacts (and thus the Exxon Valdez oil spill counted economically as a benefit in terms of the dollars it brought into the local community).

Within anthropocentrism, the primary goal for enterprise is the efficient accumulation and allocation of wealth through pursuit of economic activity.[20] Moral consideration is granted first to organizational survival and to the welfare of shareholders, and the time horizon against which results are measured is short-term (often in terms of quarterly financial reports and profits). The dominant social worldview is deeply rooted in Western religious and philosophical thought. However, people from all of the world's major religions also have begun finding common ground in the need to preserve biological diversity (creation) and ecological integrity. One example is an interfaith report defining the spiritual principles that undergird an Earth Charter (see Box 5.3).

5.5.3 Deep Ecology as a Worldview

Deep ecology was introduced in the 1970s as a biocentric (life-centered) philosophical foundation for living. It is positioned here in direct opposition to anthropocentrism[21] (see Table 5.1). It seeks to provide a normative moral and spiritual foundation for environmentalism, one that tends to be influenced by Eastern religion and philosophy. Nature is valued intrinsically rather than for its use to human beings, and the appropriate role for human beings is to live in harmony with nature. Humans are simply one species among many, and are no longer viewed as having the right to dominate nature in order to fulfill their many desires. In fact, deep ecology proposes that human beings ought to live in ways that exert minimal impact on other species and the earth in general. Bio-species equality is a central objective within this worldview—the belief that all species deserve equal moral consideration. "All things in the biosphere have the right to live, blossom, reach their own forms of self-

> **Box 5.3   Spiritual Principles of an Earth Ethic**
>
> For the past several years, a religiously-based environmental ethic has been evolving to stress the goodness of nature as God's creation. It sometimes goes beyond this to include nature as an object of God's redemption along with human beings. In preparation for the environmental summit in Rio in June of 1992, an interfaith group called the International Coordinating Committee on Religion and the Earth published a document entitled, "*An earth charter*".[19] The "Spiritual Principles" that preface the concrete recommendations for environmental responsibility reflect religion's embrace of the essential goodness of nature and its characteristic blend of concern for nature with concerns for social justice and human well-being:
>
> **Interdependence**
> The earth is an interdependent community of life. All parts of this system are interconnected and essential to the functioning of the whole.
>
> **The Value of Life**
> Life is sacred. Each of the diverse forms of life has its own intrinsic value.
>
> **Beauty**
> Earth and all forms of life embody beauty. The beauty of the Earth is food for the human spirit. It inspires human consciousness with wonder, joy, and creativity.
>
> **Humility**
> Human beings are not outside or above the community of life. We have not woven the web of life; we are but a strand within it. We depend on the whole for our very existence.
>
> **Responsibilities**
> Human beings have a special capacity to affect the ecological balance. In awareness of the consequences of each action, we have a special responsibility to preserve life in its integrity and diversity and to avoid destruction and waste for trivial or merely utilitarian reasons.
>
> **Rights**
> Every human being has the right to a healthy environment. We must grant this right to others: both those living today and generations to come.
>
> Each of these principles have as their source an essential respect for the intrinsic value of nature and they incorporate a deep understanding of the appropriate role of humanity within an interdependent ecosystem. In articulating both responsibilities and rights, the principles above also begin to reflect the way in which relationships between human beings on a global scale impact other species and the integrity of the ecological systems on which they both depend. We are responsible in our actions not only to those human beings living now, both within our own communities and those in communities far distant, but to future generations whose environmental legacy and quality of life depend on the decisions we now make. The call to " . . . avoid destruction and waste for trivial or merely utilitarian reasons" is a vital part of what it may mean to act on a fundamental ethic of "planetary responsibility," superordinate to an ethic of success or disposition.[20]

realization . . . " according to deep ecologists.[22] Human beings may only use resources as necessary to satisfy vital needs, living frugally and simply rather than relying on materialism, consumerism, and economic growth. This is necessary, it is argued, partly so as to live in harmony with nature, but also to achieve a second goal central to deep ecology: self-realization. This goal is defined as the search for one's own spiritual and biological identity, through identification first with humanity, and then through identification beyond humanity to the non-human world.

The type of economic system most conducive to biocentrism would be spaceship economics.[23] Unlike a frontier, spaceships do not have unlimited reservoirs or resources for extraction, or as sinks for pollution. Rather, spaceships require a cycling capacity that can be maintained in perpetuity. The essential measure of success is not expansion of production and consumption, but the nature, extent, quality, and complexity of the total capital stock and its self-maintenance, including the state of human bodies and minds supported by the system.[23] Thus, the economy envisaged in this worldview is anti-growth, anti-materialistic and anti-consumptive. Biocentric economic activity also is of a smaller scale, emphasizing self-sufficiency and local products rather than transport of goods from distant locations. Small scale and ecologically appropriate technologies are emphasized in order to reduce the potential for pollution or environmental degradation, and to elevate human craftsmanship in labor. The role of technology is to minimize human impact on nature.

Within this new environmentalism, or deep ecology, the primary goal of enterprise would be to achieve an integrated global ecosystem representing all stakeholders. Emphasis would be less on productivity and efficiency, and more on craftsmanship, quality, and sufficiency. The primary moral consideration of business would be socio-ecological responsibility at local, national, and global levels.

5.5.4 Sustainable Development, an Emerging Worldview

We have suggested already that a third worldview, sustainable development, is concerned with balancing resource use, environmental quality, and the equity interests of all stakeholders, including present and future generations. In important ways it incorporates elements of the two previous views (Table 5.1). Sustainable development argues that nature must be valued both intrinsically and instrumentally. Human beings are the only species capable of understanding the complexities of ecosystems and the effects human activity has on them. Thus, humans must assume full responsibility and accountability as stewards of nature and its renewable resources. Stewards are more than managers in that they are obligated to hold something in trust for another. Thus, stewardship implies a fiduciary responsibility to all stakeholders, including future generations.

As was seen in Chapter 3, the basis for a sustainable economy requires a shift from neo-classical economics to a full-cost accounting or ecological economics.[24] Sustainable development emphasizes finding sustainable forms of enterprise that neither extract resources at a rate faster than they can be renewed, nor dispose of wastes in amounts beyond the capacity of natural systems to assimilate them. Industrial systems are emerging now that close the loop by having the wastes of production processes or consumer end use used as inputs for production processes elsewhere (see the case studies). The full costs of resource use and other externalities are factored into pricing, and the new measures of macroeconomic activity (see Chapter 3) reflect the state of both the environment and the economy.

Thus, in sustainable development, the natural environment can be seen as a key stakeholder in business decisions, along with shareholders, customers, and others. Production processes that consume less energy and use fewer resources are being adopted (see the Case Studies), achieving what can be called eco-efficiency. Life-cycle analysis dominates choices about product design, engineering, and packaging of products. Management of business has become subtly ecocentric.[25] While organizational survival is still a dominant moral consideration, it is joined equally by social responsibility and sustainability. The time horizon for considering outcomes is becoming medium- to long-term, five to twenty-five years, rather than quarterly.

To function as a gyrocompass for corporate decisions that facilitate sustainability, an environmental ethic should incorporate a sense of the intrinsic value of nature. This is needed to get beyond approaches motivated by corporate survival or business expediency. If nature has only instrumental value, and the interest of human beings is the measure of all things, then there is no moral constraint to protect nature from degradation. An environmental ethic imposes new *a priori* obligations toward nature. An environmental ethic requires that we value and care for nature even as we value and care for humankind. Such an ethic requires business to go beyond the moral minimum that the law requires and develop an environmental conscience founded on the intrinsic value of the non-human world.[26]

5.6 INTEGRITY OF NATURAL SYSTEMS: A GLOBAL ETHIC FOR THE ENVIRONMENT AND BUSINESS

Because the definition of sustainable development has been imprecise, necessarily, sometimes entirely losing its ecological center, this concept alone tends to be weak as a moral compass. A need exists for an ethic that recognizes the integrity of natural systems as central, a "true north." One of the most respected arguments for the value of nature is the need to maintain the functioning of natural systems that have supported the human species and human values throughout human evolution. Some others have defined a system as " . . . a set of interrelated elements, each of which is related directly or indirectly to every other element, and no subset of which is unrelated to any other subset."[27] Each system exists in dynamic interaction with an environment, and every system is therefore part of a larger system. An ecological view of nature considers the web of relationships between organisms in a particular type of habitat or ecosystem. If we also see global ecological systems as a network of interrelated natural systems, then one concludes that all life is interrelated. This finding then leads to a shift from simplistic, linear, cause and effect thinking to a need to understand the interdependencies needed to sustain life.

The integrity of natural systems, as an important argument for the intrinsic value of nature, has emerged recently with a prominent place in environmental ethics literature. The philosopher and environmentalist Rolston argues that, starting from the dirt beneath our feet, and moving on through the biotic and abiotic entities that constitute our environment, we find the stuff of creativity. Nature continually makes and remakes things that wear out. Nature's intrinsic value, then, can be seen in its being our source, not merely our resource. (The latter would reduce it to an instrumental value alone.) Rolston writes, "Nature is a fountain of life, and the whole fountain—not just the life that issues from it—is of value. Nature is genesis."[28]

Others, such as Leopold, advocated an ethic that granted moral consideration to natural systems. In calling for a land ethic, he argued for expanding the idea of community to include the land and its soils, waters, plants, and animals. As a guide to ethical decision-making, he suggested the need to " . . . examine each question in terms of what is ethically and aesthetically right, as well as what is economically expedient. A thing is right when it tends to preserve the integrity, stability, and beauty of the biotic community. It is wrong when it tends otherwise."[29]

Recently, Westra has proposed that ecological integrity be viewed as a new categorical imperative for moral behavior in regard to nature. It provides a cohesive social value beyond the satisfaction of individual preferences that the market allows. She argues that we need a new starting point, *Homo sapiens* as inescapably a member of and dependent upon native ecosystems. Accordingly, the first principle is that nothing can be moral that contradicts this reality, or that is not consistent with nature's laws. If humans fail to act consistently with universal natural laws and processes, the species could prove unfit, eventually, for long-term adaptation and survival.[30]

We suggest here that the integrity of natural systems be at the center of moral consideration, an idea that is missing in sustainable development, leaving it open to charges of only lightly modified anthropocentrism. Without such a categorical imperative, sustainable development is a worldview open to constant drift toward forces of degradation. In business, decisions ordinarily shift in response to cost/benefit analyses centered around what is most convenient, expedient, or profitable in the short term.

Thus, the elements of integrity as an environmental ethic can be listed as follows:

1. Decisions and actions are moral when they sustain or enhance the fundamental integrity of natural systems on which all species depend. Where there is uncertainty about long-term systematic impacts, the principle of prudence should be applied.
2. Decisions and actions are moral when they value nature, first intrinsically, and then instrumentally.

3. Decisions and actions are not moral when they compromise availability of the natural resources required by present and future generations to satisfy their instrumental needs and intrinsic pleasures.
4. Decisions and actions must consider long-term consequences and must also view consequences at local and global (or regional) scales whenever possible.
5. Decisions and actions with potential to affect the environment or the economic livelihood on which human beings depend for subsistence must promote inclusiveness, allowing for the participation of affected stakeholders and dialogue-based inquiry into the consequences.

These elements are inclusive of the principles of sustainable development discussed earlier, but they also place integrity as the categorical imperative at the center of the ethic. Such placement assures the ecological focus is not lost or subordinated to an economic focus. The ethic promotes a process of dialogue, inquiry, and learning on the part of local communities and affected stakeholders as a basis for participation in society. Finally, this ethic increases the possibility of generating a cooperation that transcends the national, ethnic, religious, and cultural differences that divide people.

The question that industry asks, understandably, is how these principles can be applied or institutionalized to provide guidance in decision making. For example, a corporation that wishes to evaluate alternative technologies to produce its products needs to have in place a set of principles that lead its management and employees to consistent and defensible environmental actions. Some of the cost studies included later in this book illustrate business' adoption of such principles.

Figure 5.3 summarizes the three main ethics we have developed in this chapter and lists the following six instrumental values associated with them. All would be part of an enlightened management paradigm seeking sustainable performance.

Environmental Ethic	Instrumental Values	Corporate Sustainable Performance Through
Integrity of natural systems	Holism	Systems thinking Posterity Equity Community/diversity
	Scale	Vision Smallness
Intrinsic value of nature	Compatibility	Quality Continuous improvement Appropriate technology (benign)
	Responsibility	Accountability Measurability Database—inventory
Inter-generational and intra-generational equity	Commitment	Participation Inclusion Collaboration Dialogue Justice
	Stewardship	Humility Knowledge sharing

FIGURE 5.3 Corporate sustainable performance through an environmental ethic of system integrity.

5.6.1 Holism

Managers need a vision of the world that acknowledges its interconnectedness as a living system. This implies that most decisions made by the firm have consequences on other parts of the human and surrounding environmental systems. Organizations are essentially self-adjusting organisms controlled by a network of interconnected information channels. In fact, organizations are also systems interacting with natural systems. By means of system thinking an understanding of both long-term and short-term effects can be achieved for a sustainable result.

Holistic thinking could lead to new inter-organizational arrangements that more closely parallel natural ecosystems. For example, industrial ecosystems may be created in which the wastes from production in one organization become input for the processes of another. An example of an industrial ecosystem in Norway includes a network of companies consisting of a power plant, a chemical plant, an enzyme plant, a refinery, a cement plant, a wallboard plant, and area farms[25]. The power plant sells used steam to the enzyme plant and the refinery, instead of condensing and disposing of it in a nearby fjord. The refinery supplies the power plant with treated waste water for cooling. Fly ash produced as a waste product from the power plant is sold to the wallboard and cement plants. Local farms use waste from the fishery and the enzyme plant as fertilizer. The outcomes save water, minimize the amount of waste sent to landfills, reduce pollution, and conserve energy resources. Adopting a model such as this, however, would require many changes in business management practices, strategic decision making, and the infrastructure necessary to support it.

Sustainable performance has to include an understanding of the relationships between a highly diverse set of stakeholders, which resemble nature's own characteristics. Within the ecosystem, diversity can exist because all parties implicitly accept their niche. Similarly, the creation and maintenance of a diverse set of relationships in the corporate world must be based on moral codes that serve as a barometer by which behavior is measured. Hence, ethics becomes the basis for community life and should penetrate completely the fabric of corporate decision making.

5.6.2 Scale

A need exists, also, for businesses to determine a clear vision of who they are and what they want to become. We have seen, historically, that a need to control or dominate nature often defined a corporation's sense of power. Understanding and articulating the place an organization has within the larger forces of nature may introduce and sustain a sense of scale that, in turn, leads to responsible definition and use of its power. This also is consistent with recent management prescriptions that suggest a smaller scale enhances entrepreneurial spirit and innovation.

Replacing nationally and internationally produced items with products that are created locally and regionally, or do not require what Hawken[31] calls, "exotic" sources of capital to support unrealistic (unsustainable) growth objectives, are but two ways for an organization to support a newly defined vision of the future and its role in it. Because of nature's regenerative capacities, small scale operations tend to have small scale effects. Hence, businesses have to focus on reducing, recycling, and reusing, locally, nationally, and globally.

Recognizing the limitations of nature to sustain certain practices puts limits on the size and essence of a business. Prices and costs should reflect the limits of the earth's resources. It is, therefore, important that a complete accounting be made of all costs associated with the production and marketing of an organization's products and services.

5.6.3 Compatibility

In the past, quality used to be measured in terms of the percentage of defective products/services. More recently, however, interactions among all stakeholders have caused a shift in the definition of quality. Definitions of quality now focus on the overall perception of what an organization's products/services should be, and include not only the quality of its products or services, but also the quality of work and life within the organization. A corporation must secure a high degree of compatibility with an environment that is now demanding higher standards of quality.

Also, as indicated previously, the process by which products and services are produced needs to be compatible with the environment's ability to deal with the by-products and consequences of such activity. Quality is increasingly coming to mean reduction of wastes from production processes and increased yield with less use of resources. Compatibility means that companies remain responsible for their manufacturing processes and products even after the sale; the entire life-cycle of every product as it influences the environment must be considered.

The underlying philosophies of Total Quality Management and Total Quality Environmental Management could become the backbone of the systematic efforts that an organization expends to improve itself continuously. The development of support systems throughout the organization, and the empowerment of all those who are involved in these efforts, become critical to the successful implementation of sustainable practices.

Sustainable performance, then, is partly a process that depends on technology. The development of partnerships with governments, academe, and other businesses will aim at discovering and implementing ways to improve sustainable performance. Since no one company is likely to be able to address all environmental challenges, collaborative initiatives that facilitate and reward information transfer will be encouraged.

5.6.4 Responsibility

A basic sense of responsibility and accountability for all corporate actions provides the foundation of a corporation's decision making. Achieving this requires adequate levels of documentation, which, in turn, depend upon detailed, and auditable data. These data also need to be used to develop clear, measurable goals. Development of objective databases that form the bedrock for inclusive reporting is critical for assuring compliance with both internal and external standards. The ultimate objective is to reduce risk to all stakeholders and to obtain improved environmental efficiency and performance.

Changes in accountability may well involve the sharing of power. Whereas in the past most control of corporate actions has been with the manufacturers themselves, recent experience by a number of organizations supports the need for all stakeholders (including private citizens, governments, and special interest groups) to evaluate strategies and actions jointly.

5.6.5 Commitment

Developing equitable, community-building sustainable business strategies requires a commitment to involving all stakeholders. This is a commitment to a basic sense of equity that is needed in all strategic planning, information, and financial management systems. Cooperation between the company and its stakeholders needs to become the means by which goals for sustainable performance are implemented. Powerful, timely, inclusive dialogue aims at improving the organization's sustainable performance by educating all stakeholders about commitments.

5.6.6 Stewardship

As described earlier, stewardship denotes an awareness of and care for all affected stakeholders. It suggests a management approach that replaces economic value-driven decision making with a commitment to achieve option-driven, responsible, long-term consensus. Technology and information are less and less simply proprietary; indeed, knowledge that can contribute to industry-wide sustainable performance (such as product technology, training and education programs, and guidance documents) are shared and, in turn, are used responsibly for the betterment of all. A previously-cited example of stewardship is the self-regulation of the Chemical Manufacturers Association through their Responsible Care® Codes of Practice; members are required to help one another by sharing information that will help each firm live up to its responsibilities for product stewardship under Responsible Care®.

The *raison d'être* for the corporate steward is based on a sense of humility, and aims at replacing the organization's negative impact on natural, social, and economic environments with the positive effects of its products or services. Stewardship as a management philosophy offers a new exchange covenant that is based on mutual respect and trust. In doing so, stewardship involves sustaining the capital stock of both the organization and the earth by building knowledge and establishing ground rules for interaction.

In summary, instrumental values underlying the integrity of natural systems are characterized by high flexibility. To work with nature's broad and dynamic processes, business organizations may need to replace linear analysis with a more holistic approach that is less mechanistic. A new vision of the future can be based on values that rethink the place of "profits" generated from a drawdown of natural capital, replacing them with highly-valued social and economic services. These new values involve shifts from objects to relationships, from the parts to the whole, from domination to partnerships, from structures to processes, from self-assertion to integration, and from growth to sustainability.[32]

By means of example, we have seen why and how organizations can develop strategies incorporating a sustainability ethic based on the integrity of natural systems. The successful translation of principles into implementation is difficult. Perhaps no characteristic other than sheer willpower (or, as some argue, mere prudence) has fueled some companies' conviction that it could, indeed, should be done. Still, an important question remains: How can we choose among or incorporate from the vast array of conflicting values that permeate economic exchanges and are present in our integrity ethic? Or, said differently, how does one establish the authority of arguments for the foundation of an environmental ethic such as we have just examined? The remainder of this chapter will look at how we can do that.

5.7 THE NEED FOR DIALOGUE

The array of ideas students encounter during a four-year college program shows, clearly, that diversity is more characteristic of our national and global society than is consensus. Our pluralistic world displays diversity, ethnically, economically, politically, religiously, and ethically. In this context, consensus as to what is to be sustained seems impossible to achieve. In fact, the task is more daunting than the conditions of cultural diversity alone might suggest. The hope of the Enlightenment to provide the rational grounds for moral consensus has not been realized. The traditions that once provided moral authority seem to have been undermined by the arguments of reason, but reason has not succeeded in establishing new forms of authority for ethics.[33]

If there is a way toward greater consensus in the midst of diversity on an issue of such importance as an environmental ethic, or on sustainability of social and economic systems, it is likely to come through improved sharing of information and expectations and through dialogue. Thus, in the section that follows we elaborate fully on Principle 12, Information, Dialogue and Expectations, a key process:

5.7.1 PRINCIPLE 12

> Sustainable systems enable all participants to engage in effective dialogue, bringing their own traditions and perspectives to the decision-making process in an atmosphere of mutual respect.

Dialogue is more than communication about information. It is a discipline of collective thinking and inquiry; a process of creating shared meaning through reflective listening.[34] The dialogue needed to understand sustainability enables all who participate to bring their own traditions and perspectives to the process in an atmosphere of mutual respect. It recognizes that, in a pluralistic world, no one can claim sole possession of truth.

In this chapter we have seen that foundations of environmental ethics and sustainability can be explored through the sharing of diverse perspectives and, through that interchange, points of view, convictions, and expectations can be refined and reshaped. In the openness that dialogue requires, and that is itself required by the nature of pluralism, leaders in science, philosophy, public policy, environmental advocacy, and religion can work with leaders in business and industry toward a more comprehensive consensus on issues. This prospect is preferable to the frequently unproductive forms of confrontation that have characterized many such interchanges in the past.

Dialogue and the development of shared expectations with diverse partners helps provide a more comprehensive view of what is sought in sustainable development, and discloses the systemic character of the changes required. In this regard, then, it is important for business to see that its own economic, marketing, and manufacturing systems are organically connected to other global forces affecting the environment and the future. At the same time, those outside industry, the media, shapers of public opinion, and political leaders in public policy, are coming to see that they, too, are integral to the systemic change taking place, and bear equal responsibility with business.

In the final analysis, dialogue itself has ethical significance. The rules of dialogue require respect for other points of view and an ability to understand and articulate other views equally well. Such a requirement engenders respect for all participants and for the justice that comes with equal participation. Taking responsibility for understanding one another's positions, and the relevant facts that go with them, also fosters truthfulness. In the heat of confrontation, and in an effort to sway public opinion on what is sustainable through persuasive rhetoric, it is easy to distort an opponent's position and even misrepresent or ignore pertinent facts. This is less likely to happen when parties are committed to dialogue.

A general commitment to integrity of the environment also draws on dialogue about the intrinsic value of nature, as well as concern for justice, in order to sustain resources for present and future generations. This goal needs to be interpreted in terms of the complex options presented in different situations. Even very specific guidelines still require discussion and interpretation when used in decisions.

The dialogue between ethical principles and the demands of specific situations becomes an appropriate framework for making choices when the trends toward increased public understanding is considered. Discerning the appropriate responses and resolving vexing conflicts are a primary activity of ethical reflection. However, dialogue also is where rationalization of interests becomes expressed, an unfortunate prospect for many organizations with conflicting loyalties or interactions. For that reason, the foundations of environmental ethics need to be strong and clear; anything less than a sense of the intrinsic worth of nature can render nature vulnerable to rationalization. To be sure, some entities may have higher intrinsic value than others, as we observed earlier, and this assists in resolving certain competing claims.*

In applications of environmental ethics to management and sustainable development, moral management may need to be thought through to reflect interests of multiple stakeholders (both within and outside the organization), as well as the uncertainty of conflicting norms, ambiguous facts, and unpredictable outcomes. A new model of moral management through dialogue and sharing of expectations will involve not only the bringing of diverse norms into the context of situations (facts about businesses and communities) to determine their appropriate application or the priorities that must be set in the face of conflicting norms and claims. But this sharing of expectations includes, as well, the dialogue among stakeholders and decisionmakers. Indeed, businesses may begin to practice this process by establishing scenario dialogue situations as a tool for teaching managers about communicative ethics.

---

*Nash, for example, recognizes that "biotic rights" which we accord to all living things in the environment are *prima facie* rights, rights that may be overridden by other claims in a situation where there are conflicting obligations.[35]

5.8 CONCLUSION

This chapter has examined the ethical and value-related pressures felt by many corporations and stakeholders, all of which help to shape an organization's or society's response to environmental rights and concerns. We have also explored the view, that, absent an ethical stance toward nature as a gyrocompass for strategic decision making, business response to the environment will be idiosyncratic, shifting in relation to competing priorities. While sustainable development may be an approximate vehicle by which businesses assume their role in promoting a more ecologically viable future, we have asked here whether it can or should suffice also as an ethic for the environment. We conclude that unless sustainable development is understood as incorporating a basis for valuing nature intrinsically, it does not suffice as an ethic for the environment. When sustainable development is supported by the categorical imperative of integrity, and infused with the process for social learning through dialogue, it becomes viable as an ethic. We have termed this the global integrity ethic.

An ethical inventory of an environmental situation is but one approach for an organization to value the environment for corporate decision making. In the next chapter we will examine other means by which managers may measure and value the environment. Indeed, new means by which values can be used to quantify the costs and benefits of using resources and releasing wastes are developing, and their use will be central to competitive business outcomes.

5.9 REFERENCES

1. Sharplin, A., Union Carbide Limited and the Bhopal Gas incident: issues and commentary, in *The Corporation, Ethics, and the Environment,* Hoffman, W. M., Frederick, R., and Petry, E. S., Jr., Eds., Quorum Books, Westport, CT, 1990, 119.
2. Williams, H. E., Medhurst, J., and Drew, K., Corporate strategies for a sustainable future, in *Environmental Strategies for Industry,* Fischer, K., and Schot, J., Eds., Island Press, Washington, D. C., chap. 4.
3. Hunt, B. C. and Auster, E. R., Proactive environmental management: avoiding the toxic trap, *Sloan Manage. Rev.,*31(2), 7, 1990.
4. Costanza, R., Daly, H. E., and Bartholomew, J. A., Goals, agenda, and policy recommendations for ecological economics, in *Ecological Economics: The Science and Management of Sustainability,* Costanza, R., Ed., Columbia University Press, New York, 1991, 1.
5. Shrivastava, P., *Greening Business: Profiting the Corporation and the Environment,* Thomson Executive Press, Cincinnati, OH, 1996, 55; 1996, 153.
6. Milbrath, L. W., *Envisioning a Sustainable Society, Learning Our Way Out,* State University of New York Press, Albany, 1989.
7. World Commission on Environment and Development, *Our Common Future,* Oxford University Press, London, 1987.
8. Green, R., Intergenerational distributive justice and environmental responsibility, *BioScience,* 27(4), 260, 1977.
9. Boulding, K., *The World as a Total System,* Sage, Beverly Hills, CA, 1985.
10. Buchholz, R. A., Corporate responsibility and the good society: from economics to ecology, *Bus. Horiz.* 34(4), 9, 1991; Buchholz, R. A., *Principles of Environmental Management: Greening of Business,* Prentice-Hall, Englewood Cliffs, NJ, 1993.
11. Gladwin, T. N. and Krause, T., *Business, Nature & Society: Towards Sustainable Enterprise,* Richard T. Irwin, Burr Ridge, IL, in press.
12. Schmidheiny, S., *Changing Course: A Global Business Perspective on Development and the Environment,* MIT Press, Cambridge, MA, 1992.
13. Schmidheiny, S., The business logic of sustainable development, *Columb. J. World Bus.,* 27(4), 19, 1992.
14. Colby, M. E., *Environmental Management in Development: The Evolutionary Paradigms,* World Bank, Washington, DC, 1990.

15. Gladwin, T. N., Kennally, J. J., and Krause, T., Shifting paradigms for sustainable development: Implications for management theory and research, *Acad. Manage. Rev.,* 20(4), 874, 1995.
16. Pauchant, T. and Fortier, I., Anthropocentric ethics in organizations: how different strategic management schools view the environment, in Hoffman, W. M., Frederick, R., and Petry, E. S., Jr., Eds., *The Corporation, Ethics and the Environment,* Quorum Books, New York, 1990.
17. Catton, W. R., Jr., and Dunlap, R. E., Environmental sociology: a new paradigm, *Am. Sociol.,* 13, 41, 1978.
18. Colby, M. E., Environmental management in development: the evolution of paradigms, *Ecol. Econ.* 3, 193, 1991.
19. U.S. Citizens Network Working Group on Ethics, The earth charter, *Earth Eth.,* 3, 11, 1991.
20. Küng, H., *Global Responsibility: In Search of a New World Ethic,* Crossroad, New York, 1991.
21. Naess, A., The shallow and the deep, long-range ecology movements: a summary, *Inquiry* 16, 95, 1973.
22. Devall, W. and Sessions, G., *Deep Ecology: Living as if Nature Mattered,* G. M. Smith, Salt Lake City, UT, 1985.
23. Boulding, K., The economics of the coming spaceship earth, in Jarrett, H., Ed., *Environmental Quality in a Growing Economy,* Johns Hopkins University Press, Baltimore, MD, 1966, 3.
24. Egri, C. P. and Pinfield, L., Organizations and the biosphere: ecologies and environments, in *Handbook of Organization Studies,* Clegg, S. R., Hardy, C., and Nord, W., Eds., Sage, London, 1996.
25. Shrivastava, P., Ecocentric management for a risk society, *Acad. Manage. Rev.,* 20(4), 118, 1995.
26. Hoffman, W. M., Business and environmental ethics, *Bus. Eth. Quart.,* 1(2), 169, 1991.
27. Ackoff, R. L. and Emery, F. E, *On Purposeful Systems,* Aldine-Atherton, Chicago, 1972.
28. Rolston, H., III, *Environmental Ethics,* Temple University Press, Philadelphia, 1988.
29. Leopold, A., *A Sand County Almanac and Sketches Here and There,* Oxford University Press, New York, 1949.
30. Westra, L. S., The principle of integrity and the economy of the earth, in *The Corporation, Ethics and the Environment,* Hoffman, W. M., Frederick, M., and Petry, E. S., Eds., Quorum Books, New York, 1990, 277; Westra, L., *An Environmental Proposal for Ethics: The Principle of Integrity,* Rowman & Littlefield, Lanham, MD, 1994.
31. Hawken, P. *The Ecology of Commerce, A Declaration of Sustainability.* Harper Collins, New York, 1993.
32. Callenbach, E., Capra, F., Goldman, L., Lutz, R., and Marburg, S., *Ecomanagement, the Elmwood Guide to Ecological Auditing and Sustainable Business,* Berrett-Koehler, San Francisco, 1993.
33. MacIntyre, A., *After Virtue,* University of Notre Dame Press, South Bend, IN, 1981.
34. Isaacs, W., Taking flight: dialogue, collective thinking and organizational learning, *Organ. Dyn.,* 22(2), 24, 1993.
35. Nash., J. A., *Loving Nature: Ecological Integrity and Christian Responsibility,* Abingdon Press, Nashville, 1991.

6 Valuation and Reporting

Raymond F. Gorman

CONTENTS

6.1 Introduction 105
6.2 Valuation 106
    6.2.1 Different Meanings of Value 106
        6.2.1.1 Use Value 106
        6.2.1.2 Existence Value 108
        6.2.1.3 Intrinsic Value 108
    6.2.2 Valuation Methods 108
        6.2.2.1 Direct Methods 108
            6.2.2.1.1 Discounted Cash Flow Method (DCF) 108
                6.2.2.1.1.1 Example: A Case Study of the Evergreen Pine Tree Company 109
            6.2.2.1.2 A Critique of DCF 110
        6.2.2.2 Option Pricing Method 111
            6.2.2.2.1 Examples: Copper Mines and Oil Reserves 111
        6.2.2.3 Indirect Methods 112
            6.2.2.3.1 Replacement or Restoration Cost 113
            6.2.2.3.2 Hedonic Analysis 114
            6.2.2.3.3 Travel Cost Method 114
            6.2.2.3.4 Contingent Valuation 115
        6.2.2.4 Some Comparisons among the Indirect Methods 116
        6.2.2.5 Principle 7 117
        6.2.2.6 Public Policy and Valuation Methods 117
6.3 Accounting and Reporting 118
    6.3.1 Accounting Information Systems (AIS) 119
    6.3.2 Internal Aspects of Financial Reporting 120
    6.3.3 Environmental Audit and Compliance Measurements 122
    6.3.4 Activity Based Costing 125
6.4 External Reporting for the Environment 128
    6.4.1 Principle 13 129
    6.4.2 Risk Assessment 129
    6.4.3 Taxes and Environmental Accounting 130
6.5 Conclusion 131
6.6 References 132
6.7 Appendix I: Rubenstein's Environmental Trust Accounting 134

6.1 INTRODUCTION

If businesses, individuals, and governments are to adopt policies that promote long term sustainability, and develop policies that address their sometimes conflicting interests, then those trained in business schools need to develop linkages with those trained as scientists. Two interrelated concepts that help to develop these linkages are valuation and reporting. These concepts are represented by two additional principles: valuation and auditing and reporting. If a company's managers are to make rational decisions that reflect the corporation's goals and objectives, including inter-generational equity and sustainable development as outlined in the previous chapter, then these decision-makers need a procedure for representing the true monetary values and costs according to established norms. This valuation process must reflect not only the expected revenues and costs, but also their variability (risk). The reporting for this valuation process must ensure that the information generated meets the needs of all of the company's stakeholders.

Corporations face a difficult problem in reporting the costs and liabilities associated with environmental issues. The reports must allow for the divergent interests of all of the company's stakeholders: stockholders, creditors, employees, suppliers, customers, government, and the public at large. The scientist must play a role in these valuation and reporting processes. In addition to the physical dimensions of the problems addressed in measurement issues, scientists must assist in assessing the likelihood of future claims and costs to the company attributable to its past and present policies.

A common understanding of terminology and methodologies between the scientific and business sectors is necessary not only within corporations, but also across corporations. This allows for comparability of financial statements and technical reports across firms. Financial statements will need to be expanded from their present forms to include not only accounting for environmental liabilities, but also the recognition of depreciation of the company's and society's natural resources. An accurate determination of the values of the liabilities and depreciation expenses will be dependent on accurate measurement and risk assessment mentioned above.

We saw in Chapter 3 the difficulty of incorporating environmental costs into the national accounting system, as well as some suggestions for overcoming these difficulties. In this chapter, we are going to look at the problem from more of a financial perspective as we consider several interrelated topics: different meanings of value, discounting, valuation, auditing, and reporting, and their relationship to sustainability. To make decisions consistent with sustainable development, a firm must have data that can be used to value assets and to report the values of these assets to the company's stakeholders.

Valuation of natural resources to society, discussed in Chapter 3, also is important, especially from legal and public policy perspectives.[1] The valuation of natural resources:

- allows courts to assess damages resulting from environmental harm.
- deters future pollution.
- helps to protect existing ecosystems.

Accurately assessing the value of natural resources is important to aid in the preservation of nature. Of course, not everyone assesses the value of goods, services, or natural resources the same way. Consider the plight of the gray wolves from the perspective of environmentalists, policy makers, and people owning property in the wolves' habitat as reported by various news sources in 1995. See Box 6.1.

6.2 VALUATION

6.2.1 DIFFERENT MEANINGS OF VALUE

As we saw in the previous chapter, there are several different ways to view the meaning of value. One author identified three types of values: use value, existence value, and intrinsic value.[1]

6.2.1.1 Use Value

Use value refers to the value that humans assign to goods or services when the goods or services are used for practical reasons, such as camping or fishing. The use value may measure a person's will-

Box 6.1 Wolves: Reintroduction Sparks Wide Coverage, Sierra Suit

Greenwire 1/12/95
With a goal of returning the wolf to its original range and controlling local game populations, the U.S. Fish and Wildlife Service plans to bring up to 150 wolves over five years from Alberta to the United States. *(Globe and Mail 1/5/95)*

Legal Action

Last week, a federal judge rejected a motion by farmers and ranchers to halt the program, asserting that they had shown no evidence the animals will kill livestock rather than game. *(Greenwire 1/4/95)* The Sierra Club Legal Defense Fund went further, filing a federal lawsuit in Boise, ID, to prevent ranchers from killing any wolves that may be migrating to the area. Although gray wolves are listed as an endangered species and may not be harmed, the relocation plan includes a compromise allowing ranchers "to defend their livestock with lethal and non-lethal means." But ranchers call the program "an example of big government and urban environmentalists pulling the economic rug out from under rural Americans." *(AP/Baltimore Sun 1/8/95)*

Greenwire 1/25/95
Animal-rights groups oppose the reintroduction owing to the likelihood that some of the predators will be killed. Jane Schwerin, of People for Animals in the Prevention of Cruelty and Neglect, stated: "The wolves will be persecuted, slaughtered, and tortured to death exactly as they always were. Man's nature has not improved."*(AP/Denver Post 1/22/95)*

Greenwire 3/22/95
Rep. Barbara Cubin (R-WY) criticized the federal government's "arrogance" in the wolf-reintroduction plan, saying the "blatant disregard shown by the Interior Department toward the people of Wyoming is symptomatic of an uncaring bureaucracy." *(Cubin release 3/21/95)*

Conservationists Celebrate

Defenders of Wildlife has established a $100,000 fund to reimburse ranchers at fair market value for livestock lost to wolves. *(Defenders of Wildlife release 3/21/95)*

Fighting over Wolves or the West?

Larry Bourret, executive vice president of the Wyoming FarmBureau Federation, a group that has strongly opposed wolf reintroduction, stated: "In reality, this thing is not so much about wolves." Rather, it is about "limits on the land, the defining feature of the West." Ranchers remain convinced that wolf reintroduction is another step toward land-use restrictions. *(Chicago Tribune 3/20/95)*

"Frontier Justice"

A Canadian wolf found dead less than two weeks after federal agents released her in Idaho's Frank Church River of No Return wilderness *(Greenwire 1/31/95)* now appears to have been "a victim of frontier justice." The wolf's body was found next to a newborn Angus calf whose stomach had been torn open. But a U.S. Fish and Wildlife Service autopsy has found the calf died of natural causes. *(Newsweek 3/27/95)*

Wolf: Idaho House Rejects Greater State Involvement

Greenwire 2/2/95
Local "resentment" of the Interior Department's wolf reintroduction program prompted the Idaho House yesterday to reject by a 2 to 1 margin a state wolf-management plan. "Several rural lawmakers said they did not dare vote for anything connected to the wolf project because their constituents are so strongly opposed." Others "complained" the federal effort "was planned and carried out without state involvement."

Wolf: Newborns Are "Doing Fine" in Yellowstone

Greenwire 5/5/95
One of 14 Canadian gray wolves transplanted to Wyoming's Yellowstone National Park has given birth to seven or eight pups. According to wildlife officials, the newborns are "doing fine" near Red Lodge, MT. The wolf pups' father remains missing. His collar was found along a roadside and "officials fear someone killed the animal." *(USA Today 5/5/95)*

ingness to pay to use these resources or to avoid their loss. The use value may be reflected in licensing fees paid or travel costs for camping or hunting.

Use value is particularly attractive to economists because a measurable monetary amount is associated with the use, and because the use value is based on the actions of consumers rather than some stated willingness to pay. Others object to use value because it assumes that the value of a good is measured only by consumption value and that use value is affected only by those with sufficient income to pay for the resource. Reliance solely on a use value would assign a zero value to any resource or species that cannot be used by humans to improve their lot.

6.2.1.2 Existence Value

The existence value refers to the value that humans assign to a resource solely for the benefit of acknowledging its existence, even if no use is intended. The existence value recognizes that resources may have value (which may be captured by the option valuation model considered below), because of their future use, perhaps in ways that we presently do not recognize.

The knowledge that a preserved resource helps some people may provide a vicarious benefit that adds to the existence value of the resource. For example, many people contribute to organizations such as the World Wildlife Fund to preserve the natural habitat of animals that they have no intention of ever seeing in the wild.

The existence value, like the use value, is still essentially anthropocentric.[2] Both values are defined in terms of the importance of the resources or species to humans because of their use or existence. Without human interest, the resource or species would have no value.

Using existence value to determine the value of a resource or species can create some perverse incentives. Suppose a new and/or endangered species, having little or no use value, but considerable existence value, is found on a company's private property. The company knows that acknowledging the presence of the species may prevent it from using the land in the manner the company intends. Hence, the company is likely to withhold this information from the public in order to use the land in a profit maximizing manner unless there is some compensation mechanism for the discovery and stewardship of new species. And, as critics of the Endangered Species Act have noted, humans have a tendency to overemphasize the value of large, intelligent mammals to the detriment of insects and plants. Methods for determining the existence value of resources will be explored below under valuation methods.

6.2.1.3 Intrinsic Value

The intrinsic value refers to the value that resources may have completely independent of any value that humans may assign to the resource, and is attributable solely to their being part of the earth's ecosystem. Although this type of valuation may have some appeal, it is difficult to allocate resources based on this concept. There is no obvious way to assign a monetary value under intrinsic valuation, and without monetary values, it is impossible to allocate resources among alternative ends. With intrinsic values, the economy could come to a standstill because all natural resources were assigned an infinite value, or the economy could explode as all resources were instantly exploited because all resources were assigned a value of zero.

6.2.2 VALUATION METHODS

6.2.2.1 Direct Methods

6.2.2.1.1 Discounted Cash Flow Method (DCF)
The most commonly used method for valuation in finance and economics is the discounted cash flow method. As we saw in Chapter 3, this method involves expressing the value of assets (which could

Valuation and Reporting

be a forest or a fire truck) as the discounted present value of the cash flows from the asset. This can be represented by the following equation:

$$PV = CF_1/(1+r) + CF_2/(1+r)^2 + CF_3/(1+r)^3 + \ldots + CF_n/(1+r)^n \qquad (1)$$

where

PV represents the (present) value of the asset.
CF_i represents the (expected) cash flow from the asset in period i.
r is the discount rate which represents the opportunity cost investors bear (and demand compensation for) by postponing consumption and facing the risk of inflation and default when investing in the asset.
n is the life of the asset.

6.2.2.1.1.1 Example: A Case Study of The Evergreen Pine Tree Company

To see how this technique might work in practice, consider the following example adapted from a popular corporate finance book.[3] The Evergreen Pine Tree company is growing pine trees on 200 acres of land. Evergreen knows that it could cut all of its trees this year, sell them as Christmas trees, and net $80,000 (4,000 trees at $20 per tree). Evergreen knows that its land is worth $100 per acre. It expects that the price of the trees will grow at a rate 14 percent per year for the next 5 years, 8 percent from years 6 through 10, and 2 percent thereafter. The price of land should grow at a rate of 5 percent in perpetuity. Evergreen has just received an offer from a large cleaning products company that sells many pine scented products. The company wants to purchase the entire 200 acre tract for $120,000. Evergreen believes that a 10 percent discount rate is appropriate. All cash flows are net of taxes. Should Evergreen sell its land? If not, when should Evergreen harvest its trees?

The standard DCF solution to this problem requires a comparison between the present value of the cash flows for each possible harvesting strategy to the offer from the cleaning products company. This is shown in the table below.

	Cash Flows			
Year Sold	Land	Trees	Total	Present Value
0	$20,000	$80,000	$100,000	$100,000
1	$21,000	$91,200	$112,200	$102,000
2	$22,050	$103,968	$126,018	$104,147
3	$23,152	$118,523	$141,676	$106,443
4	$24,310	$135,116	$159,425	$108,890
5	$25,525	$154,032	$179,558	$111,491
6	$26,801	$166,355	$193,156	$109,031
7	$28,141	$179,663	$207,804	$106,636
8	$29,548	$194,036	$223,584	$104,303

We can see from the table that although the future values of the land and trees increase steadily each year, the discounted present values rise from years 1 through 5, then fall steadily thereafter. The reason for this pattern of the present value of the cash flows is that, for years 1 through 5, the growth rate in the cash flows exceeded the discount rate; therefore, the present value rose. After year five, the growth rate in the total cash flow was less than the discount rate, and so the present value fell. This points to what some people feel is a disturbing aspect of the discounted cash flow method.

Whenever the discount rate is greater than the expected growth rate of the cash flows, the wealth maximizing strategy is to exploit the natural resource today rather than preserve it for the future.

In this example, the present value of the proceeds from the sale of the trees and land never exceeds the $120,000 offer from the cleaning products company. Therefore, Evergreen should sell its assets to the cleaning products company. The trees and land are worth more to the cleaning products company than they are to Evergreen.

The discounted cash flow technique is widely used in industry and strongly advocated by the finance profession as witnessed by its prominence in most corporate finance textbooks. It is deceptively easy to use. We simply make an estimate of the cash flow, choose a discount rate, and put them all together in the appropriate fashion to produce the value of the asset. However, in spite of its relative simplicity (or perhaps because of it), the DCF method, as applied to natural resource valuation, in particular, has been attacked by people both inside and outside of the finance profession.

6.2.2.1.2 A Critique of DCF

The critique of the DCF method from inside the finance profession is exemplified by the arguments of Brennan and Schwartz.[4] Brennan and Schwartz' primary problem with the DCF method is the choice for inputs for both the numerator and denominator of equation (1). The numerator is assumed to be the *expected* cash flow from the project, while the denominator, the discount rate, is assumed to be both known and constant. The use of expected cash flows and a constant discount rate turns, what should be and at first glance appears to be, a dynamic valuation process into a static one. The valuation process implicit in equation (1) provides no room for managerial discretion for resource management throughout the life of the project. By fixing the cash flows and discount rates today, we do not recognize the possible options that managers have, such as terminating the project if cash flows are less than expected, or finding other uses for the resource.

Because future decisions, such as project abandonment or expansion, would change the risk inherent in the future cash flows, we can see that the use of a constant discount rate also does not allow for this type of managerial discretion, and is hence a weakness of the DCF technique. Inasmuch as the concept of sustainability has profound inter-generational implications, we should be somewhat leery of a technique that ignores the possibility of decision-making at a future date. As we shall investigate in more detail later, Brennan and Schwartz advocate that an option-pricing technique is more appropriate for valuing natural resources.

A second source of controversy regarding the DCF method is the choice of the discount rate. In a typical valuation done in a corporate setting, the discount rate is the company's cost of capital, or opportunity cost. However, some environmental economists[5,6] contend that the discounting process gives too little weight to distant cash flows (e.g., when the interest rate is 10 percent, one dollar received in 20 years is worth about fifteen cents today). This, in turn, causes a short run approach to resource management and leads to a rate of resource exploitation greater than what is believed to be socially optimal.

These economists also argue that the discount rate used by the government in public policy decisions should be substantially lower than that used by private interests because society would choose to save more collectively than the sum of individuals' savings decisions. Using a lower discount rate would at least ensure that the value of government-held resources would increase compared to privately held resources. Consequently, private industry would be less willing to purchase or rent natural resources from governments, and the world-wide stock of natural resources would be enhanced.

Some feel that the entire discounting process is invalid because it presumes that those of us alive today are somehow more valuable than those who will be living in the future. Certainly, if we are to achieve the type of inter-generational equity called for in the previous chapter, we must think carefully about how we value future generations compared to the present one. However, using a zero discount rate may be inappropriate in that it presumes a certain indifference between the well-being of forthcoming generations in the near future and those to come in the distant, unforeseeable future.

To the extent that people feel a close bond between their generation and the next two or three generations in the future, they will naturally behave in a way that takes these near-term generations' well-being into account to a greater extent than they will the well-being of people yet-to-be born for thousands of years. Implicit in this type of behavior is a positive discount rate. The actual level of this discount rate still remains an unresolved and controversial issue.

6.2.2.2 Option Pricing Method

In 1973, two articles on the pricing of options appeared in the financial economics literature that caused a small revolution in the valuation process. One was by Black and Scholes;[7] the other by Merton.[8] In these articles, a new option pricing formula was developed that has since proven to be been very useful to practitioners and theoreticians. Equally important, these authors have shown how option pricing can be used to value assets and securities other than the simple stock option.

An owner of a stock (call) option has the right (option) to purchase a specified number of shares of common stock at a specified price (the exercise price) within some specified period of time (the time until expiration). When someone purchases a call option, the buyer is hoping that the stock price will rise above the exercise price (the higher the better) before the expiration date. The call option owner can then exercise the option—purchase the stock at the low exercise price—and later sell the stock at the high market price. The result is a nice profit for the option buyer.

A put option gives the owner the right to *sell* a specified number of shares of a security at a specified price within a specified period of time. A put option owner makes money from the drop in the price of the underlying security. Both types of options typically cost much less than the underlying security. Their low cost can result in huge profits from the purchase of an option. However, if the underlying security price increases, then the option will expire worthless and the investor will have lost the entire investment. Options are very risky securities. More information about option pricing is available in Box 6.2.

6.2.2.2.1 Examples: Copper Mines and Oil Reserves

Brennan and Schwartz[4] and Siegel, et al.[9] have used the option pricing model to value natural resources. These authors contend that the option pricing methodology is most appropriate when the quantities and values of future cash flows are unknown, (e.g., the cash flows from a copper mine), and when it is likely that the owners of the resource may be making decisions about the project (e.g., abandonment) throughout the life of the project as new information becomes available.

In the case of a copper mine, considered by Brennan and Schwartz, the owner of a mine has the right to acquire copper (analogous to buying common stock in the case of a stock option) at an exercise price equal to the variable cost of production. As with any option, the owner has the right to walk away without exercising the option if the exercise price (variable cost) is too high. With the risk-free rate, variance of returns to a copper mine, and the expected life of the mine (time until expiration) the option pricing methodology can be used to value the mine.

The "simple" Black and Scholes formula given in Box 6.2 would not be used in the mine case because this problem contains a few complexities such as whether the mine is currently open or closed, and the amount of unexploited inventory in the mine. These complexities make the mathematical solution even more formidable than just plugging into the Black and Scholes formula. Numerical solutions do exist, but they are far beyond the scope of this book. The important point is to see the option-like characteristics of some natural resource valuation problems and to be aware of methods available for finding these values in a manner some feel is superior to the DCF method.

An undeveloped offshore oil reserve, considered by Siegel, et al. provides its owner with the option of purchasing a developed oil reserve for the price of its development cost (exercise price). Again, the measures of the variance of the rate of change of the value of a developed reserve, the risk-free rate, and the length of the lease must be determined to value the option. Siegel et al. tested the option-pricing model against the DCF method for some actual Federal sales of offshore leases. They found the values obtained from the option pricing formula to be closer to the winning bid than

> **Box 6.2 An Example of Option Pricing**
>
> As we can see from the call option pricing formula below, the mathematics behind the Black and Scholes option pricing formula is rather the foreboding, but the success of the empirical tests of this formula and its variants have been successful enough to induce most option traders to keep the formula programmed into their calculators.
>
> $$C = SN(d_1) - Ee^{-r\tau}N(d_2)$$
>
> where
>
> $$d_1 = \frac{(S/E) + r\tau + \sigma^2\tau/2}{\sigma\tau^{1/2}}$$
>
> $$d_2 = d_1 - \sigma\tau^{1/2}.$$
>
> S is the stock price, $N(\cdot)$ is the cumulative normal density function, E is the exercise price, e is the base of the natural logarithm (e=2.7183), r is the interest rate, τ is the time until expiration, σ^2 is the variance of the stock price, and is the natural logarithm.
>
> In spite of its complexity, we can make a few basic inferences from the formula:
>
> 1. The higher the **stock price** (or, in general, the underlying asset price), the higher the value of a call option. Certainly this should be obvious. An option buyer will make more money from selling the stock, after having purchased it for the exercise price, when the stock price is high.
> 2. The higher the *exercise price,* the lower the value of the option. Since the exercise price is the price a call option holder must pay to buy the underlying stock, the higher the exercise price, the more costly the purchase, the less valuable the option.
> 3. The longer the *time until expiration,* the higher the value of an option. The longer the time until expiration, the greater the chance for the stock price to rise during the life of the option.
> 4. The higher the *variance of returns* on the underlying common stock, the higher the value of an option. Stocks with a high variance of returns are likely to rise or fall in price by a bigger amount than a similar stock with a low variance.
> 5. The higher the *interest rate,* the higher the value of an option. Higher interest rates reduce the present value of the exercise price which increases the net return from holding the option.

those obtained from the DCF formula. The DCF method would not have won any of the six tracts available for bidding, but the option-pricing method would have won two of the six. In most cases, the DCF formula produces a lower estimate of the value because it typically ignores many of the future opportunities for decision-making that are captured by option pricing.

6.2.2.3 Indirect Methods

A common weakness of both the DCF and option pricing methods of valuation is that they require measurable cash flows and physical outputs. Both methods are most often used to calculate the values of *privately* held goods. Both also assume that it is the use value that determines the worth of natural resource, but note the conflicts that arise from a narrow view of property, Box 6.3. Quite often, however, we wish to know the value of goods that are less easily measured or whose benefits are shared simultaneously by many people for reasons other than their use in providing a good or service. These goods, such as air quality or national parks, are usually referred to as *public* goods. There are also situations in which the government has determined that property presently under private ownership must be regulated for the public good, and so these property owners demand compensation under a "takings" law (see the following two boxes). Finding the values of such goods through the DCF or option pricing methods is problematic.

> **Box 6.3   Property Rights: GOP Moderates' Victory "Up in the Air"**
>
> *Greenwire 2/16/95*
> As the House Judiciary Committee this morning began marking up the GOP proposal to make federal agencies pay property owners for regulatory "takings," it was impossible to confirm whether GOP moderates had managed to hang on to the victory they apparently had won yesterday afternoon. *(Greenwire Sources)* As of yesterday evening, House Republicans had "rolled back" their property-rights proposal. Under the GOPers' Contract With America, landowners would have qualified for compensation if any "final agency action" devalued their property by at least 10 percent. But the revised version would have required compensation only when property lost more than a third of its value. Further, the new version would have covered only losses due to the Endangered Species Act, wetlands provisions of the Clean Water Act, or the "swampbuster" program created by the 1985 Farm Bill. *(Greenwire Sources)* The revised bill also would have required compensation funds to come from the regulating agency's budget. *(Congress Daily 2/15/95)*
>
> **Alliance Opposes "Takings" Bill**
>
> *Greenwire 2/10/95*
> The National League of Cities, National Council of Churches of Christ, and American Community Protection Association yesterday came out against a GOP plan to compensate landowners for regulatory property "takings." Speakers at the announcement included landowners who said their property values fell when their neighbors broke pollution and wetlands laws. The announcement came as a House Judiciary panel began the first congressional hearing on the plan. *(Coalition release 2/10/95)*
>
> **Regulation: Dole Toughens His "Takings" Bill; USFWS Order**
>
> *Greenwire 3/23/95*
> Senate Majority Leader Dole (R-KS) has revised his "takings" bill to require federal compensation to landowners whose property loses value owing to regulation. His previous bill "merely" required that the federal government evaluate how a proposed rule would affect property values. *(Greenwire 3/6/95)* The plan would pay landowners when federal action lowers the value of any portion of their land by more than 33 percent. That's "less generous to property owners" than the House bill, which set the threshold at 20 percent. "Overall, however, the Senate bill may prove more expensive than its House counterpart" because it would apply to all regulations, not just the wetlands, endangered-species, and water-use rules covered by the House measure.
>
> **A Chilling Effect?**
>
> *Greenwire 3/23/95*
> Congress' "mere consideration" of property-rights and other regulatory-overhaul measures "may already be inhibiting enforcement of environmental laws." Earlier this month, the U.S. Fish and Wildlife Service sent enforcement officials a memo stating, "Effective immediately, no criminal actions under the [ESA] may be taken without prior notification to the Director or Deputy Director." *(Wall Street Journal 3/23/95)*

Even if we could determine the market values of these goods through one of the above valuation techniques, and thus find the value of the resource that would allow for its most efficient use, it is not clear that society would accept the value. As we can see from the laws governing the admission to national parks and monuments, we have chosen to make the public goods available to a wider than optimal range of people from an efficiency perspective. Implicitly, and sometimes explicitly, society has used other techniques to assess the value of these resources. We now turn our attention to several techniques that may be more appropriate for the valuation of public goods.

6.2.2.3.1   Replacement or Restoration Cost
One simple way to assess the value of a natural resource is to consider the replacement or restoration cost. This technique works best for water systems or natural habitats rather than for a particular species, because once a species is lost, the replacement or restoration costs would be infinitely high

given existing technology. (Jurassic Park notwithstanding!) One advantage of the restoration cost method is the relatively wide availability of cost data. Most biological supply firms can provide price information for many natural resources.[1] A second advantage to the restoration cost method is that it implicitly considers the unknown benefits that a species may provide in the future.

One problem with the restoration method is that it does not consider the demand for the resource. An ecosystem could cost millions to replace; but if no one would use it or otherwise care about its existence if it were to be restored, does it still have that multi-million dollar value? Some contend that restoration cost is inappropriate as a valuation technique, because they believe that no amount of restoration can return an ecosystem back to its original value.

6.2.2.3.2 Hedonic Analysis

Hedonic valuation of natural resources is based on the assumption that goods are valued for their utility-bearing attributes or characteristics. Hedonic prices are the implicit prices of differentiated products. These implicit prices are derived by comparing assets having observable market values and varying levels of public good (or bad).

The classic examples of hedonic analysis involve real estate and air quality.[10, 11] Because everyone enjoys the benefits of unpolluted air, it is difficult to assign a private property right to clean air. Hence, it is difficult to directly observe the value of clean air. However, since clean air should raise property values, it may be possible to infer the value of clean air by comparing the value of real estate for similar properties, but with different air qualities. Presumably the differences in the real estate values would reflect the value of clean air.

Hedonic valuation involves the use of an econometric technique that determines the value of environmental amenities such as clean air. This involves formulating a hedonic price equation in order to estimate the marginal implicit price of the environmental variables. From these implicit prices one can estimate marginal willingness to pay functions for groups of households. From these functions, one can compute the value of the amenities.

However, some economists are reluctant to use the hedonic methods because of some troubling assumptions that are implicit in this method.[12] For example, if differences in housing prices are caused by differences in air quality, consumers must recognize differences in air quality when they determine housing prices. Some economists would argue that consumers are unlikely to be aware of such differences and so any observed relationship between housing prices and air quality is likely to be either spurious correlation or the measurement of some other variable affecting housing prices.

Another criticism of the hedonic valuation technique is its implicit assumption that households have full information on all housing prices, and that their moving costs and other transaction costs are zero. A third problem is the determination of housing prices as the discounted present value of expected cash flows and amenities. If consumers are expecting an improvement in the air quality in one or two years, the expected improvement would be reflected in housing prices today. Yet today's housing price would then be regressed against today's higher air pollution levels. This misspecification could result in a less positive relationship between air quality and housing prices.

6.2.2.3.3 Travel Cost Method

Another commonly used valuation technique is the travel cost method. This method[13] considers travel costs, opportunity cost of time lost while on vacation, entrance fees, and other expenses from travel to determine the consumers' willingness to pay for a public good. As a use-based valuation method, the travel cost method is appealing to economists because the data needed to calculate value is easily obtained (with the possible exception of the value of the time spent while traveling). In addition, the valuation method is based on actual behavior rather than on a hypothetical willingness to pay.

However, the travel cost method is not without its critics. Several such critics have noted that the travel cost method does not adequately measure incremental changes in the quality and value of natural resources. It may take years for the public to become aware of the decrease in the quality of a national park or an increase in its user fees and then adjust the public level of consumption. The travel cost method of valuation is also sensitive to the price changes in underlying commodities, such as the price of gasoline. If the price of gasoline rises, the value of the national parks may increase as it becomes more costly to travel. However, if fewer consumers travel because of the higher cost of travel and a decrease in perceived real income caused by the gasoline price increase, then the value of national parks may decrease. Linking the value of national parks to the price of gasoline seems rather unappealing and tenuous to some. A final criticism of the travel cost method is that it does not consider the value of natural resources to those with insufficient income to travel or to those who would rather appreciate them from afar.

6.2.2.3.4 Contingent Valuation

For some environmental amenities (e.g., the solitude of the deep woods), neither hedonic analysis nor the travel cost method of valuation is appropriate. The reason is that, in the deep woods, few, if any, nearby assets with readily observable market values exist. In fact, in this case, the value of the amenity is a direct result of the absence of any assets with observable market values. Presumably the cost of travel to the deep woods is considerable, and limited access is essential to preserve the solitude of the woods. Consequently, an alternative to hedonic analysis and the travel cost method is needed to determine value. One such alternative is contingent valuation.

Contingent valuation involves surveying the public's willingness to pay (WTP) for various environmental amenities.[14] The survey involves a bidding procedure to elicit the highest possible WTP. When the WTP has been determined, the surveyor also compiles site specific (e.g., miles traveled to the site) and socio-economic data (e.g., income) about the respondent. These data are combined with the WTP data and are used to estimate a regression equation in which the WTP is the dependent variable and the socio-economic variables are the independent variables. Upon determining the WTP from the survey, the individual bids are aggregated on an equally weighted basis to assess the total value of the resource.

Researchers have found numerous problems with the contingent valuation method, but its defenders claim that most of the problems can be controlled by appropriate survey design and implementation. We shall consider several of these potential problems below.

1. *Strategic behavior*—If people who favor the restriction on the use of resources are surveyed, they may intentionally overstate their WTP in hopes of raising the inferred value. User fees based on this computation will be set artificially high, thus restricting demand and eventual use. Heavy users of the resource may intentionally report their WTP as less than the true value in order to keep user fees down.
2. *Instrument bias*—Since the respondents are oftentimes surveyed through a bidding process, the initial price suggested by the surveyor may have an undue influence on the respondents' answers. This influence may be particularly true if the respondent is pressed for time or if the surveyor presses the respondent for an answer. Whittington et al.[15] have shown that giving respondents more time has a significant negative effect on the WTP.
3. *Framing effects*—Kahneman and Knetsch[16] provide evidence that the response that is obtained in a contingent valuation can be influenced by the way the question is framed. They find that when they ask people to value a list of assets, people will respond to the order of the list which will, in turn, affect the value of the asset. Assets that are listed first receive higher values than those listed last.
4. *Embedding effects*—Kahneman and Knetsch[16] also show evidence of an embedding effect in contingent valuation. As an illustration of an embedding effect, they report that the expressed willingness of Toronto residents to pay increased taxes to prevent the drop in fish populations in all Ontario lakes was only slightly higher than the willingness to preserve fish in the Muskoka area (a much smaller area) of Ontario. It appears that the WTP

for the whole is much less than the sum of the WTP for the parts. Kahneman and Knetsch attribute the anomalies to the likelihood that the contingent valuation method is not so much an example of the WTP for the protection of a public good as it is a willingness to acquire a feeling of moral satisfaction.

5. *Willingness to buy or willingness to sell (WTS)?*—Most contingent valuation surveys explicitly ask respondents for their WTP for (buy) a public good rather than the WTS that public good. Some contend that the surveys should ask for the WTS.[17] It certainly makes a difference which is used. Most studies show that the WTP is less than the WTS.[18] This difference is true for most goods not just for natural resources. For example, the price people will pay to have someone mow their lawn is usually substantially less than the price these same people would accept to be induced to mow their neighbor's lawn. Reasons for this difference include both income and substitution effects. Paying for an asset may not change peoples' wealth, but it does reduce disposable income; selling an asset raises disposable income. In addition, consumers do not feel that private goods are easily substituted for public goods, making consumers unwilling to sell the public good.[19]

One consequence of this difference is that if court damages are based on WTP rather than to WTS, court damages will be substantially lower. Cross argues for using the WTS (at least in damage cases) since the public already owns any natural resources that have been damaged.[1] The price the public would pay to preserve the asset is, in some sense, the price the public would sell the resource for to someone who would damage the resource.

By not requiring any observable market values, contingent valuation allows for the calculation of existence value rather than the use value. However, unlike the travel cost method and hedonic methods, this method is based on a hypothetical situation rather than observable behaviors. To some, this is a strength of the contingent valuation method, but to others it is a weakness.

6.2.2.4 Some Comparisons among the Indirect Methods

There have been several studies that have attempted to value natural resources by comparing two or more of these valuation techniques. Brookshire et al.[20] compare the contingent valuation method with the hedonic techniques for valuing air in various communities around Los Angeles. They present a theoretical analysis showing why the computed value of the air should be higher when using the hedonic technique compared to the contingent valuation method. Their empirical analysis supports their hypothesis, and they conclude that this is evidence for the validity of the contingent valuation method.

Bishop and Heberlein[21] find that, when comparing contingent valuation and travel cost methods, the value based on the WTS is higher than any of the travel cost estimates. Travel cost estimates, in turn, are generally higher than the WTP (depending on the estimate of the value of time). Schulze et al.[22] examine the evidence from Bishop and Heberlein, as well as other studies that attempt to assess the values of environmental attributes. They conclude that individuals provide contingent valuations comparable to what actual market behavior implies individuals are willing to pay for those attributes. Finally, Seller et al.[23] demonstrate that a closed-ended form of contingent valuation, in which respondents are allowed to choose among some precise estimates of the value of a resource and the travel cost method is superior to an open-ended contingent valuation method.

It should be noted that these different estimates of valuation are not necessarily either/or propositions. More than one valuation method can be used to value a single resource in order to obtain a range of values. Cameron[24] shows how travel cost and contingent valuation methods can be combined to improve the point estimate of the value of non-market goods.

Consideration of the many methods of valuation leads back to our seventh principle of sustainability, the Valuation Principle:

6.2.2.5 Principle 7

To ensure an equitable and efficient allocation of resources across space and time, the values of goods and services must reflect all of the costs and benefits from their consumption. Inherent in this valuation process is the internalization of externalities and appropriate discounting of future costs and benefits.

6.2.2.6 Public Policy and Valuation Methods

As noted at the beginning of the chapter, the valuation of natural resources is more than just an academic exercise. Government agencies must arrive at estimates of value in order to set usage fees and determine sale prices. In addition, the courts utilize assessed values in order to determine and apportion damage among injured parties, a difficult proposition around Superfund sites. See Box 6.4. However, government agencies and the courts have not always been consistent in their use of the previous definitions of value and valuation techniques.

Early court cases relied on use value to uphold the regulation of public resources by the States[1] (Geer vs. Connecticut 161 U.S. 519, 534, 1895). However, when Congress passed the Wilderness Preservation Act of 1964, existence value was implicitly used in that the lands were being protected from human use. This intent for the land has been affirmed by the U.S. Supreme Court. Similarly, the Wild and Scenic Rivers Act, which directs that "environments shall be protected for the benefit and enjoyment of present and future generations," recognizes the likely increase in the future value of resources beyond their present use value.

Although neither the courts nor government agencies have shown any explicit recognition of intrinsic value, the Endangered Species Act appears to provide some evidence for the acknowledgment of intrinsic value. However, as Ehrenfeld[25] and Brown[26] have noted, the use of the term

---

Box 6.4 Superfund: New Orleans Site Threatens to Embarrass EPA

Greenwire 2/14/94

Since the EPA declared New Orleans's Press Park and Gordon Plaza subdivisions a Superfund site in December 1994, "stress levels" among the 1000 residents have "skyrocketed," while home values have "plunged," report Christopher Cooper and Coleman Warner of the *Times-Picayune*. But a $4 million EPA study of the mostly black neighborhood concludes it "is not a particularly dangerous place to live." The "most dangerous acute hazards" are chloroform in tap water and benzene in auto emissions—"toxins found everywhere in New Orleans and in most other cities."

Enviro Justice Order Led to Superfund Listing

When EPA first considered the site in 1986, it scored a 3 out of 100 on the agency's danger ranking, "far short" of the 28.5 needed to qualify for federal clean-up dollars. But after the EPA in 1990 began counting soil as a "pathway" for human exposure to toxics, and President Clinton issued an executive order on enviro justice, EPA scientists upgraded the score to 50.

Residents Want a Buyout

Now, with the Superfund designation making the area's houses "unsaleable," residents "are hoping the government will use its most drastic tool for protecting people from chemical hazards: a buyout of all homes in the neighborhood." The agency "officially refuses to discuss a buyout," but an official who requested anonymity said it seems the site "is just not dirty enough." Gordon Plaza resident Nathan Parker, who tried to convince the EPA to reconsider its position: "You have to be in my position to understand. Everything I saved for is gone."

"resource" connotes some usefulness to humans, while the term "value" implies the presence of those who can make assessments of the utility of a resource.

For public policy purposes, market valuation methods do not capture any of the existence value of resource; consequently, the courts and legislatures have been reluctant to adopt them. Market value methods have been used extensively in the measurement of damages to private natural resources. In addition, many more examples of the use of the indirect methods of valuation of public natural resources exist.[1]

Because of the readily available information about the prices of natural resources from biological supply firms, and in spite of the lack of a true relationship between restoration cost and true social value, the restoration and replacement cost method of valuation has been used by both courts and legislatures. This method was the valuation method implicit in the original Clean Water Act, and the First Circuit Court of Appeals awarded Puerto Rico six million dollars in damages based on the cost of restoring beaches, forests, and wildlife after an oil spill.

The difficulty of obtaining complete and accurate data has made the courts reluctant to employ the travel cost method in assigning damages. The hedonic valuation method primarily measures the value of the resources to private landowners. Since private landowners often represent only a small proportion of those who actually value a particular natural resource, the hedonic valuation method's use by the courts has been limited.

The contingent valuation technique has also had limited usefulness to the government. While one EPA report noted the consistency of the cost estimates provided by the different valuation methods, including contingent valuation, was strong evidence in support of their usefulness,[27] another concludes that values based on contingent valuation methods are so unreliable that they should not be used.[1] In his assessment of these techniques, Cross[1] finds that the difficulty in their use further supports the argument that public natural resources should be privatized. Privatization would allow for the use of market values in court cases, and more importantly, would avoid the "tragedy of the commons."[28] However, given many of the considerations of equity addressed in the previous chapter, this issue is far from being resolved.

6.3 ACCOUNTING AND REPORTING

In Chapter 3, we looked at the macroeconomic aspects of accounting for the environment. In this chapter we look at some of the same issues from the viewpoint of the corporation and its stakeholders. If a company's stakeholders are to make decisions about a company in a manner consistent with sustainability, the company's financial reports must reflect the true (private and public) value of resources and any liabilities resulting from environmental risks.

Unlike the valuation of public goods which has received a great deal of attention within the economics profession for the past several decades, the treatment of environmental costs on the financial statements has only recently begun to attract the attention of the accounting profession.[29] The Statement of Financial Accounting Standards No. 5 (FAS 5) directs companies to recognize liabilities on their balance sheet if a significant likelihood that the cost from the liability will be realized by the company and the amount of loss can be reasonably estimated.

However, the Financial Accounting Standards Board (FASB, the governing body of accountancy) has provided few details as to the precise recognition of the liabilities. As Rubenstein[30] notes, part of the problem with the recognition of environmental liabilities is that such liabilities are oftentimes shared with other corporations or individuals (e.g., a landfill site and its associated costs) and entail a great deal of uncertainty about the size of the environmental liability leading to valuation problems that can take a very long time to settle.

Methods that place environmental costs onto the asset side of the balance sheet have been almost totally ignored by FASB. However, some proposals for the treatment of these costs by the Big Five accounting firms, as well as by several academic accountants, have been devised. Their suggestions will be treated below. The treatments are divided into two types: external and internal (sometimes referred to as financial and managerial, respectively). The external reporting is done on behalf of stockholders, creditors, customers, and the government. Calls for financial statements that fully reflect all environmental costs are becoming increasingly common.[31]

The internal reporting system is designed for the benefit of the company's management. The principle is the same, however. The cost accounting system should be designed to allow mangers to see the full costs of the products that their company produce and the activities that produce these products. We will argue below that an activity-based accounting system will best serve management.

Before considering the details of environmental accounting, it is best to review the basics of the accounting information system, and note the differences and similarities between accounting information systems for corporations and those for economies. We should note at the outset that the output of a corporate accounting information system is an input to the national accounting system. The net income of corporations is part of the gross domestic product of the nation. The assets of corporations are included in the national economy's capital stock.

However, some fundamental differences exist between corporations and the government as to how each accounts for its stocks and flow. Many of these difference can be attributed to how each accounts for its resources. These have do to with conventions such as operating on an accrual basis or a cash basis, or the recognition of spending today rather than amortizing over the life of an asset. For example, the purchase of an Air Force bomber goes into this year's government spending even though it may have a useful life of ten years. A corporation would include the purchase of such an asset on the balance sheet and then amortize the purchase over the life of the asset.

6.3.1 Accounting Information Systems (AIS)

An AIS (see Figure 6.1) is an entity or a component within an organization that processes financial transactions to provide score keeping and decision-making information to information users.[32] The AIS is part of the overall management information system (MIS) (see Figure 6.2A, Figure 6.2B) which has the same purpose as the AIS, but the MIS is designed to process much more than just the financial information (e.g., personnel reports). An accounting system is a relatively closed system[33] whose inputs are economic events that become accounting transactions. Information to and from the accounting systems flows both vertically and horizontally through the organization structure.

The recording of the transactions and the production of the output (the financial statements) are the processes of the system. Within the accounting systems are accounting subsystems that provide control mechanisms for the overall system. The design of an accounting system that provides the company's stakeholders with meaningful information is a critical component of sustainable development. As we can see from Figures 6.1 and 6.2, the accounting information system provides information for both internal and external use, and its inputs and outputs come from both inside and outside the organization.

Both types of accounting systems, internal and external, as well as the MIS as a whole, must be designed to promote a sustainable use of resources. Such a design ensures that all relevant parties are receiving timely and accurate information. Ditz et al.[34] describe how a well designed accounting system can be used in six different decision-making processes of an organization:

1. Product mix decisions.
2. Choosing manufacturing inputs.
3. Assessing pollution prevention projects.
4. Evaluating waste management options.
5. Comparing environmental costs across facilities.
6. Pricing products.

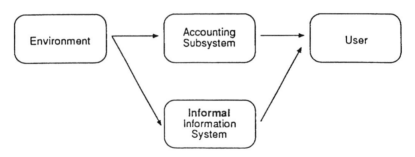

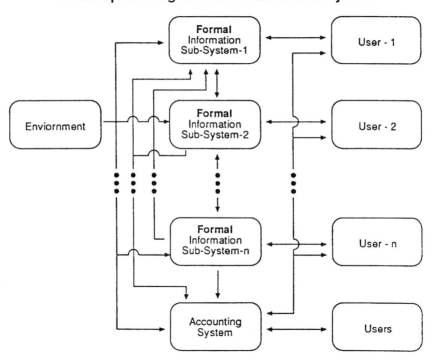

FIGURE 6.1 Two models of the relationship between the MIS of the firm and its external environment.[36]

The importance of the AIS and how it can aid decision-making can be found later in the activity based costing section.

6.3.2 INTERNAL ASPECTS OF FINANCIAL REPORTING

The Business Charter for Sustainable Development, launched in April, 1991, by the Second World Industry Conference on Environmental Management, included a listing of their own principles concerned with improving scientific assessment of industrial impacts and minimizing risks or other hazards.[35] Their 16th principle states the business commitment as follows:

Valuation and Reporting

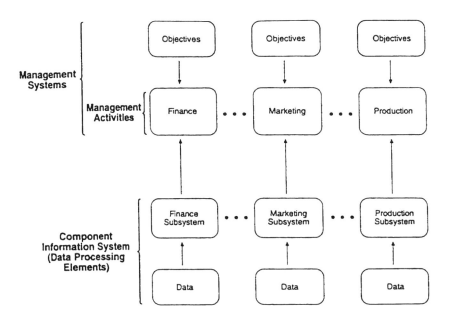

FIGURE 6.2A A component approach to system formulation.[36]

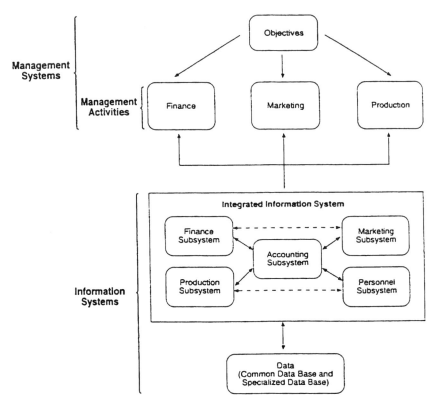

FIGURE 6.2B The systems approach to system formulation showing the relationship between the AIS system and the MIS.[36]

Compliance and Reporting—To measure environmental performance; to conduct regular environmental audits and assessments of compliance with company requirements, legal requirements, and these principles; and periodically to provide appropriate information to the Board of Directors, shareholders, employees, the authorities, and the public.

For an industrial firm to adopt this principle, its audit staff, beginning as students in business schools, need an understanding of each element in the principle: environmental audits, company requirements, legal (regulatory) requirements, and appropriate information which we interpret as financial liability information for the Board of Directors and public. Also relevant are the joint contribution of science and technology and its measurement of environmental performance in the corporate setting.

Good business practice requires that access to essential information from environmental reports be maximized, and costs, risks, and liabilities evident in each would be minimized. However, quantification of the measurements for each report is technologically complex and requires state-of-the-art contributions by all the individuals involved. All participants need to understand and accept their obligations to all stakeholders. The following sections explore the technical and business issues involved, and the training needed to make the various computations, facilitate the business-science linkages, and report the results in financial statements.

6.3.3 Environmental Audit and Compliance Measurements

The term "environmental audit" is being applied here to several related activities in environmental management. The main use of the term is to describe a programmatic review or audit of the measures underway at a site to reduce waste, air emissions, or other discharges that incur disposal charges or potential liability.[37] A second form of environmental audit relates to the review of detailed mass balance records of hazardous substances brought into a plant (e.g., mercury) and requiring that 100 percent of its ultimate use or disposition be documented and accounted for. A third form is the compliance audit, often conducted by government regulators, but critical for business as well. A waste reduction audit[38] can be a simple walk-through survey of the plant to visually identify opportunities for waste reduction. It can also be a highly technical process involving waste stream sampling and a process or engineering analysis. The evaluation may focus on only one waste stream or process, or it can become an extensive business-wide analysis.

When conducting an audit, the scientific staff will note that wastes come in many different forms including gases, liquids, solids (trash), and mixed forms. One process may produce several types of wastes in a variety of forms. For example, the parts cleaning process can produce vapor wastes, liquid wastes, and solid waste (tank bottom sludges). Other wastes include air emissions, evaporative wastes, hazardous wastes, heat and/or energy loss, maintenance and clean-up waste, obsolete or outdated stock, solid wastes (trash), spills and container leaks, system leaks, and waste water. The audit must focus on all wastes, not just regulated waste.

The protocols used in an environmental audit are typically designed around the concept of compliance rather than reporting to stakeholders. Ideally, the information obtained from the audit/compliance process should be expressed, ultimately, in terms that are amenable to decision-making by senior management and for inclusion in quarterly environmental and/or financial reports.

The ICC[37] lists many benefits from an environmental audit. Among them are the following:

- promote comparisons and interchanges of information among operations and plants.
- raise employee awareness of environmental policies.
- improve relations with authorities (e. g., regulators) by the implementation of complete and effective auditing system.
- provide an information base for internal management decision-making.

Wainman[39] has added the following to this list:

- improved efficiency and environmental cost controls.
- enhanced corporate public image.

In addition to these operations oriented-definitions of environmental audit, the term audit is also used here to describe the usual internal control system in most corporations, including audits of accounting records, projects and programs, organizational performance, and computerized information systems.[40] In some corporations, the internal accounting audit is performed in conjunction with the environmental audit, but in most companies this is not the case.

However, in those companies in which the environmental audit and accounting audit are connected, an additional benefit to the environmental audit is the creation of a data set which will provide the corporation with the means for the implementation of a full cost accounting system that includes environmental costs.

In a 1992 survey of over 1100 U.S. companies by Price Waterhouse, of the 47 percent of those companies responding to the survey, 58 percent indicated that the internal audit function of the company performed the audits in the compliance with company's environmental policies and procedures, but only 33 percent of respondents indicated that the internal audit function performed a financial audit for environmental sites. However, as environmental costs become a large part of a corporation's budget, the need to combine these activities is likely to increase.

In a 1994 follow-up survey of 445 firms, Price Waterhouse discovered that:

- 73 percent of companies have environmental auditing programs.
- 63 percent of companies have formal policies to account for clean-up costs, up from 33 percent in 1992 and 11 percent in 1990.

Not only are the internal accounting and operations audit less likely to be separate functions, they also are also less likely to be done solely for internal control purposes. Many of a corporation's stakeholders are reading the results of these audits with great interest: regulators, insurers, bankers, and suppliers. See Box 6.5. Companies are discovering the benefits of reporting these results to all stakeholders. What better way to show customers and suppliers that a company is not merely in compliance with environmental laws and regulations, but has taken a proactive stance in dealing with environmental issues.

Box 6.5   India: Two Banks to Base Loans on Enviro Assessments

Greenwire 5/2/95

The Indian Bank and the Bank of India will include environmental assessments as part of their loan approval process, the Bombay *Economic Times* reports. Because banks expect the government to alter Indian law to include lender liability provisions, adding these assessments has become "imperative," according to Bank of India Executive Director T. P. Karunanandan.

While banks were also concerned that future pollution might devalue property pledged as collateral, pressure from the World Bank also "hastened the process," Karunanandan stated, "Adoption of a clearcut policy . . . was one of the preconditions for the World Bank for sanction of soft term loans to six Indian banks."

Under the plan, the banks will assess whether the project's pollution prevention measures are sufficient. If necessary, the screening panel will recommend modifications to the project. In addition, the banks will carry out periodic checks "to ensure that pollution-control norms are not violated." The Indian Overseas Bank, the Denn Bank, and Union Bank have told the World Bank that they will create enviro-assessment procedures as well.

In spite of the emerging mandate from external stakeholders, there are disincentives regarding the self-reporting of the results from internal environmental audits. There is some incentive for corporations to attempt to hide harder-to-detect violations. Because of this, any results of an environmental audit to be reported to the public may require the same type of external certification now provided by the Big Five accounting firm in the external audits of the financial statements. To this end we are witnessing the emergence of such organizations as the National Association of Environmental Risk Auditors whose purpose is to train and "certify" environmental auditors. Presently, this organization emphasizes the operations aspect of the audit in the same way the accounting profession emphasizes the financial aspect of its audits.

The availability of the results from an environmental audit to a corporation's regulators creates another, potentially more serious incentive problem for corporations, viz., not doing environmental audits at all. Inasmuch as corporate executives can be fined and jailed for environmental violations revealed in an environmental audit,[41] these executives would rather forgo the audit altogether rather than risk having the results fall into the "wrong" hands. We hope this problem can be overcome simply by recognizing that the long-term benefits of such audits to corporations are outweighing these kinds of costs. If most of the company's stakeholders were to demand the results of an environmental audit, then the absence of such an audit would send a more damaging signal about the corporation than would any negative results from the audit.

In addition, reporting the results of the site or plant-specific environmental audit, corporations are also reporting the aggregate results of the environmental audits performed at each of their plants. For example, Dow Chemical Company issues an annual Environmental Health and Safety report in which it lists data such as combustion emissions, global fines and penalties, and the illness and injury rates per 100 employees. (See Box 6.6 for other companies.) Although their report reviews primarily the physical dimensions of environmental reporting, it does allow for inferences about the level of the company's concern for environmental protection and the likely financial obligations stemming from the physical data.

In addition to the anecdotal evidence cited above, there are also systematically gathered data indicating that companies are responding to calls for fuller reporting of environmental information in publicly available reports. Newell et al.[42] examined the annual reports of 645 of the Forbes 1000 firms. They found that 26 percent of the firms surveyed included disclosure of environmental issues. Although this may not seem like a large percentage, a closer look at the numbers reveals a more positive picture. Of those firms that included environmental disclosures in their financial statements, 38 percent came from five industries: energy, steel, chemical, pulp and paper, and utility, despite the fact that the firms from these five industries comprise only 20 percent of the original sample of firms. For each of the sample of firms examined in each of these industries, at least 45 percent of the firms in those industries included environmental disclosures in their financial statements.

The results from this study are important because they show that firms, in those industries in which environmental reporting is most likely to be important to the companies' stakeholders, have the highest incidence of reporting. These five industries have some of the more notorious past records for polluting the environment, so it is from these types of industries that the public will demand more accountability. There is less cause for alarm when only 10 percent of financial services firms make environmental disclosures than when only 10 percent of pulp and paper mills make such disclosures.

This record of reporting by these five industries also speaks to the incentives that firms have for reporting. The primary incentive is that the stakeholders of these companies want to know the environmental record of the companies in which they have a financial stake. As we saw in the case of the Indian banks, both stockholders and bankers need to know the size of environmental liabilities before they will be willing to commit funds into a corporation. Suppliers must be sure that the raw materials shipped to these companies do not become part of a toxic waste site that they (the suppliers) will be liable for.

> **Box 6.6 Corporate Reporting: Analyses Becoming More Standardized**
>
> *Greenwire 1/23/95*
>
> About 150 companies worldwide now have issued reports on their environmental performance, and the 1993–1994 reports "are leaner and more graphical" than the "first generation of reports issued just two years ago." A *Tomorrow* magazine review of several dozen reports finds their contents are becoming standardized, with most North American- and European-headquartered companies addressing topics recommended by the corporate-led Public Environmental Reporting Initiative: an organizational profile; enviro policies and management practices; releases; resource conservation; risk management; compliance; product stewardship; employee recognition; and stakeholder involvement. "Many reports also include a CEO statement espousing a belief in 'sustainable development.'"
>
> **North America vs. Europe**
>
> The recent reports "display the payoff from corporate investment in development of environmental information management systems." Many feature "credible, quantitative" data in areas such as pollution control and safety. North American companies gave the "most detailed" information on waste disposal, but European companies "appear to have taken the lead" in analyzing "upstream inputs" like water, energy, and raw materials.
>
> Indicators linking enviro performance to production levels or annual sales seem to have become more common. In the oil industry, for instance, Du Pont, BP, Neste, Norsk Hydro, and others now present energy use relative to refinery throughput or company sales.
>
> "Environmental expenditures, infrequently reported in 1991 and 1992 reports, are becoming commonplace in both North American and European reports." But the financial implications of this information "are hardly self-evident."
>
> "An increasing number of reports" also contain a statement from a consulting firm, commenting on the "acceptability of procedures used to aggregate data from facilities, the adequacy of environmental auditing procedures in place, and the commitment of top management to environmental concerns." Some statements suggest improving the data quality and a few of the boldest contain ideas for improving the auditing program or follow-up activities.

The improved public image that comes from behaving as a responsible corporate citizen is also likely to increase the likelihood of sales to those customers demanding green products. As Smart[43] notes, the companies at the forefront of environmental reporting and understanding are well-known firms whose products and processes have been hazardous and/or highly polluting. It is these companies that have first discovered the mandate for and opportunities in taking a proactive stance with environmental concerns.

6.3.4 ACTIVITY BASED COSTING

Another aspect of internal accounting is the measurement and reporting of materials, labor, and overhead costs from the production of goods and services. One of the challenges faced by the managerial accountant is the appropriate assignment of costs to individuals goods produced. Activity based costing (ABC) is an internal cost accounting system which assigns costs to products according to the impact each product makes on all of a company's resources.[44] ABC was introduced in the 1950s, but its value to the accountants has only recently been recognized.

ABC is basically a two-step process. The first step is to identify the activities (what an employee spends time doing in a company) and the related costs of the activity. The second step is to assign those activities and their related costs to products (goods and services), channels (different methods of distributing goods and services to customers, e.g. catalog, retail, or wholesale sales), and customers via cost drivers. A cost driver is a factor or event that triggers an activity to occur or causes an expense.

A more detailed list of steps for ABC has been offered by Ray and Gupta:[45]

1. Identifying activities.
2. Distinguishing between value and non-value added activities.
3. Tracing the sequence of product/service flow though activities.
4. Assigning of cost and time values to each activity.
5. Establishing linkages between activities within functions and across functions.
6. Making product/service flows more efficient; reducing non-value activities; making tradeoffs between linked activities where net savings are possible.
7. Continuing to improve the system.

The second step in this process is critical. Non-valued activities are those which do not contribute any value to a manufactured product that can be priced in the market. Examples are inspection time, move time, wait time, number of setups, and number of production orders. The overhead costs of these cost drivers are identified for calculation of activity-based overhead allocation rates to be applied to individual products. Unlike the traditional cost accounting system, ABC products that use higher proportions of non-value added activities are reported as high cost products relative to those products using a lower proportion of non-value added activities. Under the traditional method, these costs are likely to be assigned to products on a pro rata basis according to the number of units sold. Environmental cost based examples of non-value activities include filling out environmental permits, measuring and processing effluents, environmental auditing, and risk assessment.

Implementing an ABC system is a crucial component to the establishment of an information system consistent with sustainable development. In the following section, we consider an example[44] of how an ABC can differ from the traditional cost accounting system in its respective accounting for environmental costs. This example involves the costing of two types of chairs: finished and unfinished.

Overhead costs:
Total overhead costs: $30M
   Corporate administration: $12M
      Environmental Costs: $5.4M

Direct labor costs per chair:
Finished chair: $10.50
Unfinished chair: $10.00

Units sold:
Finished chair: 500,000
Unfinished chair: 500,000

Total direct labor costs:
Overhead based on direct labor costs (traditional):
Finished: $15.37M = $30M · 5.25/(5.25 + 5.00), $30.73 per chair
Unfinished: $14.63M = $30M · 5.00/(5.25 + 5.00), $29.27 per chair

Environmental overhead based on direct labor costs (traditional):
Finished: $2.76M = $5.4M · 5.25/(5.25 + 5.00), $5.53 per chair
Unfinished: $2.64M = $5.4M · 5.00/(5.25 + 5.00), $5.27 per chair

Raw materials costs per chair:
Finished chair: $15.00
Unfinished chair: $10.00

Total costs per chair (traditional):
Finished chair: $56.23 = $30.73 + $10.50 + $15.00
Unfinished chair: $49.27 = $29.27 + $10.00 + $10.00

Profit margins per chair (traditional):
Finished chair: $3.77 = $60.00 − $56.23
Unfinished chair: $0.73 = $50.00 − $49.27

Total environmental costs assigned to each chair (ABC):
Finished chair: $5.37M, $10.74 per chair
Unfinished chair: $0.03M, $ 0.06 per chair

Total costs per chair (ABC + traditional):
Finished chair: $61.44 = $10.74 + $25.20 + $10.50 + $15.00
Unfinished chair: $44.06 = $ 0.06 + $24.00 + $10.00 + $10.00

Profit margins per chair (ABC + traditional):
Finished chair: ($1.44) = $60.00 − $61.44
Unfinished chair: $5.94 = $50.00 − $44.06

Brooks et al.[44] analyze a furniture company that makes one million finished and unfinished chairs per year. Of its $30 million in overhead costs, $12 million are attributable to corporate administration. Of the $12 million, $5.4 million are environmentally related costs: permitting, compliance, risk management, recycling and reuse, and public relations. Under a traditional cost accounting method, overhead may be assigned to a product based on the direct labor costs of the product. For the unfinished chairs, the direct labor cost per chair is $10.00, while the cost for the finished chair is $10.50.

Since the company sells 500,000 of each chair, the total direct labor costs for the unfinished and finished chairs would be $5 million and $5.25 million, respectively. If overhead is based on direct labor costs, 48.78 percent of overhead ($14.63 million) would be assigned to the unfinished chairs, including 48.78 percent of environmental costs. The remaining 51.22 percent of overhead ($15.37 million) would be assigned to the finished chairs. On a per chair basis, each unfinished chair would be assigned $29.27 of overhead ($14.63 million/500,000 chairs) including $5.27 of environmental costs per chair [($14.63 million · ($5.4 million/$30 million))/500,000 chairs]. The finished chairs would be assigned $30.73 of overhead. In addition to the direct labor and overhead costs, the raw materials cost for each finished and unfinished chairs are $15.00 and $10.00, respectively.

Adding all of these costs together gives us a total cost of $49.27 (= $29.27 + $10.00 + $10.00) for the unfinished chair and $56.23 (=$30.73 + $10.50 + $15.00) for the finished chair. If the respective selling prices for each finished and unfinished chair are $50.00 and $60.00, then the corresponding profit margin is $0.73 for the unfinished chair and $3.77 for the finished chair. Under this traditional cost accounting method, while both chairs are profitable; the finished chair is substantially more profitable.

However, a different comparison would be made using an ABC system. Such a system would not naively base overhead costs, particularly the environmental costs, on the direct labor costs. In fact, the environmental costs of the unfinished chair (solid waste disposal of sawdust and residual glue) are significantly less than those of the finished chair (hazardous waste disposal of residual paints, stains, solvents, and toxic adhesives; storage permit of hazardous substances, environmental site audits, and solid waste disposal costs). Even if the other overhead costs are assigned in the traditional manner, of the total $5.4 million in environmental costs, only $30,000 are assigned to the unfinished chair for an environmental overhead cost of $0.06 per unfinished chair and $10.74 per finished chair. This changes the total costs per chair to $44.06 (=$0.06 + $24.00 + $10.00 + $10.00) for the unfinished chair and $61.44 (= $10.74 + $25.20 + $10.50 + $15.00) for the finished chair. By appropriately accounting for the environmental costs, the finished chair is now unprofitable (loss = −$1.44 = $60.00 − $61.44), while the unfinished chair is now much more profitable (net profit = $5.94 = $50.00 − $44.06).

6.4 EXTERNAL REPORTING FOR THE ENVIRONMENT

The study of methods, measurement, and external reporting of the valuation effects of low-level or chronic off-site damage, or the liability for long-term chronic site effects from emissions or waste disposal is still a developing field. The problem of unintended serious damage is diminishing, but companies seeking to meet the Business Charter principles[37] have an obligation to measure environmental performance, report compliance with company or industry requirements, and provide appropriate information to the Board of Directors. For example, the effects from gaseous discharge of mercury incineration and fossil fuel combustion are being viewed world-wide with some alarm; contributors to these discharges, even when they are in compliance with government standards, must be able to make the measurements necessary to estimate the possible amount of damage to the environment, and report that to the Board of Directors.

Clearly, all of the corporation's stakeholders have a material interest in the results of environmental audits and their implications for the financial health of the corporation. Given the sometimes conflicting interests of these stakeholders and the lack of universally recognized accounting techniques for reporting environmental impacts, corporations have been understandably slow in developing financial statements that reflect the conclusions of an environmental audit.

Green accounting is a financial reporting system that reflects the full impact of the firm's interaction with the environment. Awareness of the need for green accounting has been growing rapidly, as illustrated by examples described in Box 6.7. Calls for financial statements that more fully reflect

---

Box 6.7   Accounting: Upcoming Rule Could Spur Bankruptcies, Reform

Greenwire 4/10/95

A draft federal accounting standard set to be released next month would require fuller environmental disclosure on corporate financial statements starting on December 15. "Because of the enormity of Superfund and other environmental liabilities, some companies would show bankruptcy on their balance sheets" after the change, reports *Environment Week's* Viki Reath.

The Securities and Exchange Commission will seek public comment before issuing the final rule that was drafted by the American Institute of Certified Public Accountants. The AICPA sets the principles and procedures which accountants must follow for documents filed to the SEC.

The current standard requires firms to disclose "contingent liabilities" in guidance on how and when companies must report liabilities—"possibly earlier in the remediation process." In a 1992 survey by Price Waterhouse, 62 percent of respondents revealed that their companies had unreported enviro liabilities.

Likely to Spur Superfund Reform

The new rule could spur efforts in Congress to repeal "retroactive liability" for pollution created before Superfund was enacted in 1980. "Industry is concerned that the difficulty of assigning monetary value to natural resource damage could result in high costs to businesses," *Environment Week* reports.

But business critics see it otherwise. Jonathan Turley, director of the Environmental Crimes Project at the National Law Center, George Washington University, stated "It is very telling that just as the marketplace is moving toward greater accountability for violators . . . corporate lobbyists are pressuring Congress to relieve clients of responsibility."

Valuation and Reporting

depletion of natural resources and liabilities arising from possible cleanup costs has come from environmentalists, bankers,[46] stockholders, and other creditors. All require accurate information about the financial condition of the company, and because the financial conditions of companies are being increasingly affected by environmental considerations, it is imperative that the accounting profession comes to an agreement as how to provide this information.

Unfortunately, only modest progress has been made thus far by the accounting profession to adopt a reporting system that fully reflects the environmental costs when reporting to outside stakeholders. Great uncertainty as to the size of the environmental liabilities, combined with incentives for corporations to abstain from an accurate representation of environmental costs, have slowed the movement towards full disclosure. Rubenstein[30, 47] has taken some important steps in demonstrating how a full disclosure of environmental costs would affect a company's financial statements. We shall summarize some of his suggestions here.

Rubenstein reexamines the purchase of the Hooker Chemical Company by Occidental Petroleum in the late 1960s following the burial by Hooker of thousands of drums containing chemical waste in a landfill site by the Love Canal in upstate New York. Rubenstein argues for the creation of a natural asset trust account that would show changes in the value of the land owing to increased clean-up costs from the burial of chemical waste. Notice that this type of accounting uses the replacement cost method of valuation discussed in the beginning of the chapter. Obviously, other types of valuation could be used here. At this point we should not be so concerned as to what type of valuation to use, but rather how the changes in value will affect the company's financial statements. The journal entries and T-accounts that emerge from Rubenstein's model are presented in Appendix I.

The type of green accounting suggested by Rubenstein is not likely to be welcome by all business people in general or accountants in particular. Accounting for the environment as presented by Rubenstein is based more on non-traditional concepts such as shared resources and implicit contracts with stakeholders (such as future generations) rather than the more conventional concepts of private property and explicit contracts between the firm and its owners and creditors. The guiding ruling by the Financial Accounting Standards Board, FAS No. 5, must be revised if the accounting profession is to have explicit guidance as to the recognition of environmental liabilities.

Thus, because of the need for external reporting and the importance of environmental auditing, we state a thirteenth principle of sustainability, the Principle of Auditing and Reporting:

6.4.1 PRINCIPLE 13

The information systems of individuals and organizations must be designed so that the financial consequences of the environment are accounted for in the reporting system. These systems should reflect the true values of assets and liabilities and should be designed so that all of the company's stakeholders can utilize the information.

6.4.2 RISK ASSESSMENT

One activity that both internal auditing and external reporting requirements must address is risk assessment. As we first saw in Chapter 4, risk assessment, historically, has been a relatively focused concept, the calculation of the likelihood of a negative outcome (plant failure, disease, etc.) from some combination of low probability circumstances (human error, exposure to toxins, etc.). In corporate environmental management, the concept of risk must be carried through all forms of accidents, routine waste discharge effects, and liability for off-site damages or long-term waste depository effects.

According to the National Research Council,[48] risk assessment is largely a scientific enterprise, whereas risk management is the process by which an organization (business) decides what to do about the results of a risk assessment. The assessment of risk requires some understanding of the goals of the organization and how potential problems with inputs, outputs, and by-products of the organization may affect the achievement of that goal. One author has noted that "all risk assessments are subjective and judgmental to a certain degree, and the concept of reality is not so firm in the world of risk."[49]

Another risk assessment issue faced by corporations is the difficult problem of reporting the cost implications of the risks or liabilities in their annual financial reports. These reports must allow for the divergent interests of all of the company's stakeholders, many of whom benefit from the use of the product.[50] Scientists must play a major role in these valuation and reporting processes, but business executives must weigh all the relevant information. In addition to the physical dimensions of the problems addressed in measurements, scientists should assist in assessing the likelihood of future claims and costs to the company attributable to its past and present policies.

The problem of reporting costs and liabilities is compounded by the difficulty in translating the scientific results of a risk assessment into a form that allows for meaningful decision-making by corporate executives. A particularly complex part of this problem is the incorporation of risk into a financial statement. Whereas, for the scientist, risk assessment may involve determining the likelihood a toxic by-product causing a certain number of cancer cases in employees, the executive must think of risk in terms of variability in net income or cash flows and its effects on competitive position or stock price performance.[49]

A common understanding of methodologies between the scientific and business sectors is needed not only within corporations, but also across corporations. Common methodologies and common training would allow for greater comparability of financial statements across firms. Corporate reporting should be expanded to include accounting for environmental liabilities as well as the depreciation of natural resources used directly or indirectly by the company. An accurate determination of the values of liabilities and resource depreciation will depend on accurate measurement of resource stocks and flows from year to year as well as the risk assessments described above.

Annual reports that properly convey to the corporation's stakeholders the monetary value of the financial liabilities inherent in a risk assessment require communication with the internal auditors as well as the external financial auditors of the corporation. The American Institute of Certified Public Accountants (AICPA), Statement on Auditing Standards No. 9, specifies that external auditors must justify their reliance on any work done by internal auditors based on two criteria:

1. Internal auditor competence.
2. Their objectivity.[40]

Given the increasingly complex task of environmental auditing and the resulting financial liabilities, internal auditors will need to improve both their competence and objectivity to increase their interaction with the external auditors. This will be crucial if meaningful financial reports are to be made available for external uses.

6.4.3 TAXES AND ENVIRONMENTAL ACCOUNTING

As we will see again in the next chapter, tax policy plays a significant role in our government's management of environmental issues, and subtle changes continue, as outlined in Box 6.8. Environmental managers for a business need to become familiar with these tax policies when making business decisions. As reported in Stinson and Schaltegger,[29] there are many environmentally related tax laws in the U.S. tax code. Among them are:

1. A five cents per barrel tax on crude oil received at U.S. refineries to fund the Oil Spill Liability Trust Fund.
2. Various federal taxes to fund the Leaking Underground Storage Tank Trust Fund.
3. Taxes on petroleum, petrochemicals, and inorganic chemicals to fund the Hazardous Substance Superfund.

> **Box 6.8 PCSD: Panel to Ponder "Tax Shift," Other Economic Reforms**
>
> *Greenwire 10/28/94*
>
> President Clinton's advisory panel on sustainable development began consideration yesterday of the politically explosive issue of changing tax laws to deter pollution. At a meeting in Washington of the President's Council on Sustainable Development, charged with giving the president policy options by mid-1995, Under Secretary of State Timothy Wirth raised the idea of a "revenue neutral tax shift" which would impose levies on pollution and remove them from work and saving. Wirth said the move could lead to both economic and environmental gains. "There will be 'a great deal of pressure' in the next two years to discuss tax changes," Wirth said. "We're just nibbling around the edges with 'command-and-control' [environmental regulation]," he said, suggesting "this is precisely the kind of bold initiative" PCSD should embrace.
>
> After considerable discussion from the industry, enviro and citizen leaders on the panel, EPA Administrator Carol Browner said, "I think we've got to pull this one up and deal with it sooner rather than later." Ciba-Geigy Chairman Richard Barth said a fuller debate is warranted, but that he doubted if any proposal would be truly "revenue neutral" from the perspective of industry sectors that would see their tax burdens rise.
>
> Fred Krupp, executive director of the Environmental Defense Fund, said there are "a lot of minuses on tax policy, and other market mechanisms are more likely to produce results in my lifetime," such as pollution trading systems. "Tradable permits have worked in reducing sulfur dioxide emissions from power plants," Krupp noted, arguing that similar systems could be established for nitrogen oxides and greenhouse gases. Samuel Johnson, chairman of S.C. Johnson & Son, said the key thing is "to put our incentives where our objectives are."
>
> National Wildlife Federation President Jay Hair voiced a concern that dealing with politically charged tax issues in general terms might be "like what Alice Rivlin did," referring to a recently-leaked memo by the director of the president's Office of Management and Budget outlining other politically dangerous decisions that lie ahead. Nonetheless, the group decided to schedule a full council session at its January, 1995, meeting in Chattanooga, TN, to delve more deeply into policy options to integrate economic and environmental goals.

Besides these explicit taxes, The Internal Revenue Service, in conjunction with Congress, also decides on the tax deductibility of such expenses as site clean-up costs caused by adherence to the Resource Conservation and Recovery Act (RCRA) or the Comprehensive Environmental Response, Compensation, and Liability Act (CERCLA), and the funds used to establish a nuclear decommissioning trust as well as the depreciation schedule for pollution control facilities. Companies doing business overseas quickly find, too, that environmental taxes are by no means limited to the United States.[29] In various European countries, there are additional taxes on lead, packaging, and even aircraft noise.

6.5 CONCLUSION

The need for documentation of risks, costs, and outcomes described in this chapter shows clearly that the valuation processes outlined in early sections are inextricably linked to the reporting processes presented in the later sections. Without a comprehensive reporting system, inputs to the accounting and valuation process will be flawed, values will be inaccurate, and resources misallocated. A critical part of the reporting process is an accounting for the physical as well as the financial flows through an organization, while also considering potential external effects of these physical flows.

The practices described here also suggest that an important constraint on environmental management has been inadequate expression of scientific aspects of risks and liabilities, especially evident in the currently incomplete financial reporting. One author has stated, "It seems that any such accounting would have two major functions: One, to keep organizational decision-makers informed

> **Box 6.9   Management: More Companies Giving Environment Top Billing**
>
> *Greenwire 2/3/95*
>
> More U.S. companies are giving the environment "top priority," making their boards of directors and highest ranking executives accountable for environmental issues, according to a recent survey by consulting firm Price Waterhouse. The study, "Progress on the Environmental Challenge," surveyed 445 businesses "most likely to encounter environmental issues," including the manufacturing, public utilities, and extractive industries, but excluding the service, financial, and retailing sectors.
>
> More than 40 percent of the companies responding to the survey said they have elevated environmental compliance to the "board level." That's nearly double the 1992 number and three times the 1990 number. More than half said their environmental managers report directly to the CEO or another top executive.
>
> Dean Petracca, head of Price Waterhouse's Environmental Services Group, stated, "The growing cost of environmental compliance and the increasing attention of regulators, shareholders, and the general public have made companies more proactive":
>
> - 72 percent disclose their accounting policies for enviro costs, up from 26 percent in 1992 and 3 percent in 1990.
> - 17 percent of companies with "significant" liabilities, and 8 percent of those without them, issue annual enviro reports.
> - 38 percent of companies are factoring environmental performance into incentive compensation for executives and senior management.

of the extent to which their particular organization was depleting the planet's capital. Two, to keep society informed about the way its capital was being maintained."[31] Training that fortifies students with a foundation in both the science and business disciplines, and which provides an appreciation for the similarities and differences between these fields, provides industry and government with decision-makers capable of moving national economies closer to the goal of sustainable development.

6.6   REFERENCES

1. Cross, F., Natural resource damage valuation, *Vanderbilt Law Rev.*, 42(2), 269, 1989.
2. Callicott, J., Non-anthropocentric value theory and environmental ethics, *Am. Philos. Q.*, 21, 299, 1984.
3. Brealey, R. and Myers, S., *Principles of Corporate Finance,* 4th ed., McGraw-Hill, New York, 1991.
4. Brennan, M. and Schwartz, E., A new approach to evaluate natural resource investments, in *The New Corporate Finance: Where Theory Meets Practice,* Chew, D., Ed., McGraw-Hill, New York, 1993, 98.
5. Costanza, R. and Daly, H., Natural capital and sustainable development, *Conserv. Biol.*, 6(1), 37, 1992.
6. Marglin, S., The social rate of discounting and the optimal rate of investment, *Q. J. Econ.*, 77, 95, 1963.
7. Black, F. and Scholes, M., The pricing of options and corporate liabilities, *J. Polit. Econ.*, 81(3), 637, 1973.
8. Merton, R., Theory of rational option pricing, *Bell J. Econ. Manage. Sci.*, 4(1), 141, 1973.
9. Siegel, D., Smith, J., and Paddock, J., Valuing offshore oil properties with option pricing models, in *The New Corporate Finance: Where Theory Meets Practice,* Chew, D., Ed., McGraw-Hill, New York, 1993, 108.
10. Ridker, R. and Henning, J., The determinants of residential property values with special reference to air pollution, *Rev. Econ. Stat.*, 49, 246, 1967.
11. Freeman, A. M., On estimating air pollution control benefits from land value studies, *J. Environ. Econ. Manage.*, 1, 74, 1974.
12. Freeman, A. M., Hedonic prices, property values and measuring environmental benefits: a survey of the issues, *Scand. J. Econ.*, 81(2), 154, 1979.

13. Clawson, M. and Knetsch, J., *Economics of Outdoor Recreation,* Resources for the Future, Washington, D.C., 1966.
14. Cummings, R., Brookshire, D., and Schulze, W., Eds., *Valuing Environmental Goods: An Assessment of the Contingent Valuation Method,* Rowman and Allanheld, Totowa, NJ, 1986.
15. Whittington, D., Smith, V., Okorafor, A., Okore, A., Liu, J., and McPhail, A., Giving respondents time to think in contingent valuation studies: a developing country application, *J. Environ. Econ. Manage.,* 22(3), 205, 1992.
16. Kahneman, D. and Knetsch, J., Valuing public goods: the purchase of moral satisfaction, *J. Environ. Econ. Manage.,* 22(1), 57, 1992.
17. Knetsch, J., The WTP and the WTA and the role of law, paper presented at Western Economic Association meeting, 1994.
18. Rowe, R., d'Arge, R., and Brookshire, D., An experiment on the economic value of visibility, *J. Environ. Econ. Manage.,* 7(1), 1, 1980.
19. Haneman, W., Willingness to pay and willingness to accept: how can they differ?, *Am. Econ. Rev.,* 81(3), 635, 1979.
20. Brookshire, D., Thayer, M., Schulze, W., and d'Arge, R., Valuing public goods: a comparison of survey and hedonic approaches, *Am. Econ. Rev.,* 72(1), 165, 1982.
21. Bishop, R. and Heberlein, T., Measuring values of extramarket goods: are indirect measures biased?, *Am. J. Agric. Econ.,* 61(5), 926, 1979.
22. Schulze, W., d'Arge, R., and Brookshire, D., Valuing experimental commodities: some recent experiments, *Land Econ.,* 57(2), 151, 1981.
23. Seller, C., Stoll, J., and Chavas, J., Validation of empirical measures of welfare change: a comparison of nonmarket techniques, *Land Econ.,* 61(2), 156, 1985.
24. Cameron, T. A., Combining contingent valuation and travel cost data for the valuation of nonmarket goods, *Land Econ.,* 68(3), 302, 1992.
25. Ehrenfeld, D., *The Arrogance of Humanism,* Oxford University Press, New York, 1978.
26. Brown, T. C., The concept of value in resource allocation, *Land Econ.,* 60(3), 231, 1984.
27. Smith, V. K., Desvouges, W. H., and McGivney, M. P., The opportunity cost of travel time in recreation demand models, *Land Econ.,* 59(3), 259, 1983.
28. Hardin, G., The tragedy of the commons, *Science,* 162, 1243, 1968.
29. Stinson, C. and Schaltegger, S., Environmental accounting, 1993.
30. Rubenstein, D. B., Lessons of love, *CA Mag.,* 124(3), 34, 1991.
31. Gray, R., Accounting and environmentalism: an exploration of the challenge of gently accounting for accountability, transparency and sustainability, *Acc. Organ. Soc.,* 17(5), 399, 1992.
32. Wu, F., *Accounting Information Systems: Theory and Practice,* McGraw-Hill, New York, 1983.
33. Boockholdt, J. and Li, D. H., *Accounting Information Systems,* Irwin, Homewood, IL, 1991.
34. Ditz, D., Ranganathan, J., and Banks, R. D., Eds., *Green Ledgers: Case Studies in Corporate Environmental Accounting,* World Resources Institute, Baltimore, MD, 1995.
35. ICC, Conference report and background papers, presented at World Industry Conference on Environmental Management, International Chamber of Commerce, Rotterdam, 1991.
36. Leitch, R. and Davis, K., *Accounting Information Systems,* Prentice-Hall, Engelwood Cliffs, NJ, 1983.
37. ICC, International Chamber of Commerce position paper on environmental auditing, in Williams, J.-O., Ed., *The Greening of Enterprise,* International Chamber of Commerce, Paris, 1990, 241.
38. Wigglesworth, D., *Profiting from waste reduction in your small business,* Alaska Department of Environmental Conservation, Alaska Health Project, Fairbanks, 1988.
39. Wainman, D., Balancing nature's books, *CA Mag.,* 124(3), 6, 1991.
40. Ratliff, R., Wallace, W., Loebbecke, J., and McFarland, W., *Internal Auditing: Principles and Techniques,* Institute of Internal Auditors, Altamonte Springs, FL, 1988.
41. Hall, R. M. and Case, D. R., *All About Environmental Auditing,* Federal Publications, Washington, D.C., 1992.
42. Newell, S. J., Kreuze, J. G., and Newell, G. E., Environmental commitments and liabilities of U.S. corporations: disclosures and implications, *Mid-Am. J. Bus.,* 9(2), 15, 1994.
43. Smart, B., *Beyond Compliance: A New Industry View of the Environment,* World Resources Institute, Washington, D.C., 1992.

44. Brooks, P., Davidson, L., and Palamides, J., Environmental compliance: you better know your ABCs, *Occup. Haz.*, 55(2), 41, 1993.
45. Ray, M. and Gupta, P., Activity based costing, *Intern. Audit.* 49(6), 45, 1992.
46. Saxe, D., Caveat creditor, *CA Mag.*, 124(3), 64, 1991.
47. Rubenstein, D. B., Bridging the gap between green accounting and black ink, *Acc. Organ. Soc.*, 17(5), 501, 1992.
48. National Research Council, *Issues in Risk Assessment*, National Academy of Sciences, Washington, D.C., 1993.
49. Sheridan, P. J., Risk assessment: the path of uncertainty, *Occup. Haz.*, 52(10), 68, 1990.
50. Maxey, M., Managing environmental risks: what difference does ethics make?, No. 99, Center for the Study of American Business, Washington University, St. Louis, MO, 1990.

6.7 APPENDIX I: RUBENSTEIN'S ENVIRONMENTAL TRUST ACCOUNTING

In his article on an accounting approach to liabilities at Love Canal, Rubenstein sets up the following accounts:

1. Natural resource trust account:
   a. Natural resource asset account (market value if restored) less accumulated depreciation
   b. Amount due from company (receivable)
   c. Natural resource capital account
2. Income statement:
   a. Natural resource expenses
3. Balance sheet:
   a. Amount due to the natural asset trust account (liability account)
   b. Natural resource deferred expense (asset account)
   c. Cash (asset account)

He next considers the following transactions to see how they are reflected in the preceding accounts.

1. In 1945, Hooker Chemical, before being acquired by Occidental Petroleum, establishes a $50 million natural asset trust account for land it has purchased some time ago that will now be used as a landfill. Although the land may be on Hooker's balance sheet at a much lower price, (e. g., $15 million), its existing value (natural resource asset account) has been determined to be $50 million, also equal to the natural capital account.

   Journal entries:
   Natural Resource Trust Account:
      Natural asset account: $50 million
         Natural capital account: $50 million

2. In 1946, it has been determined that the landfill site has sustained $10 million in damages. On the natural asset trust account, a $10 million addition to accumulated depletion would exist, as well as a $10 million receivable established by Hooker owed from Hooker to the trust account to restore the asset back to its original $50 million value. On Hooker Chemical's books, a $10 million natural resource expense and a $10 million liability to the natural asset trust account would be recorded. Eventually, as the income statement is closed out to the balance sheet, the $10 million natural resource expense would be reflected as a reduction in retained earnings and/or taxes payable.

Valuation and Reporting

Journal Entries:
Natural Resource Trust Account:
   Owed from Hooker Chemical: $10 million
      Accumulated depletion: $10 million

Hooker Financial Statements
   Natural resource expense: $10 million
      Owed to natural asset trust account: $10 million

3. In 1968, just before the acquisition by Occidental Petroleum, the landfill site has been degraded by 90 percent. The accumulated depletion on the natural asset account is now $45 million as is the receivable owed from Hooker to the trust account. An additional $35 million is added to the natural resource expense and liability owing to the natural asset trust accounts on Hooker's books.

Journal Entries:
Natural Resource Trust Account :
   Owed from Hooker Chemical: $35 million
      Accumulated depletion: $35 million

Hooker Financial Statements:
   Natural resource expense: $35 million
      Owed to natural asset trust account: $35 million

4. In 1969, Occidental Petroleum purchases Hooker Chemical and, consequently, the assets of Hooker have been revalued. It is determined that the landfill, if restored to its original condition, would be worth $100 million. Thus, $50 million is added to the natural asset and natural resource capital accounts. Since the landfill is 90 percent degraded, $45 million is added to the accumulated depletion. The receivable now owing from Occidental to the trust account is also increased by $45 million. On the Occidental's books, an additional $45 million expense and liability to the trust account are recorded. Occidental's balance sheet also reflects the current market value of the land.

Journal Entries:
Natural Resource Trust Account:

   Natural asset account: $50 million
      Natural capital account: $50 million

   Owed from Occidental Petroleum: $45 million
      Accumulated Depreciation: $45 million

Occidental Financial Statements:

   Natural resource expense: $45 million
      Owed to natural asset trust account: $45 million

5. In 1988, a court has ruled that the cost to restore a similar landfill site that was 100 percent depleted was $260 million. The natural resource asset and capital accounts of Occidental's site that is 90 percent depleted need to be increased by $160 million. Occidental must also record an additional $144 million of accumulated depreciation to reflect the 90 percent degradation of the land. The amount owing from Occidental is also

increased to $234 million, providing natural resource asset and capital accounts of $260 million. Since Occidental decides to amortize the extra $144 million liability owing to the natural resource trust account, over five years, it records a natural resource expense on the income statement of $28.2 million (= 144/5) and the unamortized portion is placed in the deferred natural resource expense account.

Journal Entries:
Natural Resource Trust Account:

Natural asset account:	$160 million	
Natural capital account:		$160 million
Owing from Occidental Petroleum:	$144 million	
Accumulated depletion:		$144 million

Occidental financial statements:

Natural resource expense:	$28.2 million	
Deferred natural resource expense:		$115.8 million
Owing to natural asset trust account:		$144 million

6. In 1989, Occidental begins clean-up of the landfill site. It spends $50 million in clean-up costs. Thus, cash is reduced by $50 million as is the liability owing to the natural resource trust account. In addition, because the land has been restored, there will be a $50 million reduction in both the accumulated depletion and the amount owing from Occidental Petroleum. Also, during this year Occidental would write off another $28 million in natural resource expenses owing to last year's revaluation.

Journal Entries:
Natural Resource Trust Account:

Accumulated depletion:	$50 million	
Owing from Occidental Petroleum:		$50 million

Occidental financial statements:

Owing to natural asset trust account:	$50 million	
Cash:		$50 million
Natural resource expense:	$28.2 million	
Deferred natural resource expense:		$28.2 million

7. In 1993, the landfill site has been completely restored after having spent an additional $184 million (= 234M − 50M) to fully restore the landfill site to its $260 million value. Over the past five years the $144 million deferred natural resource expense (incurred because of the revaluation in 1988) has been fully amortized.

Journal Entries:
Natural Resource Trust Account

Accumulated depletion:	$184 million	
Due from Occidental Petroleum:		$184 million

Occidental Financial Statements:

Owing to natural asset trust account	$184 million	
Cash:		$184 million

Valuation and Reporting

These journal entries are reflected on the companies' (first Hooker, then Occidental) financial statements (in million of dollars) as follows:

1. Natural Resource Trust Account:

    a. Natural resource asset account:

(1)	50		
(4)	50		
(5)	160		

    less accumulated depreciation

		10	(2)
		35	(3)
		45	(4)
		144	(5)
(6)	50		
(7)	184		

    b. Amount due from company (receivable)

(2)	10		
(3)	35		
(4)	45		
(5)	144		
		50	(6)
		184	(7)

    c. Natural resource capital account:

		50	(1)
		50	(4)
		160	(5)

2. Income statement:

    a. Natural resource expense:

(2)	10
(3)	35
(4)	45
(5)	28
(6)	28

3. Balance Sheet

    a. Amount owing natural asset trust account (liability):

		10	(2)
		35	(3)
		45	(4)
(6)	50	144	(5)
(7)	184		

b. Natural resource deferred expense (asset account)

(5)	116		
		28	(6)

c. Cash (asset account)

	50	(6)
	184	(7)

By the end of 1993, the natural resource trust account should reflect natural resource asset and capital accounts with $260 million values with no accumulated depletion or amount owing from Occidental (receivable). There would be no liability on the balance sheet owing to the trust fund because the land has been fully restored. The cash account has been reduced by $234 million as have retained earning and/or taxes payable (depending on the tax code and the tax deductibility of these natural resource expenses).

7 Sustainability and Business Management Systems

O. Homer Erekson, Timothy C. Krehbiel, and Alison Ohl

CONTENTS

7.1 Introduction 139
7.2 Environmental Externalities and Government Regulation 140
    7.2.1 Legal Remedies 141
    7.2.2 Command-and-Control (Regulation) 141
    7.2.3 Market-Based Incentives 143
7.3 Government Regulation and Business Self-Regulation of Environmental Externalities 146
    7.3.1 The Paradigm of Self-Regulation: Values and Value-Based Management ... 147
    7.3.2 Environmental Management Systems and a Natural-Resource Based View of the Firm 148
7.4 Continuous Improvement Systems in Business 149
    7.4.1 Total Quality Environmental Management 149
    7.4.2 The History of Total Quality Management 150
    7.4.3 TQM in Environmental Management 151
    7.4.4 Examples of TQEM and Environmental Management Systems in Business ... 154
        7.4.4.1 The Council of Great Lakes Industries and TQEM 154
        7.4.4.2 Responsible Care® and the Chemical Manufacturers Association ... 155
        7.4.4.3 The CERES Principles 159
        7.4.4.4 ISO 14000 International Environmental Standards 159
    7.4.5 Characteristics of Self-Regulating Approaches 161
7.5 Conclusion 164
7.6 References 165

7.1 INTRODUCTION

S. C. Johnson & Son, Inc., the wax and polish company, has been recognized widely as a leader in promoting sustainable business practices. *Fortune* magazine has referred to its Chief Executive Officer, Samuel C. Johnson, as "corporate America's leading environmentalist." Since 1990, S. C. Johnson & Son has made several aggressive environmental changes in its operations, including cutting worldwide manufacturing waste and emissions by half, reducing virgin materials used in packaging by one fourth, and removing chlorofluorocarbons from its aerosol products (three years before government bans). Samuel C. Johnson has stated, "We believe that being environmentally conscious makes good competitive sense. For example, we aggressively seek out ecoefficiencies—ways of doing more with less—because waste is lost money. By developing a reputation as a company that produces products that are as safe as possible for people and the environment, we also improve our market share."[1]

Is the aggressive posture taken with respect to sustainable business practices by this and other companies sound? Or does it represent a strategic decision that results in environmental improvements, but with a loss of value to the firms involved? Economists and analysts who have debated these questions have used very different conceptual approaches and reached different policy conclusions. In this chapter, we will explore competing views as to the desirability of sustainable practices as a response to dealing with externalities from business activity. After proposing a framework that may bridge these differing views, we will present perspectives on environmental management systems and how they relate to sustainable business practices.

7.2 ENVIRONMENTAL EXTERNALITIES AND GOVERNMENT REGULATION

As described in earlier chapters, externalities occur when the actions of one person or company impact the welfare of another person, company, species, or other system component in a way that is not transmitted by market prices.* In Chapters 2 and 4, we introduced the notion of the tragedy of the commons, where common property assets such as water, air, land, and biodiversity are not responsibly managed with respect to their use and preservation. Negative external effects often arise in the case of common property. There is a strong tendency toward over use and degradation because there are no well-established property rights. If a ship discards waste or fuel into the ocean, there is no direct accountability for the damage that may be caused. Here, preservation of clean oceans can be seen as a public good that is under provided without some legal or social mechanism to encourage responsible behavior.

Similarly, externalities may arise as companies fail to take into account the full social effects of their actions. Consider Figure 7.1. In a competitive market, a profit-maximizing firm will equate price with marginal private cost (MPC) and produce q_0 of the product, charging P_0. The MPC includes the cost of land, labor, and capital (machinery and equipment). However, if in the produc-

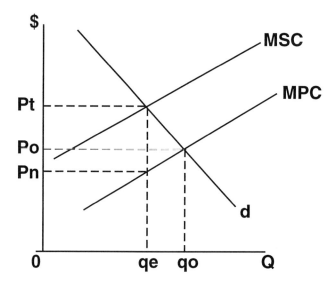

FIGURE 7.1 Externalities and social cost of production.

---

*Some externalities, called pecuniary externalities, are captured in market prices and are not a cause of environmental problems. If a community establishes attractive public parks adjacent to a neighborhood, property values may be bid up automatically. Because these property value adjustments occur automatically, they do not result in a divergence between private and social cost as discussed later on. Here we will reserve the term externalities only for non-pecuniary externalities.

Sustainability and Business Management Systems

tion process, the firm emits effluents or has other external effects on society, then the full cost of production is given by the marginal social cost (MSC) curve that includes both the private costs of production and the external costs to society from that production. The *efficient* level of production for the firm is q_e, where P_t = MSC. However, the firm does not choose to produce at this quantity, because it has no incentive to take into account the full cost of production.

The problem facing society, then, is how to internalize the externality. In effect, we must ask what mechanism can be used to have individuals and businesses confront the full social cost of their actions on others. Traditionally, there have been three approaches to dealing with environmental externalities: legal remedies, command-and-control (regulation), and market-based incentives.*

7.2.1 Legal Remedies

One means of addressing damages from environmental externalities is to pursue remedies through the courts. The options vary significantly depending upon jurisdiction and the type of externality. A frequent form of remedy is applied through common law, particularly negligence and strict liability law. Under negligence, the victim must demonstrate that the defendant did not exercise due care. Often this involves showing that the loss caused by the externality, multiplied by the probability of occurrence, exceeds the cost of preventing the externality. Under a strict liability approach, negligence is not germane; the victim need only prove that damage has occurred. In addition to common law, however, environmental externalities may be prosecuted under criminal law in some situations. In criminal cases, victims are compensated directly. Thus, there is no direct linkage to the value of damage from the externality.[2]

Clearly, legal remedies to environmental problems are expensive, both for victims and for companies accused of creating the externalities. From 1983 to 1991, the Department of Justice reported 838 criminal indictments against polluters, with 73 percent resulting in guilty pleas or convictions. Federal penalties for environmental violations increased from $340,000 in 1983 to more than $18 million in 1991.[3] In the case of negligence, the connection with economics is clear, as a benefit-cost calculus is inherent in the process. However, the linkage is not as clear with strict liability law or with criminal law. From a company's standpoint, there is an economic advantage in avoiding the costs of litigation, as well as the monetary implications of direct financial penalties, possible prison sentences, and a significant effect on the reputation of the firm.

7.2.2 Command-and-Control (Regulation)

Since 1970, the most widely-used method for dealing with environmental externalities has been direct regulation by a government agency. In particular, the regulation of environmental externalities is the primary responsibility of the U.S. Environmental Protection Agency (EPA). The EPA derives its authority from a series of major environmental laws passed by Congress, including the Clean Air Act and its Amendments (1972, 1977, and 1990), the Resource Conservation and Recovery Act and Amendments (1976 and 1984), the Toxic Substances Control Act (1976) and the Superfund Amendments and Reauthorization Act (1988).

In addressing environmental problems across a variety of resources (including, but not limited to air, water, solid waste, toxic substances, and ocean dumping, as discussed in Chapter 4), the EPA has used two primary command-and-control approaches. First, the EPA has established physical standards of a particular pollutant that may be emitted into the environment, and secondly, it has required companies to adopt "end-of-the-line" best abatement technologies or alternative production processes that minimize environmental waste and degradation of resources.

Direct regulation is useful in the case of especially hazardous materials, where the risk of the agent being carcinogenic or threatening to natural resources is clearly demonstrated, or where the

---

*For a detailed discussion of these incentive schemes, see Tietenberg.[2]

externality in question is truly global in nature. For example, chlorofluorocarbons (CFCs) that were once used as propellants in spray cans and as refrigerant agents have been shown to induce depletion of the stratospheric ozone layer. In 1977, the United States adopted regulations limiting the use of CFCs as a propellant in hair spray containers and other consumer products. Later, as substitutes for CFCs became available (and thus the cost of elimination became lower), and as scientific evidence with respect to the damages caused by CFCs became conclusive, world-wide negotiated agreements with manufacturers to end production became desirable. In 1990, an international agreement called the *Montreal Protocol on Substances that Deplete the Ozone Layer* was signed setting the timetable for reduction (and eventually cessation) of emissions of CFCs and other ozone depleting chemicals.[4]

Even though command-and-control techniques are appropriate in cases where the consequences of the externalities are acute, direct regulation has many shortcomings. First, there typically are large administrative costs in the monitoring and enforcement of standards. Second, while some would argue that establishing uniform standards provides for equality among all the regulated parties, this does not necessarily result in an efficient level of pollution emissions, as shown in Figure 7.2.

In the figure, we consider two firms A and B with different levels of emissions of hydrocarbons without regulation, 50 tons and 100 tons, respectively. The marginal abatement cost curve (MAC) shows the cost of reducing each ton of emissions.* Suppose the EPA mandates a command-and-control rule requiring each company to reduce emissions by 10 percent. The shaded area for each firm shows the total cost of reducing emissions by 10 percent. Simple calculations show the total cost to firm A to be $102.5, while for firm B the cost is $110. Thus, although firm B has twice the emissions level of firm A, their total cost of emissions reduction is approximately the same. Not only do the uniform standards fail to address vertical equity concerns (treating unequal violators differentially), they also do not provide an efficient outcome. The marginal damage cost curve (MDC) shows the value of incremental damage from increasing emissions. From society's standpoint, it is optimal for a firm to reduce emissions as long as the marginal abatement cost is less than the marginal

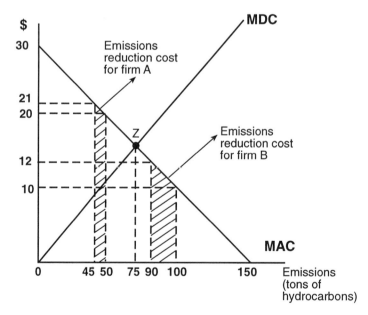

FIGURE 7.2 Externalities and uniform enforcement standards.

---

*The marginal abatement cost curve (MAC) drawn here is linear, with a slope of $-1/5$. Although this is a simplifying assumption used here for expositional purposes, the most common MAC curves in environmental applications are convex curves, where the marginal cost of reducing emissions increases at an increasing rate. That is, the cost of mitigating environmental externalities becomes increasingly more costly as the emissions fall.

damage cost for that level of pollution. The efficient level of emissions is at point Z, where MAC = MDC. In this instance, firm B would reduce emissions from 100 tons to 75 tons to reach efficiency, while firm A would make no reduction, as it already had emissions less than the efficient level. If it were possible to know with certainty the value of the damage from a level of emissions and the abatement costs facing a particular company, the regulator could require all companies to reduce emissions to such an optimal level. However, since neither of these values is known to the regulator, it is highly unlikely that any fixed emissions established as a standard would be efficient. Moreover, direct regulation does not give companies clear incentives to self-correct for externalities.

7.2.3 MARKET-BASED INCENTIVES

Most economists believe in the power of the market as an effective institution for achieving efficiency. In this view, prices serve as the important signaling tool for consumers and businesses in allocating resources to their most highly-valued use. However, free markets work efficiently only when prices reflect the full social costs of production. As noted above, the fundamental economic problem with environmental externalities is that there is no direct incentive for businesses to take into account the external social costs to the environment from their production activities. In considering that direct regulation is unlikely to assure an efficient outcome, economists have suggested that market-based incentives are a more useful way to stimulate the internalizing of externalities.

Market-based incentives have the goal of using quasi-market-based mechanisms to internalize the environmental costs through the decision-making processes of the firm. There are several types of market-based approaches, including the establishment of property rights for those affected, subsidies for abatement equipment, and deposit-refund schemes. Here we will focus on the three most widely used market-based mechanisms: taxes on products, taxes on emissions, and tradeable permits. Table 7.1 shows the adoption of these market-based incentive instruments as used in OECD countries.

Economists have long maintained that properly designed taxes can be used to internalize externalities in an efficient way. However they are designed, these taxes are intended to confront a firm with the full cost of its operations. Taxes may be imposed on the product whose production is resulting in the environmental externality. Referring back to Figure 7.1, if a tax equal to the difference between marginal private costs (MPC) and marginal social costs (MSC) is charged, the firm would reduce consumption to the efficient level q_e. The burden of the tax would be shared by consumers and producers, with consumers paying a higher price, P_t, and producers receiving a lower price, P_n. An alternative is to apply an emissions (or effluent) charge directly on the pollution-producing activity of the firm. As shown in Figure 7.3, the firm would reduce emissions as long as the tax rate was above the marginal abatement cost, in this case reducing emissions from P_0 to P_e.

Let us consider the advantages of these approaches. First, when faced with full costs, there is a direct incentive for firms to be responsive to these costs in their production activities and decision-making. This leads to more efficient outcomes. Second, either of the approaches encourages *dynamic efficiency* as well. Dynamic efficiency involves the incorporation of innovations into the production process that will lead to improved production over time and lower levels of pollutant emissions. This long-run adaptation could result from improved design of the production process, where pollution is prevented in the process itself, or in improved ability to mitigate emissions (thus lowering the marginal abatement cost). Hawken has stated[5]:

> "The purpose of integrating cost into pricing is not to provide a toll road for polluters [a polluter would be forced to pay for his harmful actions, but would still be able to do them], but a pathway to innovation. The incentive to lower costs is the same one that presently operates in all businesses, but in this case the producer's most efficient means to lower them is not externalizing these costs onto society, but implementing better design."

In fact, the tax revenues generated could be used to subsidize improved technology to combat environmental externalities, to conserve goods and exhaustible resources, or to compensate victims of the remaining level of pollution.

TABLE 7.1
Market-Based Incentives Used in OECD Countries as of January 1992

Country	Charges on Emission	Charges on Products	Tradeable Permits
United States	5	6	8
Sweden	3	11	
Canada	3	7	2
Denmark	3	10	
Finland	3	10	
Norway	4	8	
Australia	5	1	1
Germany	5	3	1
Netherlands	5	4	
Austria	3	4	
Belgium	7	2	
Portugal	2	1	
France	5	2	
Switzerland	3	2	
Italy	3	2	
Iceland	1	1	
Japan	3	1	
Ireland	2	1	
Greece		2	
Spain	3		
United Kingdom	1	1	
New Zealand	1		
Turkey			

Source: Adapted from Barde[37], as cited in Parikh.[38]

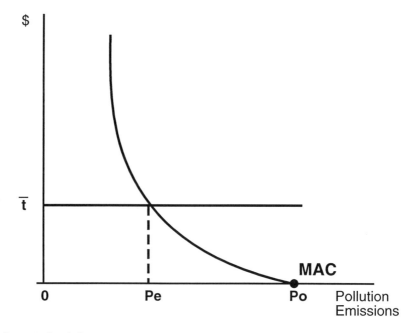

FIGURE 7.3 Impact of emissions taxes.

A third advantage claimed for taxes on polluting activities or products is the prospect of a double dividend. Not only can the tax reduce excessive levels of pollution, but the tax can improve the overall efficiency of the tax system by reducing potentially distortionary effects of taxes on income, sales, and property, and by replacing them with a tax directed at reducing an inefficient situation within the economy. Although this latter case is compelling at first glance, the evidence for it in the economics literature shows conflicting views. Depending upon how the pollution taxes are designed, the structure of the industry in question, the existing degree of distortion from the taxes that would be replaced, and other factors, pollution taxes may be more distortionary than other forms of taxation.[6] General agreement exists, however, that pollution taxes are most effective if levied directly on the pollution-producing activity, rather than on the product that is being produced.

Another market-based incentive mechanism involves "selling the right to pollute." Consider Figure 7.4. Suppose in a particular geographic area, the EPA or other government agency established an ambient air quality standard for sulfur dioxide at a level of $\bar{s}$, where the initial level of emissions in the area is S_0. Businesses in the area would have a need for some pollution, to the extent that the emission of sulfur oxide contributes to the firms' ultimate sales and profits. This is shown by D_0. If the government were to issue permits that fixed the amount of sulfur dioxide emissions at the ambient air quality standard and auctioned off the rights to pollute, firms with the highest demand for these rights would purchase the rights to pollute at the price, p_0, shown. These rights to pollute are sometimes called *tradeable permits* because they can be purchased and sold. If a firm wants to operate a plant whose sulfur dioxide emissions level results in higher profitability for the firm (e.g., because of improved pollution abatement equipment), it would have lower demand for the permits, and, indeed, have an emission advantage it can sell. Similarly, if government wanted to reduce the level of pollution even further, it could purchase these rights. In this case, the demand for the tradeable permits would be higher, D_i, and the price would be increased to p_i. The 1990 Clean Air Act Amendments established a large-scale system of marketable pollution permits that has been applied in many geographic areas, including for example, Southern California. A clear advantage of tradeable permits is that they give polluters the incentives to improve their production processes. Ultimately, however, the permits internalize externalities to the efficient level only when the geographic area affected by the emissions can be clearly identified, as well as the source of emissions within that area.

Although each of these approaches to internalizing environmental externalities is used to some degree and has important applicability, there has been a growing literature in recent years on the

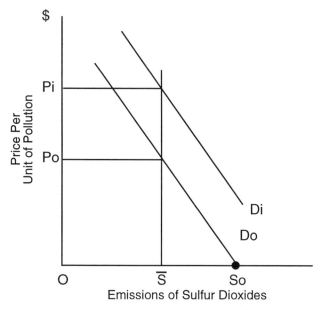

FIGURE 7.4 Tradeable permits.

desirability of firms and industries taking self-initiated actions to deal with externalities. In the following section, we will present the lively debate occurring as to the practicality and usefulness of government regulation and its relationship to self-regulating behavior.

7.3 GOVERNMENT REGULATION AND BUSINESS SELF-REGULATION OF ENVIRONMENTAL EXTERNALITIES

Historically, business response to environmental issues has been reactive as opposed to being proactive, focusing on compliance with regulations established by law or through an implementing regulatory agency. More recently, however, there have been claims from both the research literature and from corporate testimonies suggesting that sustainable business practices make good economic sense. The argument is that the pursuit of sustainable business initiatives will advance profitability, consumer acceptance, and competitiveness.[7] While a strong case has been made for this approach, others have disagreed.

One of the lessons from reviewing the traditional approaches discussed previously for dealing with environmental externalities is that efficiency is encouraged where firms are provided clear incentives. Porter and van der Linde[8] have made the case that the relationship between environmental goals and industrial competitiveness need not involve a trade-off between social benefits and private costs. They recognize that in a static world, "where firms have already made their cost-minimizing choices, environmental regulation inevitably raises costs and will tend to reduce the market share of domestic companies on global markets." However, their arguments are rooted in a dynamic framework, where companies can make adjustments that will improve their profitability and competitive advantage.

Thus, Porter and van der Linde are arguing that "competitive advantage rests not on static efficiency nor on optimizing within fixed constraints, but on the capacity for innovation and improvement that shift the constraints." They propose that environmental regulation can lead companies to adopt innovative offsets that reduce the cost of compliance with pollution control and increase industrial competitiveness. They divide "innovation offsets" into two types: First, *product offsets* occur when "environmental regulation creates better-performing or higher-quality products, safer products, lower product costs (perhaps from material substitution or less packaging), products with higher resale or scrap value (because of ease of recycling or disassembly) or lower costs of product disposal for users. *Process offsets* occur when higher resource productivity such as higher process yields less downtime through more careful monitoring and maintenance, materials savings (owing to substitution, reuse, or recycling of production inputs), better utilization of by-products, lower energy consumption during the production process, reduced material storage and handling costs, conversion of waste into valuable forms, reduced waste disposal costs, or safer workplace conditions."

Imbedded in these definitions is a clear role for regulation. Porter and van der Linde claim that regulation of companies both through command-and-control mechanisms and market-based incentives can spur innovation and increase competitiveness. Koretz[9] reports the results of a World Resource Institute study that suggest productivity growth, as conventionally measured, is seriously understated by ignoring the negative impact that pollution causes via health problems, corrosion, and environmental degradation. For instance, in the electric-power industry, multifactor productivity growth (measured as the difference between the growth of output and the growth of inputs) fell by approximately 0.35 percent per year from 1970 to 1991. However, when the value of output relative to emissions was included, total productivity increased by 0.38 percent per year.

Nonetheless, the view that regulation could spur innovation has historically been an anathema for businesses who have viewed regulation as increasing the costs of production and reducing competitiveness in the marketplace. Palmer et al.[10] and Palmer and Simpson[11] have taken issue with this approach. They find it difficult to believe that business has overlooked profitable opportunities for innovation or that enlightened regulators have sufficient information to induce cost-saving and quality-improving innovation. Moreover, they cite other research that suggests that the social costs of environmental regulation are greater when viewed in a dynamic general equilibrium context, rather than a static partial equilibrium context because of the way that environmental regulation "depresses productive investment and the consequent reduction in the rate of economic growth."

What should we make of these competing claims? We cannot arbitrate this debate here. Actually, the empirical evidence for and against environmental regulation of business and its relationship to profitability is based upon case studies and anecdotes. However, no matter which position above has the most validity, the trend of business towards voluntary self-regulation with respect to environmental externalities is quite clear. But in what way can self-imposed regulation lead to greater self-interest? Can resources devoted to minimizing externalities result in increased profitability? Or are there other motives that lead to this self-regulation?

7.3.1 THE PARADIGM OF SELF-REGULATION: VALUES AND VALUE-BASED MANAGEMENT

In reflecting on the trends of business towards self-regulation, Evan O. Jones, Director of Issues Management at International Paper Company wrote, "Evolving paradigms of business social responsibility were not spontaneously created out of the ether by self-appointed guardians of business ethics. . . . Business is changing the way it does things today because the West's entire socioeconomic context is undergoing an evolution of historic proportion." Clearly, many businesses and industries are moving increasingly towards voluntary, self-regulation in responding to environmental externalities resulting from business operations. In GM's 1994 *Environmental Report,* CEO Jack Smith stated, "We must recognize that, in order to realize this vision (*of being the world leader in transportation and product services*), we must operate in a way that is compatible with society's goals, including the goal of environmental protection. . . . We view caring for the environment as more than a responsibility: We view it as a critical factor in GM's success."

These statements from the business community have been repeated in numerous settings. Schmidheiny[12] has gathered together many of these statements, as well as several early case studies focusing on proactive response by business to environmental problems. For instance, he cites Frank Popoff, President and CEO of Dow Chemical as saying, "If we view sustainable development as an opportunity for growth and not as prohibitive, industry can shape a new social and ethical framework for assessing our relationship with our environment and each other. Seeking cooperative dialogues and partnerships among and between industry, government, special interest groups, and the public will move the process of sustainable development much further, much faster."

Reviewing the quotations from business leaders cited above, as well as insights from various academic and professional sources, at least four major forces may be identified that contribute to business self-regulation. First, there is an economic advantage to avoiding government regulation, whether it be command-and-control or market-based. Although Porter and van der Linde have posited credible circumstances where innovation offsets spurred by regulation may be economically justified, there is still strong evidence that much regulation is quite expensive for business. Various studies by the EPA and other organizations showed that the total costs associated with complying with federal government environmental regulation in 1992 were in the range of $100 to 135 billion.[10] In his book, Schmidheiny has noted a recurring concern of the business community with respect to command-and-control regulation. Such regulation is inflexible and requires adherence to specific requirements, rather than encouraging continuous improvement and innovation. Where government regulation has often centered around application of specific end-of-pipeline technologies, business operating on its own can consider multiple venues that may reduce pollution and minimize the creation of waste as by-products of production.

A second force is related to the growing globalization of business. Shrivastava and Hart have noted that "increasingly stringent national laws and new international treaties" have created a "political imperative" for business to respond aggressively to environmental challenges, and an economic imperative to avoid the likelihood of rapidly escalating compliance and liability costs.[13] Moreover, as public awareness of international environmental problems increases, multinational corporations face added pressures to establish consistent standards across operations in all countries in which they operate.

The third significant force leading to self-regulation involves the influence of public pressure. Schmidheiny[12] argued that industries operate within an implied contract with the public; loss of confidence in a firm or industry affects market viability. Kelly[14] hypothesized that business has gone through a maturation process. Kelly argued that business has turned to self-regulation to preserve self-interest and freedom by responding to social pressure, fear of punishment, good role models, and personal values. In responding to the need for public esteem, companies and trade associations have turned to public advisory panels to share information and, in some cases, to monitor their performance. An interesting example of stakeholder involvement with a company was the relationship between Foron, a German appliance maker, and Greenpeace. Greenpeace had taken a strong position against the use of ozone-damaging chlorofluorocarbon refrigeration technology in Germany. When Foron introduced an environmentally-friendly "Clean Cooler" refrigerator, using an ozone-safe technology (which was called "Greenfreeze"), Greenpeace advertised the new product and sought media attention to build product demand.[15]

Gro Harlem Brundtland, Prime Minister of Norway, has stated, "With greater freedom for the market comes greater responsibility." The final driving force leading business to active self-regulation is the personal values of its leadership. As described in Chapter 5, business leaders operate to varying degrees from an ethical frame of reference. Many business executives believe that they have a moral responsibility to minimize the impact of their companies on the environment and to preserve the natural resource base for future generations. Fundamentally, one might argue that the main responsibility for a company's business leaders is to maximize shareholder value for its stockholders. Walley and Whitehead[16] argue that businesses must operate in the "tradeoff zone," where environmental benefit is weighed against destruction of shareholder value, rather than focusing on compliance, emissions, or abatement costs. Gokcek and Lyons[17] add an interesting twist to this perspective. Using a Cornell-Shapiro stakeholder model,[18] they argue that the market value of a firm depends not only on explicit contracts with factor input suppliers, but also on implicit contracts with customers, suppliers, employees and communities. These implicit contracts would include expectations of continuing supply, timely delivery, product enhancement, and job security. Voluntary environmental self-regulation may not be valued by all stockholders, but to the extent that environmental responsibility matters to other stakeholders, the effect on shareholder value can be negative if implicit contracts with the various stakeholders are ignored. As discussed in Chapter 5, sustainable systems enable all stakeholders to engage in constructive dialogue and to have an effective voice in decision-making processes.

7.3.2 Environmental Management Systems and a Natural-Resource Based View of the Firm

In considering the significance of the preceding discussion for the design of effective environmental management systems, we must go well beyond the simple economic model based upon profit-maximization. The traditional economic models do not readily lend themselves to full incorporation of values or influence from external stakeholders. Moreover, they are, more often than not, static models that fail to capture important dynamic, continuous improvement perspectives.

Hart[19, 20] has utilized a different framework, a resource-based view of the firm, to offer a different way of considering environmental management challenges. A resource-based view of the firm builds upon notions of competitive advantage as developed in the strategic management literature. In this approach, firms gain competitive advantage when they match their distinctive internal capabilities to external circumstances. The resulting competitive advantage can arise from a least-cost position, product differentiation, or first mover advantage (i.e., preempting competitors).

Hart adapts this framework to a natural-resource-based view of the firm. Fundamentally, he argues that businesses will be increasingly constrained in their future decisions by environmental considerations. To the extent that businesses can facilitate environmentally sustainable economic activity, he argues they can gain competitive advantage in the marketplace. Table 7.2 summarizes the basic structure of his model. At the first level, firms focus on pollution prevention. The objective at this level is to minimize emissions, effluents, and waste so as to gain a cost advantage over other competitors. This involves an internal continuous improvement process where firms review and improve upon existing processes and products. The second level involves the concept of product stewardship, incorporating the views of external stakeholders into product design and development. It is at this stage where insights from suppliers and buyers, as well as community residents and other concerned stakeholders, provide feedback and monitoring of business activities. Hart terms the final stage of his model sustainable development. Here a firm looks to the future to develop products and production processes that minimize the "footprints of the company," that is, the environmental impact of firm growth and development. At this level, companies may consider the integrity of ecosystems, the rights of non-human species, and the importance of providing consistent standards of operation in developing countries and in emerging markets.

There are several important elements in Hart's approach to highlight. First, it involves an evolutionary, maturing process for business. Starting from a basic compliance and waste reduction position, it calls for a firm to move to a more proactive life-cycle-based approach. Second, the focus of the firm turns from an internal focus to an external focus, integrating the views of external stakeholders, ultimately to a global perspective. This continuous improvement process reflects well the quality-based environmental management systems that more and more companies are utilizing. As shown in Figure 7.5, Wever[21] has effectively illustrated a similar framework in her model for training environmental, health, and safety (EHS) managers.

Up to this point in this chapter, we have moved from the role that government has played in addressing environmental externalities to a framework based in continuous quality improvement where businesses take a proactive role in promoting sustainable business practices through self-regulation. In the next portion of the chapter, we will take a closer look at total quality environmental management, present a framework that categorizes different approaches to self-regulating behavior by firms and industries, and consider a few examples of self-regulating systems.

7.4 CONTINUOUS IMPROVEMENT SYSTEMS IN BUSINESS

7.4.1 TOTAL QUALITY ENVIRONMENTAL MANAGEMENT

Total quality environmental management (TQEM) is the application of total quality management (TQM) to an organization's environmental needs and concerns. This systemic approach to

TABLE 7.2
A Natural-Resource-Based View: Conceptual Framework

Strategic Capability	Environmental Driving Force	Key Resource	Competitive Advantage
Pollution Prevention	Minimize emissions, effluents and waste	Continuous improvement	Lower costs
Product Stewardship	Minimize life-cycle cost of products	Stakeholder integration	Preempt competitors
Sustainable Development	Minimize environmental burden of firm growth and development	Shared vision	Future position

Source: Hart.[19]

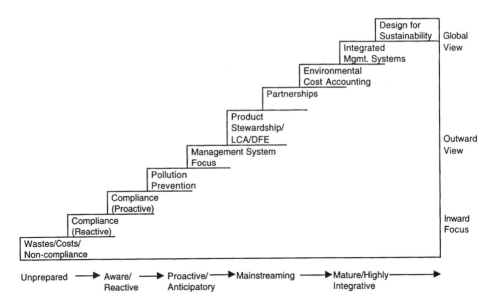

FIGURE 7.5 Stages in implementing an EHS management system; adapted from Reference 21.

management has a very rich history. The early roots of TQM can be traced to the 1940s. It is believed that TQM was first applied to environmental issues in Japan, during the 1970s. During the 1980s American industry greatly expanded its knowledge and use of TQM, as well as its application to environmental issues. In 1991, the Global Environmental Management Initiative (see below) formally coined the phrase total quality environmental management, or TQEM.

7.4.2 THE HISTORY OF TOTAL QUALITY MANAGEMENT

The philosophy which governs modern quality management can be traced back to the work of American statisticians in the field of statistical quality control. While working for Bell Laboratories in the 1920s, Walter Shewhart developed the control chart. His protégé, W. Edwards Deming, greatly broadened control chart usage and scope over the next two decades. During the same period, Feigenbaum[23] developed what might be called the first systematic approach to quality management. In 1951, Feigenbaum's book, *Quality Control,* discussed the holistic role of the industrial statistician throughout the product life cycle: from customer requirements, to product design, to material selection, to the production process, to inspection, and, finally, to sales. American industry, which was thriving in a seemingly endless economic boom, paid little attention to Deming and Feigenbaum. Their warnings of the disastrous consequences to be paid when quality is not assured fell upon deaf ears.

Deming was working for the Census Bureau when he received orders from General Douglas MacArthur to help rebuild Japan's industrial base. In June of 1950 Deming traveled to Japan to meet with Kenichi Koyanagi, head of the Japanese Union of Scientists and Engineers (JUSE). Deming was invited to teach a series of seminars concerning the basic principles of statistics and quality control to Japanese industrialists, including laborers, middle managers, and top executives. Over the next thirty years, Deming continued to teach and consult in Japan. During this time, Japanese industry flourished and the phrase "made in Japan" became a symbol of world class quality, particularly in the electronics and automotive industries.

Deming's knowledge and philosophy of management grew immensely during his years in Japan. His ideas of total quality control were being developed into a managerial model that could be used to implement a whole new philosophy of management. Methods of quality improvement were being applied to all aspects of the organization. During the 1970s, the first applications of quality management to environmental issues occurred in Japan. Although the original projects were relatively small in scope and impact, the framework for managing environmental issues within a management model was being laid.

Deming returned to the United States in the 1970s to consult and teach part time at New York University. Little attention was paid to his work until a 1980 NBC documentary entitled "If Japan Can . . . Why Can't We?" captured the fascination of American industry. In 1981, Deming began working with Ford Motor Company, the first of a long line of major American corporations to receive his consulting services. Although he was in his eighties, Deming started teaching vigorous four day seminars. These seminars were immensely popular with industry. He continued to teach eight hours a day to within a month of his death in 1993 at the age of 93.

The Deming philosophy of management was being adopted by numerous organizations in both the private and public sectors. During the mid-1980s the U.S. Navy became deeply interested in the Deming philosophy. In 1986, a behavioral psychologist in the Navy named Nancy Warren attended a four day Deming seminar. Understanding that the Deming philosophy was much more than just *quality control*, Warren suggested that it be called *total quality management*. Today, total quality management, or TQM for short, has become an important part of business vocabulary worldwide. It is interesting to note, however, that Deming himself never used the phrase TQM.

In 1986, Deming's classic book, *Out of the Crisis,* was published.[24] Besides being given to all seminar participants, the book was widely purchased and read by individuals around the world in academia, government, and industry. The book focused on fourteen points for management. (See Box 7.1.) Deming believed that the fourteen points could transform America by returning quality to industry and government. The points focused on the involvement of top management, management by fact, the understanding of systems and processes, continual improvement, and teamwork.

Although not all institutions use the phrase TQM or advertise themselves as strong advocates of Deming, many organizations have implemented to some degree the concepts encompassed in Deming's fourteen points. Quality management and TQM are evolving concepts. A study group composed of leading quality experts at the 1992 Total Quality Forum defined TQM as follows:

" . . . a people-focused management system that aims at continual increase in customer satisfaction at continually lower real cost. TQM is a total system approach (no separate area or program), and an integral part of high-level strategy. It works horizontally across functions and departments, involving all employees, top to bottom, and extends backwards and forwards to include the supply chain and the customer chain."[22]

7.4.3 TQM in Environmental Management

TQM starts at the top of the organization and impacts all aspects of the organization. TQM also recognizes the needs of all the organization's stakeholders. Therefore, its application to environmental issues is a logical extension of the models put forth by Deming and other quality experts. During the 1980s, several leading corporations in the field of TQM, including Dow Chemical, Eastman Kodak, DuPont, and Xerox, began integrating quality philosophy and methods into their environmental management departments.

Perhaps the most notable achievements were made by the Xerox Corporation. In 1980 Xerox formed a corporate environmental, health, and safety department. Soon after, Xerox adopted its basic environmental policy:

Xerox Corporation is committed to the protection of the environment and the health and safety of its employees, customers, and neighbors. This commitment is applied worldwide in developing new products and processes.

> **Box 7.1 Deming's Fourteen Points for Management**
>
> 1. Create constancy of purpose toward improvement of product and service, with the aim to become competitive and to stay in business, and to provide jobs.
> 2. Adopt the new philosophy. We are in a new economic age. Western management must awaken to the challenge, must learn their responsibilities, and take on leadership for change.
> 3. Cease dependence on inspection to achieve quality. Eliminate the need for inspection on a mass basis by building quality into the product in the first place.
> 4. End the practice of awarding business on the basis of price tag. Instead, minimize total cost. Move toward a single supplier for any one item, on a long-term relationship of loyalty and trust.
> 5. Improve constantly and forever the system of production and service, to improve quality and productivity, and thus constantly decrease costs.
> 6. Institute training on the job.
> 7. Institute leadership. The aim of supervision should be to help people and machines and gadgets to do a better job. Supervision of management is in need of overhaul, as well as supervision of production workers.
> 8. Drive out fear, so that everyone may work effectively for the company.
> 9. Break down barriers between departments. People in research, design, sales, and production must work as a team, to foresee problems of production and in use that may be encountered with the product or service.
> 10. Eliminate slogans, exhortations, and targets for the work force asking for zero defects and levels of productivity. Such exhortations only create adversarial relationships, as the bulk of the causes of low quality and low productivity belong to the system and thus lie beyond the power of the work force.
> 11. Eliminate work standards (quotas) on the factory floor. Substitute leadership. Eliminate management by objective. Eliminate management by numbers, numerical goals. Substitute leadership.
> 12. Remove barriers that rob the hourly worker of his right to pride of workmanship. The responsibility of supervisors must be changed from sheer numbers to quality. Remove barriers that rob people in management and in engineering of their right to pride of workmanship. This means, *inter alia*, abolishment of the annual or merit rating and of management by objective.
> 13. Institute a vigorous program of education and self-improvement.
> 14. Put everybody in the company to work to accomplish the transformation. The transformation is everybody's job.

Work regarding environmental management took on added direction and significance in 1984 when Xerox instituted the leadership through quality (LTQ) program. Xerox began extensive employee training in quality principles, quality tools, and management philosophy. Xerox stressed the use of benchmarking to assess progress and plan improvement strategies.

Teams of Xerox employees began continuous improvement projects in all areas of the corporation, including environmental issues. Xerox teams made improvements in processes throughout the corporation, including pollution prevention, designing green products, reusable packaging, and recycling. Because of the success of the LTQ program, Xerox won the second annual Malcolm Baldrige National Quality Award in 1990. The award, presented by the National Institute of Standards and Technology and named after former Secretary of State Malcolm Baldrige, represents the highest standard of quality management in the United States.

Xerox's excellence in both environmental issues and quality management were not independent events. Rather, they were the result of a systemic approach to management. Furthermore, being on the leading edge of environmental issues became a key corporate goal of Xerox, and the new management philosophy made the realization of the goal more attainable. The commitment to environmental issues was clearly stated by C.E.O. David Kearns in the June 1990 issue of *Xerox World*: "If we cannot afford to protect the environment, we should get out of the business."

In 1990, several leading corporations formed the Global Environmental Management Initiative (GEMI). The corporations represented many different industries, but had a common goal: to foster environmental excellence worldwide through publicizing the leadership and examples set by business.[25] In 1991, GEMI was the first to formally use the term total quality environmental management (TQEM) to refer to the application of the quality management philosophy to environmental issues. In 1993, GEMI published a TQEM primer which identified four basic elements of TQEM:

1. Identify your customers.
2. Continuous improvement.
3. Do the job right the first time.
4. Take a systems approach to work.

The primer also included Deming's plan-do-check-act (PDCA) cycle as the model for continuous improvement and illustrated the use of several quality tools: cause and effect diagrams; Pareto charts; control charts; flow charts; histograms; and benchmarking. By 1997, twenty-two corporations were members of GEMI.*

Just as TQM differs from one organization to another, so does TQEM. However, several basic principles are present in all TQEM initiatives:

1. TQEM involves a systemic approach.
2. TQEM requires organizations to identify and listen to their customers.
3. TQEM requires continual improvement.
4. TQEM is proactive management, i.e., it goes beyond compliance and strives to design environmental-friendly products and processes.
5. TQEM is management by data and sound science.
6. TQEM requires organizations to make environmental issues, like sustainable development and environmental ethics, key corporate goals.

The specific methods organizations stress as being vital to the implementation of TQEM also differ among organizations. Several key tools are imperative, however, for TQEM to be effective:

1. Benchmarking is required for continual improvement to take place.
2. Cross-functional teams are necessary to obtain a systemic approach to environmental management.
3. Channels for effective dialogue between different stakeholders must be developed.
4. Statistical tools are necessary to analyze data so that management by fact can occur.

---

*The members of GEMI in 1997 were: AT&T, Anheuser-Busch, Bristol-Myers Squibb, Browning-Ferris Industries, Ciba Specialty Chemical Corporation, Coca-Cola, Colgate-Palmolive, Coors Brewing, Dow Chemical, DuPont, Duke Power, Eastman Kodak, Georgia-Pacific, Halliburton, Hughes Electronics, Johnson & Johnson, Merck & Company, Olin, Procter & Gamble, Southern, Tenneco, and WMX Technologies.

7.4.4 EXAMPLES OF TQEM AND ENVIRONMENTAL MANAGEMENT SYSTEMS IN BUSINESS

Environmentally conscious organizations have an inherent tendency to adopt TQEM programs, and many corporations have.* Some of the reasons to implement TQEM, however, are purely economic. For example, increasing efficiency, reducing waste, and conserving resources lead to increased profitability. In today's environmentally-conscious society, it is also true that being perceived as "green" is good for public relations, and that being perceived as "green" increases employee morale. Another major reason for adopting a TQEM program is to attempt to be proactive towards further government mandates and regulations. By instituting a TQEM program, a corporation can argue that it is doing everything feasible to improve its environmental scorecard and that added government interference will only make environmental issues worse. In this section, we will present several examples of companies and businesses that have developed environmental management systems based on the principles of TQEM.

7.4.4.1 The Council of Great Lakes Industries and TQEM

The Great Lakes region enjoys a rich natural resource base including an extensive large lake system containing 20 percent of the world's fresh water. The region also has a substantial economic base accounting for 20 percent of U.S. economic activity and 50 percent for Canada, entirely in Ontario province. This enormous economic activity has threatened the resilience of the ecological systems in the region. Over the past three decades, intensive efforts have been directed by government and industry toward environmental improvements. As CGLI President and CEO, Tippett[26] has testified, "These efforts have greatly reduced discharges and levels of chemicals in the water column, sediments, and tissues, and have led to significant improvements in the health of aquatic and wildlife communities." According to environmental groups, however, the majority of Lake Michigan's shoreline remains unsafe for fishing and swimming. The heavy industrialization also has resulted in the heaviest concentration of superfund sites in the United States being located in Indiana, bordering Lake Michigan, and a significant portion of the land in Great Lake cities are classified as brownfield sites.[27]

The Council of Great Lakes Industries (CGLI) was founded in 1991 by several large firms from the Great Lakes area. As of November, 1996, twenty-seven firms or organizations were active members: American Forest and Paper Association, Avenor, BASF, Bayer Rubber Corporation, Browning-Ferris Industries, CN North America, Canadian Petroleum Products Institute, Clark Hill, Dofasco, Dow Chemical Canada, Dow Chemical Company, Dow Elanco, Eastman Kodak Canada, Eastman Kodak Company, Environmental Research Institute of Michigan, Falconbridge Limited, Federal Reserve Bank of Chicago, Ford Motor Company, Ford Motor Company of Canada, General Motors Corporation, GTE North, Geon Corporation, Georgia-Pacific Corporation, ITT Automotive, The New York Power Authority, Occidental Chemical Corporation, Ontario Hydro, and Xerox Corporation. Thus, CGLI is a binational, multi-industry organization that serves as a hub for industry coordination on public policy, promoting sustainable development (described as economic growth and viability in harmony with human and natural resources).

CGLI represents industries' views on environmental and economic policy positions. Members of CGLI have spoken at Congressional Hearings, meetings of the International Joint Commission (IJC), and other public forums. CGLI tries to dispel the notion that all industries are anti-environment. It publishes Environmental Stewardship Reports that document significant investments the regional industries have made in pollution prevention, product stewardship, and improved environmental systems.

Environmental issues for CGLI are headed by Grace Wever, Vice President of CGLI and Senior Manager for KPMG Peat Marwick (and formerly of Eastman Kodak). CGLI's environmental man-

---

*Barriers to initiating TQEM programs also exist in corporate America. Many corporations will not make a commitment to long-term environmental goals in the reality of short-term financial costs. Secondly, until reliable measures for the total cost of discharges and environmental remediation are available, many corporations claim that they cannot manage that which cannot be measured.

agement philosophy is well-documented in its *TQEM Primer and Self-Assessment Matrix*, which CGLI published in 1993.[28] The self-assessment matrix was developed by taking Eastman Kodak's *Total Quality Management Self-Assessment Matrix* and adapting it for environmental management generally. According to the *Total Quality Management Self-Assessment Matrix* (page 1), "The basic elements of TQEM include a high level of management commitment, a strong customer/stakeholder focus, teamwork and empowerment, continuous improvement, and a prevention approach."

The self-assessment matrix is the focal point of CGLI's TQEM effort (See Figure 7.6). It is a self-evaluation tool, composed of seven criteria adapted from those used in the Malcolm Baldridge National Quality Award. The seven matrix categories include leadership, information and analysis, strategic planning, human resources, quality assurance of environmental performance, environmental results, and customer/stakeholder satisfaction. Within each criteria, the matrix provides benchmarks for companies to evaluate themselves against. Organizations can measure their progress in each area, identify their strengths, and determine areas where the most improvement is necessary. At the bottom of each column is the lowest performance level, which indicates that the organization is just beginning its TQEM program. At the top of each column is a performance rating equal to the industry leader or "Best-in-Class."

The matrix is a framework for implementing TQEM. The TQEM Primer and Self-Assessment Matrix summarizes the advantages of this matrix approach[28]:

1. The matrix provides a building block system and guide for TQEM implementation.
2. It fosters a prevention-based approach and continuous improvement of performance.
3. It sets standards for excellence.
4. It provides a tool for economic and environmental improvement by encouraging integration of environmental goals into business plans.
5. It reinforces partnerships and encourages sharing information.
6. It fosters a consensus approach by business, government, and the public to environmental priority and goal setting, planning, and commitment to resources based on sound information.
7. It promotes environmental stewardship among all sectors of society.

Grace Wever has led many workshops for employees of CGLI members. At these workshops, Dr. Wever asks participants to rate their companies' performance in each of the seven criteria using a three level metric. A green dot indicates that the company always or almost always meets the standard, a yellow dot indicates that the organization sometimes meets the standard, and a red dot indicates that the organization's performance is not up to that level at this time. Dr. Wever notes that a company's total performance is typically represented by a diagonal across the matrix. High scores are most often seen on the left-hand side (leadership), medium scores are in the middle (planning and development), and low scores are seen on the right-hand side (performance and customer satisfaction). For companies to improve, advances in environmental results and customer satisfaction must be pre-dated by a commitment of top management, followed by advances in information, planning, and development. For a complete discussion of the self-assessment process, see Wever.[21]

7.4.4.2 Responsible Care® and the Chemical Manufacturers Association

In December 1984, a Union Carbide Plant in Bhopal, India, suffered a major leak of methyl isocyanate (MIC), killing more than 3000 people and injuring tens of thousands of others. According to Union Carbide CEO Robert Kennedy, this accident was "the single most astonishing and terrible event in the history of the chemical industry."[29] The response to this horrible incident by the chemical industry was swift and led to the development of the most comprehensive industry-wide environmental management program to date.

The Chemical Manufacturers Association (CMA) is an industry-wide trade association, consisting of over 190 companies who represent about 90 percent of the nation's basic industrial

Level	Rank	Category: Leadership Weighing: 15.0%	Information and Analysis 7.5%	Strategic Planning 7.5%
Maturing	10	Benchmarking indicates unit is "Best-in-Class" in leadership area.	Benchmarking indicates unit is "Best-in-Class" in area of information and analysis.	Benchmarking indicates unit is "Best-in-Class" in area of strategic planning.
	9	Top management proactively participates in public policy decision-making processes in environmental area.	Environmental data/analysis directly affect behavior and lead to improved environmental performance of products, operations, services (results).	Environmental improvement plans for processes, products and services are totally integrated into long-term and short-term business plans.
	8	Top management's external actions reflect commitment to unit's environmental principles; management encourages employees to do the same.	Environmental data/analysis used in strategic decision-making.	Improvement plans in place at all organization levels support unit's key environmental objectives.
	7	Top management has completed at least one full continuous improvement cycle, management performance measures based on meeting key environmental objectives, decisions based on vision.	Process in place to use environmental data to plan/design for new products/operations/services.	Strategic planning process is supported by a system of rewards and consequences based on both behavior and results.
Growing	6	Management uses reward/consequence system in all areas to reinforce commitment to environmental management.	Process in place to continuously improve environmental data collection/analysis/dissemination.	Process in place to include stakeholder contributions to strategic planning.
	5	At least half of top management is using environmental considerations as part of decision-making process.	Environmental data routinely used to improve current products, operations, services, focused on prevention.	Long- and short-term plans that include environmental management are reviewed and improved at least annually.
	4	Dialog occurs between top management and customers/stakeholders regarding environmental principles.	Environmental data analyzed for trends.	Resource allocation is consistent with environmental plan implementation needs.
	3	Unit-wide plan in place to implement environmental programs including necessary resources.	Environmental data inventory and management process established; some external environmental data collected.	Consistency exists at all levels for environmental management planning and implementation.
	2	Management directly involved in environmental quality management processes as leader/role model. Employee empowerment framework established.	Processes in place to assure validity (Quality Assurance/Quality Control) of basic environmental data.	Quality mangement process links existing and anticipated environmental regulatory requirements with the planning process.
Beginning	1	Environmental mission, vision, principles defined, published, and understood internally.	Basic internal environmental data identified and gathered.	A long-term (2–5 y) & short-term (1–2 y) planning process used that addresses environmental needs; annual operating plan includes environmental management needs.

Source: Council of Great Lakes Industries.

FIGURE 7.6 Total Quality Environmental Management implementation Guide and Assessment Matrix.

Sustainability and Business Management Systems

Human Resource Development 10.0%	QA of Environmental Performance 15.0%	Environmental Results 30.0%	Customer/Stakeholder Satisfaction 15.0%
Benchmarking indicates unit is "Best-in-Class" in levels of employee morale and attitudes toward environmental management.	Benchmarking indicates unit is "Best-in-Class" in level of QA of environmental performance.	Benchmarking indicates unit is "Best-in-Class" in environmental results.	Benchmarking indicates unit is "Best-in-Class" in customer/stakeholder satisfaction with respect to environmental quality.
Career development and education opportunities in environmental management are widely available.	Processes in place in all areas to continuously improve environmental performance of products, processes, and services.	Sustained improvement in environmental performance of processes, products, and services is evident in all areas.	Active customer/stakeholder involvement contributes to sustained improvement in environmental performance of processes, products, services.
Education and career development plans exist and are linked to environmental management goals, tactics, and strategies.	Formal process used to consider all stakeholder input to environmental performance improvement.	Benchmarking measures identified; benchmarking initiated.	Customers/stakeholders are actively involved in environmental problem-solving.
Environmental management is an essential element of reward and consequences systems.	Environmental expertise included in cross-functional teams involved in development cycle for new and existing products, processes, and services.	Rewards/consequences are used to reinforce environmental performance improvement.	Customer/stakeholder satisfaction data is integrated into the continuous improvement cycle for all aspects of the unit's functions.
Environmental management training is evaluated for improvement.	Process in place to obtain/use stakeholder input to develop environmental objectives for products, processes, services.	Measures are reviewed and updated at least annually to reflect all stakeholder input.	Customer/stakeholder satisfaction data is integrated into the continuous improvement cycle for some aspects of the unit's functions.
Measures and trends of employee attitudes toward environmental performance exist.	Evidence exists that quantitative measures of environmental performance extend fully into all aspects of unit's operations.	Improving trends of environmental performance in major areas.	Customer/stakeholder satisfaction measures indicate positive trends.
All employees have completed appropriate environmental training. Employees are empowered. System in place for periodic retraining.	Evidence exists for prevention focus, rather than reaction (e.g. pollution prevention), root-cause analysis used for problem-solving. Audit systems used to assure continuous improvement.	Improving trends of environmental performance in some areas.	Measures of customer/stakeholder satisfaction exist with respect to environmental considerations.
Appropriate environmental awareness and training/education programs developed and scheduled for all employees.	Process in place to assure goals/objectives followed for modification/production of current products, processes, services. Document control process in place and used.	Management system in place for improving environmental performance; major areas for environmental improvement identified.	Proactive process exists to identify customers/stakeholders and environmental considerations beyond measurement of questions, complaints.
Resources are allocated for developing/implementing environmental training and education.	Process in place to assure environmental principles translated into policies, practices; environmental objectives followed to develop new products, processes, services.	Baselines for environmental performance established.	Process exists to respond to customer/stakeholder environmental questions/ concerns.
Clear assignment of environmental responsibility exists.	Processes in place for ensuring precision and accuracy of measurment systems; internal standards are in place.	Measures of environmental performance are identified.	Process exists to insert existing environmental regulatory requirements for customer/stakeholder information about products, services, and operations.

chemical manufacturing capacity. In response to the Bhopal accident, CMA developed the Community Awareness and Emergency Response (CAER) program in April 1985. The CAER program was designed as a voluntary program to help assure emergency preparedness and to foster community right-to-know. At the same time that the CAER program was beginning to see promising results at the local plant level, the Canadian Chemical Producers Association was formulating a program it called "Responsible Care," involving codes of practice for transportation, distribution, manufacturing, research and development, and hazardous waste operations.

In 1987, CMA commissioned a Public Perception Committee "to evaluate current status and trends regarding the overall public image of the chemical industry; to understand the impacts of negative public opinion on the long-term prospects for innovation and industry growth; and to recommend industry initiatives which could improve the legislative, regulatory, market, and public interest climate for the industry." Out of the work of this committee came the genesis of the Responsible Care® initiative, whose objectives were "to promote continuous improvement in member company environmental, health, and safety performance in response to public concerns and to assist members' demonstration of their improvements in performance to critical public audiences."

In September 1988, Responsible Care® was unanimously approved by CMA's Board and adopted by 170 member companies as an obligation of membership in CMA. The initiative is organized around ten guiding principles (see Box 7.2) and six codes of management practice. It requires each of the member and partner companies to continually improve performance in health, safety, and environmental quality. The six codes of management practice (community awareness and emer-

---

Box 7.2   Guiding Principles of Responsible Care®

Member companies of the Chemical Manufacturers Association are committed to support a continuing effort to improve the industry's responsible management of chemicals. They pledge to manage their businesses according to these principles:

- To recognize and respond to community concerns about chemicals and their operations.

- To develop and produce chemicals that can be manufactured, transported, used, and disposed of safely.

- To make health, safety, and environment considerations a priority in our planning for all existing and new products and processes.

- To report promptly to officials, employees, customers, and the public, information on chemical-related health or environmental hazards to recommend protective measures.

- To counsel customers on the safe use, transportation, and disposal of chemical products.

- To operate our plants and facilities in a manner that protects the environment and the health and safety of our employees and the public.

- To extend knowledge by conducting or supporting research on the health, safety, and environmental effects of our products, processes, and waste materials.

- To work with others to resolve problems created by past handling and disposal of hazardous substances.

- To participate with government and others in creating responsible laws, regulations, and standards to safeguard the community, workplace, and environment.

- To promote the principles and practices of Responsible Care® by sharing experiences and offering assistance to others who produce, use, transport, or dispose of chemicals.

Source: Chemical Manufacturers Association, April 1991.

gency response, pollution prevention, process safety, distribution, employee health and safety, and product stewardship) commit the chemical industry to responsible management of chemicals from development through production, distribution, and ultimate use. Besides emphasizing continuous quality improvement, the Responsible Care® initiative encourages close relationships between participating companies and their communities through active public dialogue in the form of advisory councils and other citizen forums, and has begun to emphasize the importance of quantitative Code performance improvement data for credibility of the program.[30] To pursue these latter objectives, the Responsible Care® initiative is now undergoing a Phase II implementation stage focusing on issues such as management systems verification and mutual assistance among CMA companies.

7.4.4.3 The CERES Principles

Another major environmental incident, the *Exxon Valdez* oil spill gave rise to another set of environmental principles, the CERES Principles. These principles were developed by the Coalition for Environmentally Responsible Economies (CERES). This organization is a non-profit organization which includes in its membership corporations, social investors, major environmental groups, public pensions, labor organizations, and public interest groups. For instance, corporate members include Ben & Jerry's Homemade, Domino's Pizza Distribution Corporation, General Motors Corporation, and Sun Company, Inc. Associations and organizations include the National Audubon Society, the National Wildlife Federation, the Sierra Club, the Social Investment Forum, the Interfaith Center on Corporate Responsibility, and the AFL-CIO Industrial Union Department.

The CERES Principles are intended to serve as a model corporate code of conduct and include ten principles (see Box 7.3). Like the CGLI and Responsible Care® programs, the CERES Principles fundamentally call for a continuous improvement process. However, they are distinguished by their emphasis on two dimensions. First, the CERES Principles emphasize "citizenship" responsibilities of corporations. Companies are asked to look beyond narrow self-interest to the interests of its employees, neighbors, and other stakeholders. Although this broad sense of stewardship is present in both the CGLI and Responsible Care® initiatives, the CERES Principles clearly delineate the public duties of companies. With respect to risk reduction, safe work places and environs focus business responsibility to its employees and the surrounding communities in which it operates. In fact, stakeholders are very broadly defined within principles such as the *protection of the biosphere principle,* the *sustainable use of natural resources principle,* and the *environmental restoration principle,* all of which extend the concept of stakeholders beyond human beings to all living things and natural habitats. There is a recognition that companies must fit into a system of larger values, human and non-human based, in order to be viewed as a community asset.[31]

The second distinguishing feature of the CERES Principles is the emphasis on auditing and reporting. Companies are required to complete the CERES Report annually, disclosing information necessary to assess the real environmental impacts of their operations. The CERES Report was the first attempt to produce a comprehensive and accessible environmental reporting format for corporations.

7.4.4.4 ISO 14000 International Environmental Standards

Since 1946, the International Organization for Standardization (ISO) headquartered in Geneva, Switzerland, has become a major international agency with the objective of creating harmonized, uniform world-wide standards for business. In 1987, ISO issued standards (ISO 9000) designed to guide internal quality management and external quality assurance issues. In the years to follow, countries and corporations adopted versions of the standards. In November 1996, ISO issued the first edition of the ISO 14000 Series of Environmental Management Systems standards. These are standards prescribing requirements for environmental quality management and management systems. The environmental management system refers to the organizational structure, responsibilities, practices, procedures, processes, and resources for implementing and maintaining environmental

> **Box 7.3   The CERES Principles**
>
> *Protection of the Biosphere*—We will reduce and make continual progress toward eliminating the release of any substance that may cause environmental damage to the air, water, or the earth, or its inhabitants. We will safeguard all habitats affected by our operations and will protect open spaces and wilderness, while preserving biodiversity.
>
> *Sustainable Use of Natural Resources*—We will make sustainable use of renewable natural resources, such as water, soils, and forests. We will conserve non-renewable natural resources through efficient use and careful planning.
>
> *Reduction and Disposal of Wastes*—We will reduce and where possible eliminate waste through source reduction and recycling. All waste will be handled and disposed through safe and responsible methods.
>
> *Energy Conservation*—We will conserve energy and improve the energy efficiency of our internal operations and of the goods and services we sell. We will make every effort to use environmentally safe and sustainable energy sources.
>
> *Risk Reduction*—We will strive to minimize the environmental, health, and safety risks to our employees and the communities in which we operate through safe technologies, facilities, and operating procedures, and by being prepared for emergencies.
>
> *Safe Products and Services*—We will reduce and where possible eliminate the use, manufacture, or sale of products and services that cause environmental damage or health or safety hazards. We will inform our customers of the environmental impacts of our products or services and try to correct unsafe use.
>
> *Environmental Restoration*—We will promptly and responsibly correct conditions we have caused that endanger health, safety, or the environment. To the extent feasible, we will redress injuries we have caused to persons or damage we have caused to the environment and will restore the environment.
>
> *Informing the Public*—We will inform in a timely manner everyone who may be affected by conditions caused by our company that might endanger health, safety, or the environment. We will regularly seek advice and counsel through dialogue with persons in communities near our facilities. We will not take any action against employees for reporting dangerous incidents or conditions to management or to appropriate authorities.
>
> *Management Commitment*—We will implement these Principles and sustain a process that ensures that the Board of Directors and Chief Executive Officer are fully informed about pertinent environmental issues and are fully responsible for environmental policy. In selecting our Board of Directors, we will consider demonstrated environmental commitment as a factor.
>
> *Audits and Reports*—We will conduct an annual self-evaluation of our progress in implementing these Principles. We will support the timely creation of generally accepted environmental audit procedures. We will annually complete the CERES Report, which will be made available to the public.
>
> *Source:* "A Healthy Economy and a Healthy Environment," CERES.

management. Environmental management refers to the parts of the overall management function of an organization that develop, implement, achieve, review, and maintain the environmental policy. Delegates from more than 45 countries worked on developing the environmental management system standards.

The basis for these international standards is found first in the BS 7750, the national environmental management system developed by the British Standards Institution in 1992 for all United Kingdom companies. In 1994, the European Community established a certification process for companies as part of its Eco-Management and Audit Scheme (EMAS).[32]

The ISO environmental standards have been dubbed ISO 14000, and have been described by some as so important that they are "expected to become a *de facto* requirement for doing business in the 21st century."[33] They involve a series of management standards with the following objectives:

- Minimize trade barriers owing to inconsistent national standards.
- Promote a common process and language for environmental management based upon quality principles.
- Enhance the measurement of environmental performance.
- Establish a uniform certification and registration procedure.
- Reduce duplicative audits of environmental performance.

Although ISO 14000 addresses various aspects of environmental performance, including auditing, life-cycle assessment, and product labeling, the only element to which companies can be certified is the environmental management standard, ISO 14001.[34]

The ISO 14001 standard is concerned with the framework within which business sets environmental goals and the attainment of those goals. It does not specify particular performance goals for a business. Rather, a company must describe their current state of environmental performance and the management system in place to achieve its own specified goals. It is possible for a company performing poorly with respect to environmental matters to be certified under ISO 14001 as long as it has established an acceptable program to reach its self-imposed objectives. Thus, two companies with very different levels of environmental performance can be certified. Companies who have been ISO 9000 certified are expected to have little trouble becoming ISO 14000 certified. Moreover, companies that have pursued environmental performance improvements aggressively through diligent use of TQEM are likely to be well-beyond the ISO 14001 standards.[35] In fact, some critics argue that "ISO 14001 is more likely to help advance the 'trailing edge' in corporate environmental management than extend the 'leading edge.'"[32]. The ISO 14001 does have the advantage of specifying a uniform global standard whereby companies can demonstrate attainment with respect to its environmental goals and objectives.

7.4.5 Characteristics of Self-Regulating Approaches

Reflecting on the different self-management systems presented, there are a variety of prevailing approaches to self-regulating environmental management. The approaches include adherence to principles or codes of conduct, voluntary adoption of performance standards, and adoption of management practices or standards. Some of these mechanisms may be industry-specific, while others are broadly based and intended to cross industry lines. Some may be completely voluntary, while others may have some ties to a legally binding organization. Many of these approaches are not distinct at all, but rather parts of a larger decision-making system within the company. In other words, a management system cannot function without being grounded on some principles. However, the breadth of self-regulating mechanisms suggest that each firm has a different comfort level and need for these systems. The programs can work hand-in-hand. For example, a company may establish goals based on the CERES principles and use the ISO 14000 standards to establish an environmental quality management system.

In this section, we propose four characteristics of environmental management systems that help to categorize the various approaches previously described. The four characteristics may be conceived of as each being on a continuum from high emphasis to low emphasis. The characteristics are the following:

- Binding codes of practice.
- Systemic nature.
- Dialogical basis with stakeholders.
- Performance measurement or audit.

The first characteristic of environmental management systems is the degree to which the adoption of the program requires binding codes of practice as a criteria for participation. A program low on the *binding* continuum suggests a voluntary program without great consequences for failing to meet goals. A program high on the same continuum involves a greater risk of consequences for non-compliance. Binding here refers to either the legal obligation involved or the non-legal feeling of compulsion or sense of duty. The degree of compulsion is strongly related to the degree of consequence.

The second characteristic is the degree to which systems are emphasized, as opposed to conceptual principles. A program that is low on the *systemic* continuum focuses on broad principles or tenets. This type of mechanism typically consists of a list of abstract goals. The goals may be part of a vision statement or statement of beliefs. They may be considered a part of the business ethic. They do not suggest specific steps or changes in practice that must occur. They do provide for the beginnings of definable and measurable goals. A program high on the *systemic* continuum indicates that adoption or adherence involves facilitation of an entire environmental management system. This mechanism is more sophisticated and complex. By publicly adhering to known standards of management and/or codes of management practice, a company is committing to certain behaviors. This requires integrating environmental issues into decision making. If an effective environmental management system is in place, a firm will be able to respond to changes in political, legal, and social environmental pressures.

The third characteristic of environmental management systems is the degree to which stakeholder dialogue is emphasized. A program that is low on the *dialogical* continuum would have little expectation that a company would establish effective dialogue with its stakeholders. A program high on this continuum would emphasize the responsibility of business to its multiple stakeholders, in fact, requiring a change in corporate culture to accept a stewardship role. As suggested in our *stakeholder interests* and *information dialogue and expectation* sustainability principles, a program emphasizing dialogue would recognize that maintaining a resilient, well-functioning community system requires that all stakeholders' interests (human and non-human) must have effective voice, and that the dialogue must recognize varying traditions and perspectives and be conducted in an atmosphere of mutual respect.

The final characteristic is the degree to which measurable performance in achieving sustainable business practices is emphasized. Setting goals allows a company to assert specific quantitative accomplishments. A system low on the *performance measurability* continuum would not emphasize establishment of quantitative standards of performance. Programs high on this continuum would require establishment of quantitative performance objectives, and would recognize that credibility of a company's or industry's environmental effort is greater where achievements can be verified by an auditor or by external stakeholders. Such a program would require that financial and physical consequences be included in a corporate reporting system. Since these typically include a disclosure role, there would need to be confidence in the process of communication with stakeholders.

Figure 7.7 includes these four characteristics in a two-by-two continuum and plots our best assessment of where the four programs detailed above would lie.

As discussed above, the CERES Principles, known as the Valdez Principles until 1992, are ten statements to guide the firm to environmental responsibility and to establish a moral code of conduct. The CERES organization approaches companies to urge adoption of its principles. The CERES argument program is relatively less systematic in its effect and relatively non-binding. Although companies must pledge to live by the CERES Principles, there are not specific expectations about specific management practices. The emphasis is on the ethics of environmental responsibility. As such, the CERES approach is very high on the dialogical continuum, not only including diverse interests (business, environmental groups, social investors, and others) in dialogue, but also emphasizing the rights of non-human species and the need to preserve the integrity of the ecosystem. Interestingly, however, as noted above, one condition of membership is the preparation of the annual report on environmental performance pioneered by CERES. Although one might think at first that

Sustainability and Business Management Systems

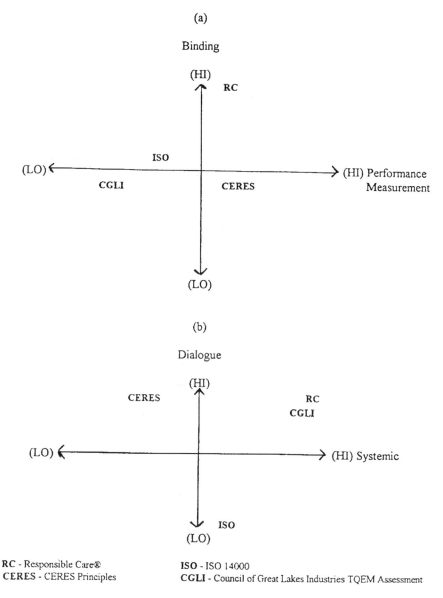

FIGURE 7.7 Characteristics of business environmental management systems. Key: **RC**—Responsible Care®; **ISO**—ISO 14000; **CERES**—CERES Principles; **CGLI**—Council of Great Lakes Industries, TQEM Assessment.

this emphasis on specificity is not congruent with the CERES approach, it actually recognizes the important role that effective communication and environmental reporting have on establishing credibility with external stakeholders.

The ISO 14000 standards offer a quite distinct approach from that of CERES. In general, ISO 14000 is to "support the corporate goals of achieving compliance with legal requirements; establishing internal environmental quality policies; and managing marketplace expectations." These goals are "accomplished by implementing environmental quality management systems, environmental audits, environmental performance evaluations, product life-cycle assessments, and product labeling." With respect to the characteristics of environmental management systems, ISO 14000 is fundamentally non-prescriptive. Even though the intention is presumably to establish international standards, the emphasis is rather on assuring stakeholders globally that companies in different countries are attentive to environmental issues. On one hand, ISO 14000 does involve a binding certification process; however, this applies only to the first level environmental management system whereby a company describes its current level of environmental quality, its target for the future, and its implementation plan. ISO 14000 does not prescribe specific environmental quality standards. Thus, ISO 14000 is low to moderate in terms of performance measurement and with only limited systems expectations. Particularly striking about ISO 14000 is the almost total avoidance of involvement with the public in any form of dialogue. It does, however, require proof that the outcomes from the systems meet the objectives of those systems.

The Council of Great Lakes Industries (CGLI) approach takes a much greater systems approach than either CERES or ISO 14000. As described above, CGLI utilizes a Baldrige-based TQEM approach. The Assessment Matrix at the heart of this process is quite thorough in describing the process that companies would follow in each of the seven categories (leadership, information and analysis, strategic planning, human resource development, quality assurance (QA) of environmental performance, environmental results, and customer/stakeholder satisfaction) to become "Best-of-Class," and thus rates very high on the *systemic* continuum. However, although companies would need to pay attention to quantitative assessment of its activities to maximize performance under this system, the CGLI approach itself does not establish specific performance measurement standards. CGLI does emphasize dialogue with stakeholders. In terms of the *binding* continuum, this is not a point of emphasis for CGLI, probably owing to its interindustry membership and lack of specific standards of performance.

The Responsible Care® initiative is similar to the CGLI approach in some ways, especially with respect to its emphasis on a systemic approach. As described above, Responsible Care® involves not only a set of guiding principles, but also a quite prescriptive set of codes of management practice. Built into these codes is a strong emphasis on dialogue with external stakeholders; in particular, significant reliance is placed upon citizen advisory groups. One of the distinctive features of Responsible Care® is that it is the only well-articulated industry-wide environmental, health and safety program, and that the Chemical Manufacturers Association requires adherence to its principles and codes as a condition of membership. In terms of measurement, this is now an area of emphasis by Responsible Care®. To this point, performance measurement under Responsible Care® has been primarily through member company annual environmental performance reports and through selected initiatives such as taking leadership in facilitating the measurement of worst-case scenarios in the event of chemical accidents.

7.5   CONCLUSION

In this chapter, we have reintroduced the notion of environmental externalities and the inefficiency that results when companies fail to take full account of the costs of their operations. After reviewing the traditional role of government in dealing with these externalities, we explored the ways that business voluntarily or through trade associations has increasingly turned to self-regulation to achieve sustainability of their growth or development. Although the approaches taken by these companies and groups are different, they all are based to some extent on total quality principles and require a change of corporate culture.

In reflecting upon the role that business must play in promoting sustainable business practices and a sustainable environment, Smart[36] has written, "Important as pressure from environmentalists and governmental direction are to stimulating change, in the end only the corporate community can efficiently provide the necessary organization, technology, and financial resources needed to design and implement change on the scale required. Companies that are trying to be leaders on a new path to a sustainable future merit our encouragement and support, just as the inevitable backsliders deserve a vigorous shove onto the trail."

7.6 REFERENCES

1. Rigoglioso, M., Emmons, G., Goodspeed, L., and Gottlieb, E., Stewards of the seventh generation, *Harv. Bus. Sch. Bull.*, April, 48, 1996.
2. Tietenberg, T., *Environmental and Natural Resource Economics,* HarperCollins College, New York, 1996.
3. Lis, J. and Chilton, K., *The Limits of Pollution Prevention,* Contemporary Issues Series No. 52, Center for the Study of American Business, Washington University, St. Louis, MO, 1992.
4. Kahn, J. R. *The Economic Approach to Environmental and Natural Resources,* Dryden Press, Fort Worth, TX, 1995.
5. Hawken, P., *The Ecology of Commerce: A Declaration of Sustainability,* Harper Business, New York, 1993.
6. Oates, W. E., Green taxes: can we protect the environment and improve the tax system at the same time?, *South. Econ. J.*, 61(4), 915, 1995.
7. Erekson, O. H., Loucks, O. L., and Aldag, C., The dimensions of sustainability for business, *Mid-Am. J. Bus.*, 9(2), 3, 1994.
8. Porter, M. E. and van der Linde, C., Toward a new conception of the environment-competitiveness relationship, *J. Econ. Perspect.*, 9(4), 97, 1995.
9. Koretz, G., A hidden source of efficiency, *Bus. Week,* October 21, 28, 1996.
10. Palmer, K., Oates, W. E., and Portney, P. R. Tightening environmental standards: the benefit-cost or the no-cost paradigm?, *J. Econ. Perspect.*, 9(4), 119, 1995.
11. Palmer, K. L. and Simpson, R. D., Environmental policy as industrial policy, *Resources,* Summer, 17, 1993.
12. Schmidheiny, S., *Changing Course,* MIT Press, Cambridge, MA, 1992.
13. Shrivastava, P. and Hart, S., Creating sustainable corporations, *Bus. Strat. Environ.*, 4, 1995.
14. Kelly, M., Capitalism grows up, *Bus. Eth.*, 9(1), 9, 1995.
15. Hartman, C. L. and Stafford, E. R., Green alliances: building new business with environmental groups, *Long-Range Plan.*, 30(2), 184, 1997.
16. Walley, N. and Whitehead, B., It's not easy being green, *Harv. Bus. Rev.*, 72(3), 46, 1994.
17. Gokcek, G. and Lyons, B., A stakeholder model of corporate environmental policy, prepublication manuscript.
18. Cornell, B. and Shapiro, A., Corporate stakeholders and corporate finance, *Financ. Manage.*, 16(1), 5, 1987.
19. Hart, S., A natural-resource based view of the firm, *Acad. Manage. Rev.*, 20(4), 986, 1995.
20. Hart, S., Beyond greening: strategies for a sustainable world, *Harv. Bus. Rev.*, 75(1), 66, 1997.
21. Wever, G. H., *Strategic Environmental Management: Using TQEM and ISO 1400 for Competitive Advantage,* John Wiley & Sons, New York, 1996.
22. Bounds, G., Yorks, L., Adams, M., and Ranney, G., *Beyond Total Quality Control: Toward the Emerging Paradigm,* McGraw-Hill, New York, 1994.
23. Feigenbaum, A., *Quality Control: Principles, Practice, and Administration,* McGraw-Hill, New York, 1951.
24. Deming, W. E., *Out of the Crisis,* MIT, Cambridge, Mass., 1986.
25. Carpenter, G. D., GEMI and the total quality journey to environmental excellence, in *Proceedings of the Corporate Quality/Environmental Management, 1st Conf.,* Global Environmental Management Initiative, Washington, D.C., 1991, 1.
26. Tippett, P. and Wever, G., Written testimony, U.S. House of Representatives, Joint Hearing, Subcommittee on Oceanography, Gulf of Mexico, and the Outer Continental Shelf and Subcommittee on Environmental and Natural Resources, U.S. Congress, Washington, D.C., 1994.
27. Wever, G. H., Managing for a clean environment and a healthy economy: a case study of the Great Lakes region, Cleaner Production Symp., Taiwan, ROC, 1994.
28. CGLI, *TQEM Primer and Self-Assessment Matrix,* Council of Great Lakes Industries, Ann Arbor, MI, 1993.

29. Wood, A., 10 years after Bhopal, *Chem. Week,* 155, 25, 1994.
30. CMA, *Improving Responsible Care® Implementation and Enhancing Performance and Credibility,* Report of the Ad Hoc Board Responsible Care® Committee, Chemical Manufacturers Association, Washington, D.C., 1993.
31. Kluth, F. J., Implementing the CERES 10 principles of environmental management, October, 1994.
32. Willson, J. S. and McLean, R. A. N., ISO 14001: is it for you?, *Prism,* First Quarter, 27, 1996.
33. Rhodes, S., International environmental guidelines to emerge as the ISO 14000 series, *Tappi J.,* 78(9), 65, 1995.
34. Powers, M. B., Companies await ISO 14000 as primer for global eco-citizenship, *ENR,* 234(21), 30, 1995.
35. Hemenway, C. G. and Hale, G. J., The TQEM-ISO 14001 connection, *Qual. Prog.,* 29(6), 29, 1996.
36. Smart, B., *Beyond Compliance: A New Industry View of the Environment,* World Resources Institute, Washington, D.C., 1992.
37. Barde, J. P., Economic instruments in environmental policy: lessons in the OECD experience and their relevance to developing economies, Technical paper No. 92, OECD Development Centre, Paris, 1994.
38. Parikh, K., "Sustainable development and the role of tax policy," *Asian Dev. Rev.,* 13(1), 127, 1995.

8 Decision-Making and the Environment: Integrating Scope and Values

O. Homer Erekson, Orie L. Loucks, Jan Willem Bol, Raymond F. Gorman, Norman K. Grant, Timothy C. Krehbiel, Allison R. Leavitt, and Nigel C. Strafford

CONTENTS

8.1 Introduction 167
8.2 Decision Option Analysis: Scoping the Steps to Sustainability 168
8.3 Scale and the Globalization of Commerce 170
    8.3.1 Principle 14 170
8.4 The Social Context of Sustainability Revisited 174
    8.4.1 Trust and Consensus as Elements of Choosing 174
    8.4.2 Sustainability as the Design of Win-Win Outcomes 175
    8.4.3 The Safe Minimum Standard: Making All the Parts Work Better 175
8.5 Conclusion 177
8.6 References 178

> "In closing, let me stress the need for all of us to view environmental problems in interdisciplinary terms, not in narrow terms of specialization. The world is replete with projects that made excellent engineering sense but were economically disastrous, or that were economically sound but environmentally catastrophic. The global environment cannot be separated from political, economic, and moral issues. Environmental concerns must permeate all decisions, from consumer choices through national budgets to international agreements. We must learn to accept the fact that environmental considerations are part of the unified management of our planet. This is our ethical challenge. This is our practical challenge—a challenge we all must take."
>
> (Gro Harlem Bruntland, Prime Minister of Norway, 1989)

8.1 INTRODUCTION

In the preceding chapters, we introduced thirteen principles (and will introduce one other here) to guide business and the public as to what matters most in seeking personal and commercial development that is sustainable. In this chapter we will consider the questions: How do these principles come together practically, and over what time frames and geographic scales should we judge outcomes? Given the perspectives from business, economics, and the sciences introduced in this book, how can these considerations be brought together to make sustainable choices?

To address these questions, we use a standard approach to problem-solving: application of the general rational comprehensive approach to decision-making, while emphasizing systems thinking and sustainability principles. As we have emphasized throughout, sustainability requires viewing the natural world and the human, political, social, and economic spheres as interwoven systems. Dysfunctional outcomes result when the underlying unity and closed nature of the larger system and its components are ignored, as when resource depletion and externalities are not considered.

Whether in decrying the ineffectiveness of regulation for controlling environmental externalities, or in remarking on the emergence of market economies throughout Eastern Europe and other places in the world, market-driven processes are increasingly touted as the most effective means for allocating resources within an economy and for promoting human welfare. While we have supported use of market-incentives and self-regulation of business to address environmental externalities where appropriate, the limits of the marketplace must be recognized explicitly. In his book, *Everything for Sale,* Kuttner[1] reminds us:

> "A limitation on the reach of the market is necessary to limit the market's self-cannibalizing tendencies within its own realm. A set of extra-market or premarket values—such as honor, trust, loyalty, decency, fairness—makes markets work better, even though market pressures keep undermining those values … If markets are not perfectly self-correcting, then the only check on their excesses must be extra-market institutions. These reside in values other than market values, and in affiliations that transcend mere hedonism and profit maximization. To temper the market, one must reclaim civil society and government, and make clear that government and civic vitality are allies, not adversaries."

Sustainability requires industry, political institutions, and the general populace to adopt policies that will "close the loop" and reconcile manufacturing, use, and disposal of goods with the need to maintain the earth's capital. This can be accomplished through waste management, minimization of chemical residues, recycling, adoption of benign technologies, and environmental conservation described here and in the case studies that follow. Such situations, and the associated decisions, are encountered by consumers, business managers, employees, and government officials nearly every day. Too often these decisions are made on the basis of near-term self-interest. Attainment of sustainable development, however, will require a long-term and much more broad-based focus. Below we will describe the integration and scope of analysis necessary to achieve sustainability in business and society.

8.2 DECISION OPTION ANALYSIS: SCOPING THE STEPS TO SUSTAINABILITY

Many people seek, generally, to apply the principles of sustainability in their decisions, but they are frustrated by the absence of any straightforward methodology. Let us understand why, while also recognizing that simple methods are inappropriate for complex questions.

Decisions, whether by the public or by business, involve problem resolution. As shown in Figure 8.1, the first step is problem definition, in our case in the context of sustainability of outcomes. Solving the problem also involves accepting that choices must be made with options evaluated fully. Systematic analysis of the options, then, is the problem and a decision is one outcome.

Problem definition involves establishing goals and objectives for the decision. In our approach to decision-making, we consider the 14 principles of sustainable development developed here and in the previous chapters. In setting a goal, one is to determine what it is that is to be resolved by a decision, and how the outcome is to be measured. The objectives are subquestions, aspects of secondary choices made in order to achieve secondary goals. The process itself is dynamic and not necessarily linear; choosing among alternatives often involves problem redefinition and reanalysis as perspectives of various stakeholders are fully considered.

The next step is to define the boundaries of the system relevant to the problem at hand. This usually is referred to as scoping, an analysis and specification of the scope of the problem. For

Decision-Making and the Environment: Integrating Scope and Values

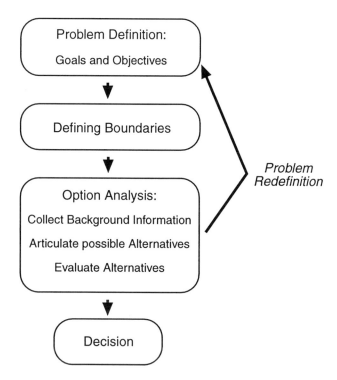

FIGURE 8.1 Decision option analysis.

example, in Chapter 4, we described the types of issues inherent in sustainability of fisheries. These include:

- *Geographical* dimensions defining the waters to be included, as well as the migration of fish in and out of the fishery.
- *Biological* dimensions describing the relationship among various fish stocks, the food base, and physical characteristics of the water (such as oxygen level).
- *Human* dimensions including the market for different fish species, employment implications for fishery communities, investment in boats and equipment, and the role of government in regulating fish harvests.*

Thus, scoping includes establishing spatial, temporal, political, financial, and stakeholder interests, all of which are needed to determine the constraints limiting the choices of action.

Consideration of options preparatory to a decision usually involves three steps: collecting background information to help identify and assess alternatives, articulating all the possible alternatives, and evaluating the alternatives. For each, the information required must allow determination of feasibility of the outcome and long-term ramifications, after considering all the constraints, and the acceptability of each outcome with respect to criteria for sustainability of development. These steps lead to selection of an option (or set of options), and the initial steps toward implementing that option. In the long run, we then need to be prepared to evaluate outcomes with respect to the sustainable development principles outlined in previous chapters.

Evaluation of the alternatives is where the principles of sustainability can be brought to bear most clearly. Not all principles have to be considered for every option. But as with the total quality

---

*For a more complete description of the interdependecies implicit in fisheries management, see Krehbiel et al.[2]

environmental management approach cited in Chapter 7, any one or more principles may be critically significant for a given decision. This approach, used systematically, shows the open loops leading to externalities that now should be closed.

Certain principles of sustainability bear critically on assessment. Especially significant is characterizing the *integrity of naturally functioning systems, the value of that integrity,* and consideration of suitable *audits for stakeholder interests.* Since no system exists without variability, it is important also to know how measures of integrity and value are related to measures of resilience and social justice (e.g., both inter- and intra-generational equity). Most importantly, quantitative information, to the extent it is available, should inform the process of multi-stakeholder assessment of alternatives imbedded in the decision using *information and dialogue.* Where possible, feedback effects may be quantitatively estimated through the use of simple material mass-balances or spread-sheet calculations.

Now, how are these steps different from other approaches to individual, public, or business decision-making? The differences in our approach lie in how spatial and temporal limits (boundaries) are used, reflecting the principles of long-term sustainability, and how these, in turn, have informed the process. Like most applications of problem-solving and decision-making, these spatial, temporal, political, financial, and design constraints determine the outcomes, but application of our principles also leads to somewhat different decisions. What is especially salient, and sometimes not emphasized in other problem-solving approaches, is identification of the full array of stakeholders and their concerns, which include both short-term and long-term equity interests. Especially important is consideration of externalities affecting stakeholders and the requisite context regarding political jurisdictions, geographical (spatial) influences, and the prospect of temporal dislocations.

8.3 SCALE AND THE GLOBALIZATION OF COMMERCE

Environmental issues have only begun to receive significant international attention in the last 25 to 30 years. In fact, more than half of the multilateral environmental treaties that have been adopted since 1921 were concluded since 1973.[3] Increasingly decision-makers in business, government, and world organizations are providing significant revenues to address environmental problems. For instance, in 1996, 68 countries were receiving financial and technical support from the World Bank for environmental policy reforms and associated investments ranging from industrial pollution and coastal-zone management to protected-areas management and biodiversity conservation. As shown in Figure 8.2, World Bank loans in support of environmental protection have increased from 1 minimal loan in 1986 to 153 loans totaling more than $11 billion in 1996.[4] Keohane et al.[3] have argued that sustaining the quality of the planet for future generations depends upon successful cooperation among international organizations to guide international behavior along a path of sustainable development. By this they mean development of "persisted and connected sets of rules and practices that prescribe behavioral roles, constrain activity, and shape expectations."

In this light, organizations such as the World Bank may establish expectations or understandings with companies or countries to provide funding for environmental projects. As discussed in Chapter 7, two of the major purposes of the ISO 14000 standards are to minimize trade barriers that may arise owing to disparate national environmental standards and to promote a common approach and language for environmental management within the international community.

Indeed, because environmental externalities know no political boundary, promotion of sustainability often becomes an international scale problem. This observation leads to our final principle for sustainability, the *globalization of commerce:*

8.3.1 PRINCIPLE 14

Sustainable processes include explicit recognition of international interdependencies among consumers, businesses, and governments, with the result that solutions to environmental problems cross cultural and national boundaries.

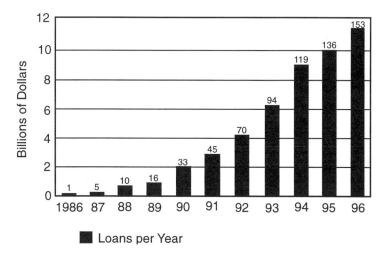

FIGURE 8.2 World Bank Financing for the Environment—The number of projects approved each year is given by the number above the bar chart. The dollar volume of loans per year is given by values on the vertical axis; adapted from Reference 4.

Two brief examples illustrate the importance of recognizing the international dimension of sustainability. First, because many environmental problems involve significant effects that cross national political boundaries, international agreement on dealing with these externalities is essential. One of the most broad-based environmental agreements was the Framework Convention on Climate Change coming out of the June 1992 Earth Summit in Rio de Janeiro. More than 100 nations have ratified the Rio treaty, which calls for reducing the total annual emissions of greenhouse gases world-wide and stabilizing the overall concentration in the atmosphere. It calls for industrialized countries to take the lead in conforming to this treaty, aiming "as a first step to return greenhouse emissions to 1990 levels by the year 2000."[5] Unfortunately, this convention has minimal compliance obligations. Countries are called to "develop inventories of greenhouse emissions, prepare national programs for mitigating and adapting to climate change, and to take climate into consideration when formulating other government policies." The vigor with which countries have pursued the Rio objectives is quite inconsistent. Some countries have moved the date for achieving 1990-level emissions to 2005; Finland did not set a base year at all; France and Japan have agreed to stabilize emissions per capita, rather than total; the United States has set its standard as allowing a 3 percent increase in carbon dioxide emissions; and Spain and Ireland have disregarded the 1990 target completely, agreeing only to slow emissions growth by some unspecified amount.[6] Thus, compliance with the Rio Climate Change Treaty has operated like a classic public good where there is an incentive for each country to be a free rider, resulting in underprovision of the good, in this case, optimal reduction of carbon dioxide and other greenhouse gases.*

A second example involves the effects of international trade on the environment. There is a significant debate as to whether trade liberalization has positive or negative effects on the environment. Those who argue for the positive effects of trade liberalization argue that free trade leads to increased economic growth, and that environmental quality is positively related to economic growth. There are two hypotheses imbedded in this previous sentence.

First, the argument that free trade leads to increased economic growth is rooted in the concept of comparative advantage. In this theory, individuals and nations gain by specializing in the production of commodities for which they have relatively low production costs, exporting these goods and

---

* The *free rider* problem emerges in the case of non-excludable public goods. It occurs where consumption of a good (such as a fireworks display) does not require a person to pay for that good. Thus, each person has an incentive to let other people pay while still enjoying the benefits. This normally leads to an underprovision of the good. For a full discussion of public goods and free ridership, see Reference 7.

services to other countries who have a comparative advantage in the production of another commodity, who trade this good or service in return. An important corollary is that countries tend to specialize in the export of goods and services for which the production processes involve a factor of production that is relatively abundant in that country.

Application of this principle to actual trading patterns is more problematic. As an example, consider the European Community's Common Agricultural Policy (CAP). This policy involves large subsidizes for the European agricultural sector. As shown in Figure 8.3, agricultural subsidies vary significantly across countries, along with the use of chemical-intensive and energy-intensive agricultural production. Johnstone has argued that the CAP has overridden the principles of comparative advantage with large subsidies that have created artificial incentives for agricultural production in regions not best suited to such cultivation. This has resulted in agricultural specialization by the first eight countries in the table, each of which practices agricultural production that is very chemical and energy intensive. Protectionism in this case has resulted in significantly greater environmental stress than would be the case should trade liberalization and removal of subsidies be introduced.[8]

The second hypothesis in this example involves the U-shaped relationship we presented in Chapter 1 which argues that environmental degradation increases with increases in income to a point, beyond which the quality of the environment improves as income rises further. As we discussed then, the usefulness of this relationship depends upon the type of environmental problem being considered, as well as the time scale involved. While the positive relationship between economic growth and environmental quality may have some validity for basic local sulfur dioxide emissions, it is not likely to be applicable in situations such as those involving transboundary deposition of acid rain or carbon dioxide emissions where economic growth may result in long-term irreversible environmental degradation. Thus, trade that promotes economic growth may result in decreased environmental quality, even for highly developed economies.

In fact, some writers have suggested that free trade will necessarily lead to environmental degradation. Daly has argued that trade liberalization distorts the international political arena by thwarting the potential for national governments to enforce environmental standards. He claims that "to

	Agricultural subsidy equivalent (%) 1979–1989	Chemical fertilizer use (kg/ha of arable land) 1987–1989	Tractors and harvesters (#/ha of arable land) 1987–1989
Denmark	39	243	0.078
France	39	312	0.087
Germany	39	405	0.212
Netherlands	39	662	0.214
Italy	39	172	0.117
Spain	39	101	0.037
Ireland	39	717	0.177
United Kingdom	39	359	0.085
Argentina	−38	5	0.007
Brazil	22	46	0.009
India	−2	62	0.005
Indonesia	11	113	0.002
Thailand	−4	33	0.006
Australia	11	26	0.008
Canada	35	47	0.020
United States	30	95	0.028

FIGURE 8.3 Agricultural subsidies and production techniques in selected countries, adapted from Reference 8.

globalize the economy by erasure of national economic boundaries through free trade, free capital mobility, and free, or at least uncontrolled, migration is to wound fatally the major unit of community capable of carrying out any policies for the common good."[9] Within an environment of free trade, he argues that multi-national companies can avoid higher production costs associated with compliance with environmental standards by moving their operations to countries with more lax environmental standards. Korten builds on this argument noting that "globalization tends to delink the fate of the corporation from the fate of its employees."[10] To the extent that Daly and Korten have valid arguments, free trade would then limit the ability of national governments to enforce environmental policies and would limit the voice of employees and other stakeholders, relative to that of corporate managers.

Another dimension of the globalization of commerce involves the connection between large-scale human-induced environmental and resource pressures and international security. Homer-Dixon has argued that atmospheric, terrestrial, and aquatic environmental pressures are likely to contribute eventually to four main, causally interrelated social effects: reduced agricultural production, economic decline, population displacement, and disruption of regular and legitimized social relations.[11] These social effects, he contends, may lead to acute domestic and international conflict, including scarcity disputes between countries, clashes between ethnic groups, and civil strife and insurgency. As an example, in almost every region of the world, conflicts and potential conflicts over the availability of water are occurring more frequently. The importance of water to basic human functioning and the fact that the water for many countries originates outside their borders (see Figure 8.4) makes water a natural focus for international conflict. Countries have an incentive to capture and use water within their boundaries, but little reason to conserve or maintain the quality of water supplies for downstream users. The inability of establishing well-defined property rights again leads to free rider effects, as well as the failure of countries needing to face the full social cost of their water use and downstream externalities.[12]

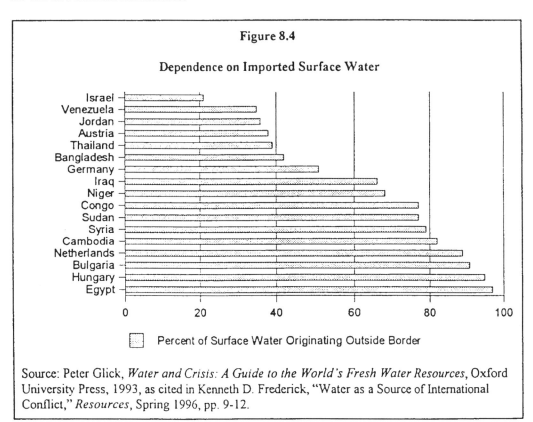

Figure 8.4

Dependence on Imported Surface Water

Percent of Surface Water Originating Outside Border

Source: Peter Glick, *Water and Crisis: A Guide to the World's Fresh Water Resources*, Oxford University Press, 1993, as cited in Kenneth D. Frederick, "Water as a Source of International Conflict," *Resources*, Spring 1996, pp. 9-12.

8.4 THE SOCIAL CONTEXT OF SUSTAINABILITY REVISITED

8.4.1 Trust and Consensus as Elements of Choosing

Although, at various points in this book, we have noted the importance of measurement, auditing, and reporting, and other seemingly formalistic processes and systems related to sustainability, we want to emphasize here that sustainability involves values at its core. Sustainability is more than analyzing the size of the packaging or the mode of material transport. There is much more, mostly having to do with how to apply the values principles in sustainability.

Living in a democracy, we recognize that political leaders, writers, and philosophers, at various times, have sought to approach utopian goals through ethical analysis and decision-making. We assert the importance of equality of opportunity and responsibility among stakeholders, balance in the use and long-term availability of resources, and balance among equity interests. An appropriate balance is expected to lead to congruence between individual and community-wide behavior.

Seen this way, there need be no long-term conflict between the survival of human beings and maintenance of the resource base that supports humans. Considering the many conflicting stakeholder interests, however, requires cooperation, tolerance, and a large-group vision of self-interest leading to trust. Have the questions posed by leaders of democratic interests changed through time in the direction of recognizing more broad-based equity interests on the part of all members of society? Are the questions posed in the past different from those asked now, and do they relate to good business practice? Yes, they are different, and sustainability, widely accepted as good business practice, is now a component in the visioning of an ideal society and a new relationship between the natural world and human values and beliefs. Evolution, and new understandings in science, have replaced medieval views of heavenly hierarchies, but this also has led society and business to see themselves as part of a complex, evolving system.

The term *economics* itself comes from the Greek *oikonomikos,* and connotes "management of a household or community system." As business leaders increasingly have taken a worldview as to their responsibility to current and future generations, their understanding of sustainable business practices has come to emphasize business' role in responsibly using natural resources, while preserving the resource base. Thus, it is common now to hear business and stakeholders refer to the concept of *stewardship,* "the careful and responsible management of something entrusted to one's care."[13] This has led business to embrace full valuation of human activity to measure both direct and indirect depletion of resources, measure the costs of externalities from industrial production, and place a value on those activities that contribute to or detract from maintenance of natural resources and the ecosystem. Thus, the notion of *product stewardship* has become an important component of responsible business strategy and calls for business to consider all environmental impacts involved from purchase of raw materials from suppliers through product development, production, distribution, and ultimate use and disposal.[14]

This discussion needs to be seen as part of a process underlying the building of consensus and trust between otherwise disparate constituency interests. The process provides a means of amending human perceptions of apparent needs (individual goals, community goals), not merely improving the efficiency of transactions, technology, and production to keep up with human demands. The process provides an over-arching center against which production and technology options are evaluated and choices are made. It draws on a succession of consensus-building steps, each providing information and filling communication gaps.

What is the means by which the principles of sustainability help build multi-stakeholder trust and consensus? In small communities where everyone's contributions can be recognized, this is not even a question. In an industrial society, however, no alternative exists to use of dialogue and communication to build consensus. Without confidence in our potential to build a broad consensus in business decisions we could lose confidence in the prospects for sustaining the diverse multi-cultural society we enjoy. Does this mean that managing for sustainability leads to a reduction in choices?

Not really, although there can be no long-term consensus (or trust) in choices that benefit a few and impoverish others. Systematic building of trust derives from awareness that every stakeholder has participated in the choosing of options for the future.

8.4.2 SUSTAINABILITY AS THE DESIGN OF WIN-WIN OUTCOMES

Reference often is made to people and businesses as winners or losers. Implicit here is the idea that their activities are a game in which the scores at the end of a day, or a year, can be characterized as zero-sum (wins are balanced by losses), or positive-sum (win-win outcomes) where all participants win and there may be no losers. It is our thesis that sustainability principles provide an explicit framework and calculus for designing and quantifying positive-sum strategies for people and business in an information-rich society.

For example, when the Chemical Manufacturers Association adopted principles and codes of management practice for establishing a new materials auditing standard for their members, society as a whole benefited in the long run, as did many individual companies (by reducing waste), even though higher costs were incurred in the short run. Individual chemical companies also were able to expand their overall market opportunities because of increased public trust in their performance and products. This is a win-win situation long-term because both producers and consumers benefit, producers through less expense in waste, more efficient production, and lowering of the risk of accidents, and consumers through less pollution and related externality costs. The short-run increase in costs for chemical products must be seen as closing a loop, incorporating the cost of a safe product into the price paid by the user, rather than the externality burden born by the public generally. The resulting expression of trust is an outcome of using the principles of sustainability in the decision process. Society as a whole benefits when the needs of multiple stakeholders are legitimized and the value of external costs are incorporated into decision-making.

Internationally, some firms may pay high wages for high productivity workers in developing countries, possibly incurring short-run increases in production costs, and consumers in developed countries may have to pay higher costs for the products. This may result is a zero-sum outcome in the short-run, a lose and win situation, as the consumers in developed countries pay higher prices, but the workers in underdeveloped countries benefit from the higher wages. In the long-run, however, this may have the potential for being a win-win opportunity, as the increased purchasing power in the developing countries will result in higher demand for products from the developed countries. The higher wages also allow developing countries to raise their standard of living, often leading to greater capacity to protect the environment and reduce risks to health.[15]

These examples illustrate the win-win perspective that may arise from a long-term intergenerational view of sustainability. As larger numbers of people share in higher and more secure disposable income (because resource supplies and air and water quality are secure), societies are more capable of accepting the price increases from internalizing pollution costs. All of the larger systems benefit. Thus, as societies develop benign product design and technology improvements, they can become more resilient and more accommodating to stakeholder justice. This is a positive sum outcome for the world economy and its peoples.

8.4.3 THE SAFE MINIMUM STANDARD: MAKING ALL THE PARTS WORK BETTER

Throughout the earlier chapters where we developed the principles of sustainability, and in the business case studies which follow, the development of new "low impact" or benign production technologies is critical to the decision process and application of sustainability principles. These applications are obvious in initiatives such as those in Europe where new manufacturing approaches are requiring products to be designed for the environment, that is, to have the capacity for complete disassembly and return to the manufacturer for recycling. A related technological underpinning is evident in the Procter and Gamble or Fetzer case studies, where life cycle analysis is central. This now widespread methodology allows precise comparison of what is or is not a benign technology or low-impact production, where the whole cycle from raw material through production to disposal (or recycling) is considered, and product stewardship may be seen in application.

If the principles of sustainability are followed, measures of performance by our society for the subsystems or for the whole, can be stimulated or designed, like our economy, so as to minimize unintended externalities from open loops. Proposals for improvements need to be examined and evaluated across time, across societal components, and across national boundaries. Individuals or firms who define their perspective narrowly and short-term, and who place constraining boundaries around their activity or business system, have little means of judging long-term risks or benefits to themselves or others with whom they interact or do business.

Toman has suggested approaching these issues through use of the concept of a *safe minimum standard* (see Figure 8.5).[16] In this conception, human activities are characterized along two dimensions. The vertical axis shows the increasing cost of ecological damages, while the horizontal axis shows the degree to which the ecological damages are irreversible. The upper left-hand corner reflects a situation where cumulative ecological damage is large and irreversible, likely resulting in a threat to the survival of human and other species. The bottom right-hand corner captures ecological damage that is relatively minor in scope and easily reversible. In this case, Toman argues, "private market transactions or corrective government policies based on comparisons of benefits and costs" are appropriate and likely sufficient. The safe minimum standard is "a socially determined dividing line between moral imperatives to preserve and enhance natural resource systems and the free play of resource tradeoffs." The inter-generational contract implicit with sustainability "would rule out in advance actions that could result in natural impacts beyond a certain threshold of cost and irreversibility." Human activity that might lead to catastrophic global climate, change or loss of genetic diversity, would be examples of actions that would not be allowed under this approach.

We find the notion of the safe minimum standard to be attractive, not just because it attempts to relate human and ecological systems, but also because "societal value judgments determine the level of safeguards," the role of the public in decision-making, and the formation of social values is

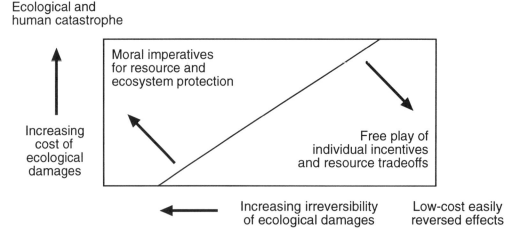

FIGURE 8.5 The safe minimum standard for balancing natural resource tradeoffs and imperatives for preservation, adapted from Reference 16.

explicit. The degree to which sustainability can be maintained is dependent upon the following dimensions:

- The productivity of the economic subsystem, a function of the stock of natural capital, man-made capital, labor, technology, and management of these factors of production; the economic subsystem must provide for information systems and business strategic planning that will lead to sustainable outcomes.
- The carrying capacity of the physical and biological systems, which determines the scale of resource use that can be maintained; this dimension focuses on the relationship of the input base to human and other species population sizes and quality.
- The assimilative capacity of the physical system, the threshold value of the environmental subsystem, which if eclipsed would threaten the vitality of the system.

The *resilience* of the system relates back to the concept of irreversibility, and reflects the ability of the overall system to accept fluctuations while maintaining itself. We believe that resiliency of the earth's aggregate system has three inseparable components: *ecological resilience* (e.g., necessary nutrients to maintain a native species or human population are available, and critical tolerance for waste is not exceeded); *economic resilience* (e.g., the structural base for the economy is sufficient to allow for efficient production); and *social resilience* (e.g., social change occurs in ways that preserve justice and equity among stakeholders). Each of these components is like one part of a three-legged stool. For the stool to support the weight of human activity, each leg must be firm.

8.5 CONCLUSION

In this book, we have introduced fourteen principles of sustainability (see Figure 8.6). At first, the number of principles and their inter-relationships may have seemed overwhelming. Hopefully, they now form a coherent whole. The eight systems principles require approaching sustainability from a unified systems perspective, including specifying appropriate scales (time and geographic), internalizing externalities within closed-loop systems, and establishing principles of system resilience. Six process principles emphasis the role of information, measurement, and communication in achieving sustainable outcomes. Finally, the six values principles include valuation of human activity and the necessity for broad-based participation by stakeholders.

At the core of these systems, linking them together is the principle of integrity of individuals, communities, and nature. Maintaining the health of each of these components requires a confluence of systems, processes, and values. Perhaps the most fundamental value is an appreciation for the wealth of nature itself. Kellert and Bormann[17] argue that:

> "perhaps what we require is an attitude more of studentship than of stewardship, a view of nature as teacher and us as apprentices to its mysteries and methodologies. This posture would not relinquish the need for greater scientific research and even environmental manipulation, but instead would temper our actions with a spirit of caring and humility. The most compelling reasons for environmental conservation are not material enhancements or altruism or the acceptance of scientific knowledge, but the personal conviction that nature is as much a place for nurturing human fulfillment as it is for raising crops or generating harvestable commodities."

Or as exhorted by Wilson[18]:

> "Humanity is part of nature, a species that evolved among other species. The more closely we identify ourselves with the rest of life, the more quickly we will be able to discover the sources of human sensibility and acquire the knowledge on which an enduring ethic, a sense of preferred direction, can be built."

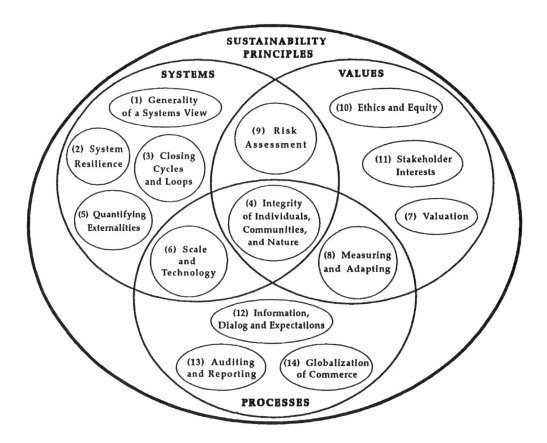

FIGURE 8.6 Sustainability principles.

REFERENCES

1. Kuttner, R., *Everything for Sale: The Virtues and Limits of Markets,* Alfred A. Knopf, New York, 1996.
2. Krehbiel, T. C., Gorman, R. F., Erekson, O. H., Loucks, O. L., and Johnson, P. C., Advancing ecology and economics through a business-science synthesis, *Ecol. Econ.,* in press.
3. Keohane, R. O., Haas, P. M., and Levy, M. A., The effectiveness of international environmental institutions, in *Institutions for the Earth: Sources of Effective Environmental Protection,* Haas, P. M., Keohane, R., and Levy, M. A., Eds., MIT Press, Cambridge, MA, 1993, 3.
4. Steer, A., Ten principles of the new environmentalism, *Finan Dev,* 3(4), 4, 1996.
5. Grubb, M., Koch, M., Munson, A., Sullivan, F., and Thomson, K., *The Earth Summit Agreements: A Guide and Assessment,* Earthscan Publications, London, 1993.
6. Flavin, C. and Tunali, O., Getting warmer: looking for a way out of the climate impasse, *World Watch,* 8(2), 10, 1995.
7. Rosen, H., *Public Finance,* 4th ed., Irwin, Chicago, 1995, chap. 5.
8. Johnstone, N., Trade liberalization, economic specialization, and the environment, *Ecol. Econ.,* 14(3), 165, 1995.
9. Daly, H. E., Fostering environmentally sustainable development: four parting suggestions for the World Bank, *Ecol. Econ.,* 10(3), 183, 1994; Daly, H. E., The consequences of global competitiveness, *Ecol. Econ. Bull.,* 2, 4, 1997.
10. Hinrichs, D. and Roodman, D., Economic globalization: an interview with David Korten, *Ecol. Econ. Bull.,* 2, 14, 1997.
11. Homer-Dixon, T. F., On the threshold: environmental changes as causes of acute conflict, *Int. Security,* 16(2), 76, 1991.
12. Frederick, K. D., Water as a source of international conflict, *Resources,* Spring, 9, 1996.

13. Barrett, C. B., Fairness, stewardship, and sustainable development, *Ecol. Econ.,* 19(1), 11, 1996.
14. Erekson, O. H., Loucks, O. L., and Aldag, C., The dimensions of sustainability for business, *Mid-Am. J. Bus.,* 9(2), 3, 1994.
15. Grossman, G. M. and Krueger, A. B., Economic growth and the environment, *Q. J. Econ.,* 110(2), 353, 1995.
16. Toman, M. A., The difficulty in defining sustainability, *Resources,* Winter, 3, 1992.
17. Kellert, S. R. and Bormann, F. H., Closing the circle: weaving strands among ecology, economics, and ethics, in *Ecology, Economics, and Ethics: The Broken Circle,* Bormann, F. H. and Kellert, S. R., Eds., Yale University Press, New Haven, CT, 1991, 205.
18. Wilson, E. O., *The Diversity of Life,* W. W. Norton, New York, 1992.

Section II

Case Studies on Sustainability

9 Ashland Chemical: Achieving Sustainability Through the Use of Responsible Management Systems—Internal Environmental, Health, and Safety Auditing Case Study Number 1

Raymond F. Gorman

CONTENTS

9.1 Introduction 183
9.2 Company Background 184
9.3 Operations Auditing—The Beginning 187
9.4 The Operations Auditing Group 188
9.5 The Audit Process 188
9.6 The Walnut Station Plant 190
9.7 Selecting and Scheduling the Site Audits 190
9.8 Preparing for the Audit 191
9.9 Conducting the Walnut Station Audit 192
9.10 Reporting and Communicating the Audit Evidence 195
9.11 Completing the Audit 196
9.12 Summary 198

9.1 INTRODUCTION

Ashland Chemical Company (ACC), a Division of Ashland Inc. (AI)* maintains an Operations Auditing Group in the Environmental, Health, and Safety (EH&S) Department. The Operations Auditing Group plays an important role in sustaining ACC's world-wide operating facilities by determining the extent to which ACC facilities are operated in compliance with applicable laws, regulations, company policies, and good management practices. ACC Management uses the information gathered from operations audits to assess the effectiveness of the company's Responsible

---

*In January 1995, Ashland Oil, Inc. (AOI) officially changed its corporate name to Ashland Inc., (AI). For the purposes of this case, references to this corporation (both current and past) will be made as "Ashland Inc. (AI)."

Care®* effort and to identify where additional resources may be needed to improve overall performance.

In early November 1994, the Operations Auditing Group was preparing for an audit at one of its larger distribution facilities in Walnut Station, PA.[†] It will take the group almost nine months to prepare for, perform, and report on the audit and then work with the facility to resolve the audit findings. An audit team consisting of individuals with expertise and experience in the areas of employee health, safety, the environment, and plant operations is carefully selected to perform the audit. The Operations Auditing Group selected a Plant Engineer from one of ACC's manufacturing facilities to be a guest auditor for the Walnut Station audit. As part of the audit participation, the guest auditor is challenged to become familiar with the history and philosophy of the Operations Auditing program. The guest auditor also must become familiar with the auditing procedures and protocols that will be used during the five day site visit scheduled for the first week in December 1994.

9.2 COMPANY BACKGROUND

Ashland Chemical is an operating division of Ashland Inc. (AI). AI is a diversified energy and chemical company with operations in petroleum refining, transportation, and wholesale marketing; retail gasoline marketing; motor oil and lubricant marketing; chemicals; construction; and oil and gas exploration and production. AI also has equity interests in two coal companies. AI, a publicly held Fortune 50 company with over 33,000 employees world-wide, achieved over $10 billion in sales in FY 1994 with a return on stockholder's equity of 14.5 percent.[‡] The company's long-term strategy is to maintain a strong financial position by investing and growing earnings from its energy and chemical businesses while keeping a competitive strength through refining and wholesale marketing.

AI strives to fulfill its responsibilities as a corporate citizen and is very active in the communities in which it operates. Through education and community relation programs, Ashland provides financial aid and leadership for many activities that enhance the quality of life in these communities. Ashland also realizes the importance of environmental considerations regarding its current and future operations. The company has recently invested hundreds of millions of dollars in environmental control capital projects. While many of the corporation's environmental improvements are the result of mandated regulations, an increasing number are the result of voluntary initiatives and cooperative agreements with regulatory agencies. For instance, Ashland was one of the first companies to participate in the Environmental Protection Agency's voluntary Industrial Toxics Project (ITP). The goal of the ITP was to reach a 50 percent reduction in the emission of 17 target chemicals by 1995 over 1988 baseline levels. Eleven of the chemicals relate to the company's refining and chemical operations. Ashland is a participant in EPA's Green Lights Pollution Prevention Program. This is a national effort to use energy-efficient lighting to help reduce air pollution, conserve energy, and lower utility costs. The company also works cooperatively with the Ohio River Sanitation Commission (ORSANCO) to support the Ohio River Sweep, a voluntary effort to clean debris from the banks of the Ohio River. During the past six years, Ashland Inc. has been the major corporate sponsor of this project which has resulted in the removal of over 60,000 tons of trash from about 2,000 miles of the river's shoreline.[§] The corporation also has worked with the Wildlife Habitat Enhancement Council to create nature reserves on nearly 800 acres of company-owned land in Kentucky and Louisiana.

Ashland Chemical Company (ACC) is a large, global company that maintains an aggressive growth strategy by operating business units that are well positioned in their respective markets. The

---

* Responsible Care® is a registered service mark of the Chemical Manufacturers Association.

[†] The Walnut Station facility is a fictitious location created for the purposes of this case. The operations audit reported here for the Walnut Station plant represents a composite of audits that may be conducted at different Ashland Chemical facilities by different auditors to illustrate a broad perspective.

[‡] Annual report, Ashland Oil, Inc., Ashland, KY, 1994.

[§] Environmental, Health, and Safety Annual Report, Ashland Oil, Inc., Ashland, KY, 1994.

company has and continues to grow through acquisition and has made over 20 acquisitions since 1990. Its sales have grown steadily from $1.5 billion in 1984 to close to $3 billion in 1994. ACC manufacturers, markets, and distributes a wide variety of chemical and plastic products world-wide. The company holds strong positions in the chemical manufacturing and chemical distribution markets by operating 38 manufacturing facilities in 9 states and 16 foreign countries and over 100 distribution facilities in 30 states and 10 foreign countries. ACC is the largest distributor of chemicals and solvents in North America. A profile of ACC's operating divisions is shown in Table 9.1.

Another key element of ACC's organization is the many resource (staff) groups to which all of the company's operating divisions have access. Some of these groups are Accounting, EH&S, Engineering, Information Systems, and Research and Development (see Figure 9.1). The resource groups are located at ACC's Dublin, OH world headquarters. The Operations Auditing Group is a part of the Environmental, Health, and Safety (EH&S) Department as shown in Fig. 9.2.

The diversity and complexity of ACC's operating divisions create a challenge regarding environmental, health, and safety issues. Despite this diversity, however, three common threads weave through the company to make it one cohesive organization. These threads are:

- ACC's Vision Statement.
- ACC's participation in the Responsible Care® Initiative.
- ACC's *Simply The Best* philosophy that merges the vision statement elements with the Responsible Care® initiative to enhance overall company growth and performance.

TABLE 9.1
A Profile of Ashland Chemical's Business Units

Division	Primary Product Line
General Polymers FRP IC&S	Markets and distributes over 3500 chemical, plastic, and related products. Market segments include the industrial manufacturing, paint and coatings, plastic processing, fine ingredients, commercial cleaning and sanitation, and automotive and transportation. All marketing and distributing activities are managed by ACC's Distribution Services Organization.
Composite Polymers	Manufactures and sells automotive and high-performance unsaturated polyester marine resins. The product line also includes a broad range of chemical-resistant, fire-retardant, and general purpose grades of unsaturated polyester and vinyl ester resins for the reinforced plastics industry.
Specialty Polymers and Adhesives	Manufactures and sells phenolic resins, acrylic polymers for pressure-sensitive adhesives, a variety of adhesives used for structural and composite bonding, induction bonding systems for thermoplastic materials, and butyl rubber tapes and adhesives for roofing applications.
Drew Ameroid Marine	The world's leading supplier of specialty chemicals for water and fuel treatment for the marine industry; Drew Marine provides shipboard technical service for over 15,000 vessels in the merchant marine fleet from 140 locations and 600 ports around the world.
Drew Industrial	The world's leading supplier of specialized chemicals and consulting services for the treatment of boiler water, cooling water, and steam, fuel and waste streams. The division also supplies process chemicals and technical services to the pulp and paper and mining industries.
Electronic Chemicals	Manufactures and sells a variety of high-purity chemicals for the semi-conductor manufacturing industry and blends and packages high-purity liquid chemicals to meet customer specifications. Maintains a marketshare leadership position in the United States and has just entered Europe.
Foundry Products	The world's leading manufacturer of foundry chemicals including foundry binders, core and mold coatings, sand additives, mold releases, and other specialties.
Petrochemical	Manufacturers selected petrochemical products (methanol and maleic anhydride) that are strategic raw materials. The division markets those products as well as a variety of aromatic hydrocarbons, aromatic and aliphatic solvents, and propylene produced at Ashland Petroleum's Cattlettsburg Refinery. (Ashland Petroleum Company (APC) is a division of AI.)

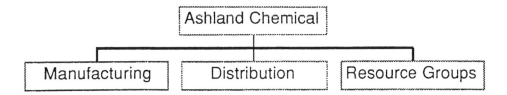

FIGURE 9.1 Ashland Chemical general organization.

ACC's Vision Statement is shown in Figure 9.3. The vision statement clearly addresses the company's commitment to its employees, customers, and the communities in which it does business.

In 1988 the Chemical Manufacturers Association (CMA) initiated a Responsible Care® initiative. The goal of the Responsible Care® initiative is to implement improved management systems throughout the chemical industry to improve environmental, health, and safety performance, and improve the general public's perception of the chemical industry. The program requires all CMA member companies to continuously improve performance in the following six areas of environmental, health, and safety:

- Community Awareness and Emergency Response (CAER).
- Process safety.
- Distribution.
- Pollution prevention.
- Employee health and safety.
- Product stewardship.

The Responsible Care® philosophy echoes ACC's Vision Statement elements regarding the value of the Company's employees and community. ACC has taken an aggressive role in integrating the principals of the Responsible Care® initiative throughout all of its businesses.

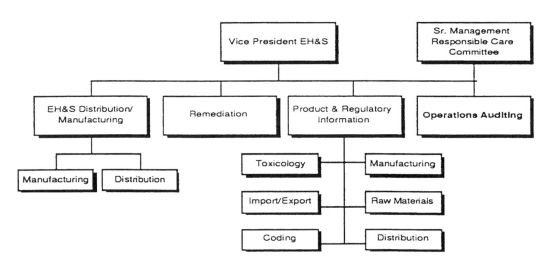

FIGURE 9.2 EH&S department organization chart.

> **Vision Statement: Ashland Chemical Company**
> Ashland Chemical is a profitable, professional, responsible, growth-oriented. customer-driven team that supplies high quality products and services while maintaining sensitivity to our employees and the community.
>
> **Vision Statement: Community**
> Ashland Chemical is an active and responsible member of the communities in which we operate. We obey all laws and regulations governing our operations and work in a manner that ensures the safety of our employees, our communities, and our environment.
>
> **Vision Statement: Employees**
> Ashland Chemical's employees are our most valuable asset. We are committed to hire, train, develop and reward people who are enthusiastic, intelligent, hardworking team players dedicated to the company's success.
>
> **Vision Statement: Customers**
> Ashland Chemical is dedicated to achieving mutually porfitable relationships with our customers and suppliers by providing quality products and services, on time and consistently, and by remain-

FIGURE 9.3 Ashland Chemical's vision statement.

9.3 OPERATIONS AUDITING—THE BEGINNING

ACC's Operations Auditing Group was formed in 1989 as the result of two significant factors:

- The company's own commitment to, and participation in, the Responsible Care® program.
- Corporate-wide changes made by AI as the result of an environmental incident involving Ashland Petroleum Company (APC), an operating division of AI.

As the Responsible Care® program was conceived in 1988, ACC developed its Responsible Care® initiative around a set of ten essential elements that commit the company to the responsible management of chemicals as part of its day-to-day operations. One of the essential elements is "Management System Audits." ACC Senior Management believed there to be great value in pro-actively assessing the EH&S performance of its facilities. The Operations Auditing (OA) program was seen as being an effective tool for providing this value. Also, in early 1988, AI experienced the first major pollution accident in the corporation's 64 year history.* Early on the evening of January 2, 1988, a 4 million gallon storage tank operated by APC ruptured and spilled about 750,000 gallons of diesel fuel into the Monongahela River in Floreffe, PA, a suburb of Pittsburgh. A diesel fuel slick proceeded to travel downriver into the Ohio River and threatened public drinking water supplies in Pennsylvania, Ohio, and West Virginia as well as the health and safety of nearby residents. AI Chairman, John Hall, took a very proactive and responsible position in responding to the crisis, apologized for the incident, and offered to reimburse communities, businesses, and residents for reasonable costs associated with the incident. However, while AI and its Chairman were commended for the compassionate and timely response, a number of factors associated with the tank surfaced in the hours and days after the incident that raised questions about the company's management systems. The issues were concerned with the tank's construction, design, testing and permits.

---

*Details of this case were extracted from Harvard Business School case #9-390-017, dated January 19, 1990.

Repercussions from these issues and the tank failure, in general, resulted in changes in the culture and organization of APC and AI. One of those changes was APC's formation of a Compliance Audit Program in December 1988. Shortly thereafter, AI directed each division to organize their own compliance audit program. Since ACC was already developing an audit program for Responsible Care® implementation, it was well prepared to meet this corporate directive, and formalized the Operations Auditing Group in June 1989. ACC Management now uses Operations Auditing as a tool for reviewing the operations of its facilities and providing feedback to management on the status of EH&S issues and programs.

9.4 THE OPERATIONS AUDITING GROUP

The Operations Auditing Group is part of ACC's EH&S Department and includes a manager, an office assistant, two auditors, and a regulatory specialist. The auditors are selected based on their expertise in 17 areas that are reviewed during a site audit. The Group also has a pass-through position where an individual from one of ACC's operating divisions will spend 18 to 24 months in the group. This individual is selected based on his or her potential to move into a management position within the company. The pass-through opportunity allows that person to obtain training, knowledge, and experience in the areas of operations, environmental, health, and safety to better prepare them for a management position. To supplement the regular auditing staff, the Operations Auditing Group also invites guest auditors to participate in the audits.

The mission of the Operations Auditing Group is to support the sustainability of ACC as a company and the sustainability of each individual facility by:

- Performing routine audits at all world-wide facilities to determine the extent of adequate management systems to assure the continued compliance and responsible facility management.
- Identifying and correcting potential problem areas.
- Increasing the awareness of plant management teams and employees of their environmental, health, and safety responsibilities, and enhancing the facilities' ability to maintain compliance with rules, regulations, and company policies.
- Informing ACC Senior Management of operational status and identifying opportunities for improvement.
- Auditing non-EH&S areas that are important to the successful and responsible management of ACC facilities.
- Providing all levels of ACC management with an increased assurance that the company's operations are:
  - In compliance with applicable laws and regulations.
  - Adhering to company policies and procedures (including the Responsible Care® initiatives).
  - Utilizing good management systems in the areas of environment, health, and safety.

9.5 THE AUDIT PROCESS

The Operations Auditing Process includes five main steps:

- Selecting audit sites and establishing a schedule.
- Preparing for audits.
- Conducting the audit.
- Reporting.
- Tracking the audit status through resolution of all findings.

Achieving Sustainability Through the Use of Responsible Management Systems

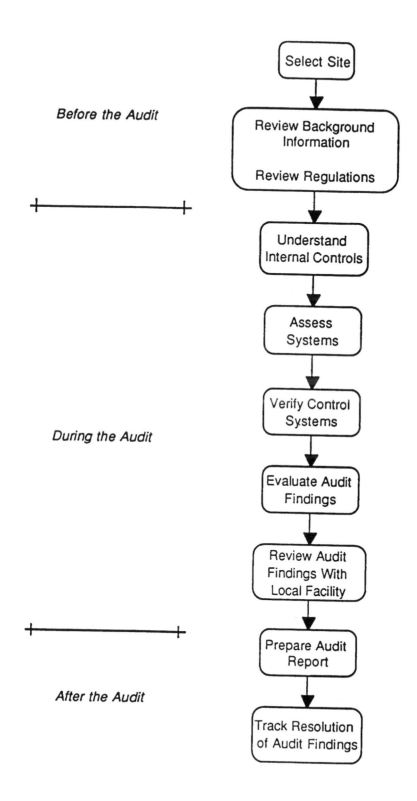

FIGURE 9.4 The operations auditing process.

9.6 THE WALNUT STATION PLANT

The Walnut Station plant is an ACC distribution facility situated on a 5 acre industrial site in southern Pennsylvania. It is located on the western bank of Chestnut Creek. The facility receives liquid chemicals by rail and tank truck, unloads them into 24 storage tanks that range in size from 5,000 to 30,000 gal, repackages chemicals and ships them to ACC customers in containers ranging in size from 1 gal pails to tank truck quantities. The facility also receives and stores packaged goods (both dry bags and drummed liquid chemicals) in a 40,000 ft^2 warehouse. Many of these chemicals are flammable and/or hazardous. Some of the operations result in hazardous waste that must be disposed of according to federal and state regulations. The Operations Auditing Team will perform a week-long audit of this facility during the first week of December 1994. The audit team will consist of a Team Leader, three Operations Auditors, and the Guest Auditor.

9.7 SELECTING AND SCHEDULING THE SITE AUDITS

All ACC domestic and international facilities are included in the Operations Auditing Program. Because of the large number of facilities ACC owns and operates world-wide, the audit schedule and frequency for each facility must be prioritized. From 1989 to 1993, all manufacturing facilities were audited on a three-year cycle and all distribution facilities on a five-year cycle. In 1994, the audit scheduling process was changed to consider the relative environmental, health, and safety risk of each facility. The Operations Auditing Group developed the Audit Priority Index (API) to determine a risk-based audit schedule for all ACC facilities. The API is calculated using a detailed formula that considers the following criteria:

- Type of facility (warehouse, distribution, blending, complex reactions).
- Number of employees.
- Community risk (location with regard to schools, waterways, or other high risk locations where an upset condition could pose a threat to the health and safety of the surrounding area).
- Performance on previous operation audits.
- Elapsed time since previous audit.
- EH&S performance (OSHA recordable injuries or illnesses, spills, regulatory deficiencies, etc.).
- Management factor (changes in facility management, high employee turnover, ISO* certification, etc.).

The numerical values assigned for each of these criteria are then put into an equation and the API is calculated. Facilities that have a higher API are perceived to have a higher relative risk and will be scheduled on a priority basis. The Operations Auditing Group used the API to determine that the Walnut Station facility should be audited during 1994.

The operations audits are announced so that each facility has enough time to prepare for the audit. The Operations Auditing Group distributes the audit schedule to division and plant management at the end of each calendar year. Therefore, facilities may have between 3 to 14 months advance notice of an upcoming audit. The Walnut Station facility had almost a year's notice prior to this audit. One of the tools that is used to help the plant manager prepare for the audit is a preaudit questionnaire. The Operations Auditing Manager sends the questionnaire to the plant managers 6 to 8 weeks before the audit to inform them about the management systems that will be verified during the audit. This helps the plant manager determine the types of documentation that will be requested by the OA

---

* ISO 9000 Quality Standards Series, International Standards Organization.

team during the audit. The plant manager returns the completed preaudit questionnaire to the Operations Auditing Manager before the audit. This document then is distributed to the Team Leader and members of the audit team to help familiarize them with the operations and issues at each site. A copy of the preaudit questionnaire is included in Attachment 1.

The facility manager, staff, and divisions also learn a great deal about the areas of environment, health, and safety from the auditing process. The Operations Auditing Group developed a "Regulatory Guidebook" for the facilities to use in preparing for an audit and arranging their files in a manner that makes the audit process go more smoothly.

9.8 PREPARING FOR THE AUDIT

About three weeks before the audit, the Team Leader contacts the Guest Auditor and the other auditors to begin audit preparations. The Team Leader reviews the audit process with the Guest Auditor and provides a list of areas that the audit team will review during the audit. Most of these audit areas are related to the fields of environmental, health, and safety. However, ACC Management has also asked the Operations Auditing Group to audit other selected non-EH&S management systems such as Contract Management and Record Retention.

Table 9.2 shows the audit area assignments for the Walnut Station Audit. The Operations Auditing Group uses written protocols that serve as a guide for auditing each management system. The protocols contain a general outline, process flow diagram, and specific questions that the auditor will ask to assess the facility's management systems. Depending on the subject matter of the management system, the protocol may include three categories of questions:

- Laws and regulations.
- ACC company policies and procedures.
- Management control systems.

The Operations Auditing Group updates the protocols as changes occur to the audit scope, regulations, or company policies. Approximately two weeks before the audit, the Team Leader sends the

TABLE 9.2
Assignments for the Walnut Station Audit

Management System	Auditor
Air Quality	Team Leader
Asbestos	Auditor #1
Contract Management	Guest Auditor
Emergency Planning and Control	Auditor #2
Employee Safety	Guest Auditor
Employment Practices/	
Americans With Disabilities Act	Auditor #2
Fire Safety and Loss Prevention	Team Leader
Food/Pesticide/Drug Control	Auditor #3
Industrial Hygiene	Auditor #1
Process Safety Management	Auditor #1
Record Retention	Team Leader
Responsible Care	Auditor #3
Solid and Hazardous Waste	Guest Auditor
Storage Tanks	Auditor #3
Toxic Substance Control Act (TSCA) and PCB Management	Auditor #3
Transportation	Auditor #2
Water Quality Control	Team Leader

Guest Auditor a copy of the three protocols that she will be responsible for during the audit (Employee Safety, Contract Management, and Solid/Hazardous Waste). A Guest Auditor tape is also provided to give an overview of the audit process.

Preparation for the audit is a team effort that is directed by the Team Leader. The Team Leader for the Walnut Station audit has primary responsibility for assuring that the audit is effectively planned, conducted, and completed. The week before the audit, the Team Leader packs all the information that the audit team will need during the audit. Some of the Team Leader's duties include:

- Reviewing available records at the ACC headquarters; these records may include permits, memos, agency correspondence, industrial hygiene surveys, previous audits, insurance inspections, citations or other legal documents, federal, state, and local regulations or codes, an equipment list, etc.
- Providing information to each auditor along with a copy of the preaudit questionnaire, a facility plot plan, and any other information that may be helpful for preparing the auditor.
- Making all travel and lodging arrangements for the entire audit team.

The Team Leader uses an Audit Preparation Handbook that was developed by the Operations Auditing Group to be sure that all preparatory activities have been completed. Having packed all these resources, the Team Leader reviews the audit preparation with the Operations Auditing Manager and the Audit Team members. The Audit Team is now prepared for the Walnut Station audit.

9.9 CONDUCTING THE WALNUT STATION AUDIT

Day One:—Upon arrival at the Walnut Station plant early Monday afternoon, the Team Leader introduces the Audit Team to the Plant Manager, the Service Center Manager, and other selected members of the plant staff. The Team Leader asks the Audit Team and plant staff to gather in the large conference room and calls the Opening Meeting to order. The Guest Auditor listens closely to pick up as much insight into the audit process and plant issues as possible.

The Team Leader begins by describing how the Operations Auditing Group was formed, the importance of the program, and how the audit process will flow over the next five days. The first thing that becomes obvious to the Guest Auditor is that open and effective communication will be an important part of this process. In addition to the opening meeting, daily afternoon meetings will be held to review the issues discovered during the day and to give the plant the opportunity to respond. The week will end with a final meeting on Friday morning to review the results. The Team Leader then describes the five-step approach that the audit team will use to review the 17 management systems (see Table 9.3).

The Team Leader also notes that the audit review period is two years unless federal, state, or other regulations require otherwise. Therefore, the audit team will review pertinent records, files, etc. for a two-year period prior to the date of the audit.

Table 9.4 shows some of the important information and items that must be packed and transported to the audit site.

After the opening presentation, the Team Leader asks the plant management team if they have any questions. One of the questions concerns audit findings that the plant may be able to correct during the week. The plant manager wanted to know if these could be excluded from the audit report. The Team Leader explains that the audit represents a snapshot in time and that any deficiencies observed during the audit will be included in the report. However, any finding that the plant resolves during the week is footnoted as "CORRECTED AT TIME OF AUDIT" in the final report. There will be no follow-up or reporting required for that finding after the audit.

Having no further questions or issues to discuss, the plant manager reviewed the plant safety rules with the audit team. Hard-hats, steel-toed shoes, and safety glasses are required in all produc-

TABLE 9.3
Five-Step Operations Auditing Process

Step	Objective	Comments
1	Understand Management Systems	The Audit Team will use information gathered from the opening meeting, a facility tour, and interviews with facility associates to understand the facility's overall management systems. It is important to understand how plant management assigns responsibilities, how decisions are made, and how the facility manages the programs that are to be audited.
2	Assess Management Systems	The overall goal of the audit is to determine if the facility, operating under the management systems described in Step #1 is in compliance with the applicable regulations, policies, and procedures.
3	Gather Audit Findings	The auditors will spend most of their time during the week interviewing associates, reviewing records, observing work practices, and testing various systems and procedures to determine if: 1. Actual operations are consistent with stated management practices. 2. Regulations, policies, etc. are being followed. The auditors will record all notes and deficiencies in the Auditing Workbooks. These deficiencies are called "findings."
4	Evaluate Audit Findings	Before any audit finding is entered into the reporting software, each auditor is responsible for confirming that sufficient audit evidence exists to justify the finding. Once this has been done, the auditor summarizes all audit observations and findings and enters them into the reporting software.
5	Report Audit Findings	At the end of each day, and at the final on-site audit meeting, the auditors present the audit findings to facility management. After the audit, the Team Leader prepares a final audit report and distributes it to appropriate ACC management. All findings are tracked until resolved.

tion and warehousing areas and goggles are required within 15 ft of any liquid transfer operation. The entire group then went on a plant tour that included the storage tank farm, truck loading/unloading rack, warehouse, drumming room, food grade storage area, and the food grade "white room" (an ultra-pure room used to transfer liquid food grade materials from delivery trucks into 55 gal drums). By this time, it was about 5:30 p.m. and the Audit Team was ready to leave the facility, check into the hotel, and meet the plant manager for dinner around 7:00 p.m. to further discuss the facility operations and the audit process.

Days Two to Four of the Audit:—The Audit Team arrived at the facility by 7:30 a.m. each day and began the intensive process of assessing the plant's management systems and gathering audit evidence. On Tuesday, the Guest Auditor asked to meet with the Team Leader to ask some last minute questions on the management systems that she would be reviewing. Of the three protocols that the Guest Auditor would review, the Team Leader felt that Employee Safety would take the most time and he recommended that she start with it.

The Guest Auditor began her audit activities by walking through the plant and observing employee safety issues. There are many federal Occupational Health and Safety Administration (OSHA) regulations that apply to this facility. Table 9.5 shows the areas of the Employee Safety protocol that the Guest Auditor evaluated on Tuesday and Wednesday morning.

TABLE 9.4
Audit Resource Checklist

Item	Comments
Notebook Computers	The computers contain a software program that generates the audit report. One computer is required to run a compact disk (CD) player used to access regulations.
CD Player and Regulatory CD	The regulatory CD contains many pertinent federal and state, environmental, health, safety, and transportation regulations that the audit team may need to reference during the audit.
Regulations and Codes	Hard copies of certain federal, state, or local regulations, codes, (including building, fire, etc.) may be required If they are not on the regulatory CD.
Audit Protocols (see Table 9.2)	The Audit Team uses the protocols to review each of the management systems.
Ashland Chemical Company and Division Policy Manuals	The company has a controlled policy manual. The Audit Team will review facility performance related to these policies. Each division may also have a policy manual.
Preaudit Questionnaire	Sent to the Plant Manager 6 to 8 weeks before the audit.
OA Workbooks for Each Auditor	This workbook is where the auditor will make all notes related to the audit. The Team Leader collects all the workbooks after the audit and manages them as controlled documents.
Facility Permit List	ACC's EH&S department keeps an up-to-date permit list for each facility.
EH&S File Information	Miscellaneous information that may be in corporate EH&S files.
Miscellaneous Reference Materials	Various reference documents that may add supporting information to the audit protocols.
Directions to the Plant and Hotel	
Opening Meeting Discussion Packages	Packets prepared for the opening meeting.
Safety Equipment	Safety shoes, glasses, etc.
Miscellaneous Supplies	Extension cords, power strips, clipboards, pens.
Clothing	Appropriate clothing for working outdoors.

Table 9.6 shows the subareas of the Solid and Hazardous Waste protocol that the Guest Auditor applied Wednesday afternoon and Thursday morning. The local waste authority has a very active inspection program, especially for facilities that handle hazardous, flammable, and combustible materials. The Walnut Station Plant maintains a very open and active relationship with this agency and had invited them to an open house in October. The Plant Manager felt that the Audit Team would find the plant's waste management system to be in very good shape.

The Guest Auditor spent the last part of Thursday reviewing the Walnut Station Contract Management system. ACC policy requires contracts for all contract work performed at company facilities and that each contract contains the following five basic elements:

- The signed contract document.
- The vendor's insurance certificate.
- A labor rate schedule.
- An EEO policy statement.
- A signed health and safety statement.

Because the Walnut Station Plant was in the process of a large capital project that involved significant demolition and new construction, a large number of contractors had performed work on site

TABLE 9.5
Subareas of the Employee Safety Operations Auditing Protocol

Subarea	Keys to Effective Management System
Aisles and Walkways	No slipping, tripping or fall hazards.
Confined Space Entry	Precautions are taken prior to, and during, any entry into a confined space (tank, pit, vessel); training is provided/documented.
Derailers	Derailers are placed on rail tracks entering the plant when no rail movement is expected.
Electrical	Electrical gear is properly enclosed and labeled; precautions are taken when working on electrical equipment.
Eyewash and Safety Showers	This safety equipment is properly placed and maintained in the event of employee contact with hazardous chemicals.
First Aid	Proper first aid training is given and documented; adequate first aid equipment is located and maintained around facility.
Forklifts	Inspections are performed on each shift; each truck has appropriate safety equipment; trucks are safely operated; operator training is provided and documented.
Housekeeping	The facility's housekeeping practices are such to provide a safe workplace.
Ladders/Platforms/Stairs	This equipment is properly secured; elevated areas have proper handrails and kick plates.
Lockout/Tagout	Electrical equipment is properly deenergized prior to performing work on it; locks and tags are available and properly used; training is provided and documented.
Material Storage	Material compatibility is considered when storing chemical products; incompatible materials are not stored in the same area.
OSHA 200 Log	The required OSHA log is maintained to document industrial injuries and illnesses; log is posted annually during February.
OSHA Poster	The OSHA poster is located so employees and visitors have access to information.
Personal Protective Equipment	Plant workers and visitors have access to and use appropriate protective equipment; protective equipment is properly stored and maintained; training is provided and documented where required (e.g., respirators).
Training	All required safety-related training is provided and documented.

over the last year. The Guest Auditor spent over 4 hours reviewing the contract files and applying the Contract Management protocol. She followed the protocol to be sure that each contract included the five elements listed above.

9.10 REPORTING AND COMMUNICATING THE AUDIT EVIDENCE

At the end of each day the audit evidence was reviewed with the Plant Manager and the Audit Team. All potential findings were discussed and some refinements were made. Once an audit finding was determined to be accurate, the auditor entered it into the computerized audit reporting system. By Friday morning, the Audit Team was ready for the final on-site audit meeting.

Day Five:—After calling the final meeting to order, the Team Leader thanked the Walnut Station management and staff for their hospitality and openness during the audit. The Team Leader then distributed a draft copy of the audit finding report. Some of the findings that were reported are shown in Table 9.7.

TABLE 9.6
Subareas of the Solid and Hazardous Waste Protocol

Sub-Area	Keys to Effective Management System
Disposal	All regulated waste is disposed of in accordance with applicable regulations; ACC-approved waste disposal facilities are used; industrial waste is not sent to a municipal landfill per ACC policy.
Inspection	Routine inspections are performed on waste storage areas as required by regulations or permits.
Labeling	Waste containers and storage areas are properly labeled; storage containers are properly dated.
Manifests	Federal or state manifests exist for all hazardous waste shipments; all required information is included on manifest for generator, transporter, and disposal site; return manifests are received from disposal site within required time.
Permits	All appropriate permits are in place for the generation, storage, treatment, and/or disposal of waste materials.
Plans/Programs	Emergency and contingency plans are in place.
Recordkeeping	All federal and state recordkeeping requirements are met for inspections, manifests, training, permits, etc.
Training	All associates who have responsibilities for the waste management program have been properly trained; training documentation exists.
Storage	Waste materials are stored in designated areas; regulatory time limits are not exceeded for storage; storage area is properly inspected and maintained.
Security	Fences, gates, signs, etc. are in place to prevent unauthorized access.
Waste Minimization	Waste minimization plans are in place; progress is being made towards Responsible Care Pollution Prevention goals.
Reports	Quarterly, semi-annual or annual reports are submitted to agencies on waste generation, shipments, etc.

After some discussion on the findings and some limited refinements, all parties agree that the information in the draft report was accurate. Then, the Team Leader reviewed the reporting process. Within four weeks after the audit, the Team Leader will complete the final audit report. All audit correspondence is distributed through ACC's Law Department and is marked "**PRIVILEGED AND CONFIDENTIAL—DO NOT DUPLICATE**". Copies of the report are sent to the Plant Manager, appropriate managers within the operating division and the Vice President of EH&S. See Figure 9.5 for an illustration of how audit information flows through the ACC organization. The Team Leader indicated that he will make the revisions to the draft report that were agreed to during the final meeting, issue the report, and begin the tracking process. Attachment 2 includes the reporting and tracking timeline for the Walnut Station audit. The Audit Team left the plant and began its journey back to headquarters. At this point, the Guest Auditor's formal participation in the audit is completed. As the Team Leader begins the process of reporting and tracking the audit results, he may have some follow-up questions for the Guest Auditor on the management systems that she reviewed during the audit. The Team Leader thanks her for doing an outstanding job and presents her with a token of appreciation—a glass paperweight engraved with the Operations Auditing logo.

9.11 COMPLETING THE AUDIT

Although the on-site portion of the audit is complete, the Team Leader still has a responsibility to follow-up with the plant, track progress towards finding resolution, and work with the facility to resolve each audit finding. Findings are "resolved" when the facility either corrects the deficiency

TABLE 9.7
Audit Findings Reported for the Walnut Station Distribution Facility

Finding	Type of Finding
The facility stored hazardous waste in containers that were not labeled. *[40 CFR 262.34(a)(2)]*	Regulatory
The facility was discharging treated sanitary effluent into Chestnut Creek without an NPDES permit. *[PA NPDES Regulations; Paragraph 12.26(a)]*	Regulatory
Aisles and passageways were obstructed with pallets in the warehouse area. *[29 CFR 1910.22(b)]*	Regulatory
The plant does not have an air permit to operate the six new 10,000 gal storage tanks. *[Pennsylvania Air Permit to Install/Operate Regulations; Section 3.07]*	Regulatory
The facility could not document that hazard communication training had been provided to employees. *[29 CFR 1910.1200]*	Regulatory
The fire extinguishers in the tank farm and truck rack were not inspected during 1994. *[29 CFR 1910.157(e)(3)]*	Regulatory
The facility sent eight shipments of hazardous waste to a federally permitted disposal site that was not approved by EH&S. *[ACC Policies and Procedures: CF 3-2]*	Company Policy
The facility had not tested the integrity of its underground lines leading from the tank farm to the drumming room. *[ACC Policies and Procedures: CF-15]*	Company Policy

The facility did not have service contracts for the following services: *[ACC Policies and Procedures: CG-3]*
    ABC Office Equipment Services Company
    Rusty Mechanical Contractors, Inc.
    No Freeze Insulating Company
	Company Policy

The insulation on piping in the drum room was crumbled and deteriorating. The facility could not document if the insulation had been tested for asbestos.	Management Control System
The north wall in the solvent tank farm has several unsealed cracks.	Management Control System

noted during the audit or develops a plan of action for resolving findings. ACC Management strongly encourages its facilities to resolve all audit findings as quickly as possible and within the two calendar quarters following the audit.

An audit tracking tool called "Form 103" is used to track the progress of each finding. The original finding and the corresponding citation are entered onto the Form 103 by the Operations Auditing Office Assistant. The facility then adds its response in the space provided. The response includes information on how the plant resolved, or plans to resolve, the audit finding. The Team Leader uses Form 103 to track each audit finding until it is resolved. The Team Leader also plays an important role as a facilitator to help the facility resolve each finding. The Team Leader may work with other resources in the EH&S Department, other resource groups, or the division/plant management to help resolve certain findings.

Attachment 3 shows a completed Form 103 for a Walnut Station audit finding that was resolved after two quarters of tracking. An air permitting engineer in the EH&S Department helped the plant obtain the permit that was needed to resolve this particular finding.

The Operations Auditing Manager forwards any audit findings not resolved within two calendar quarters to ACC's Law Department for tracking and final resolution. Some audit findings may require extended periods of time to resolve, especially if capital investment or some type of agency permit or approval is required. Once all findings are resolved, the active audit file for that facility is closed.

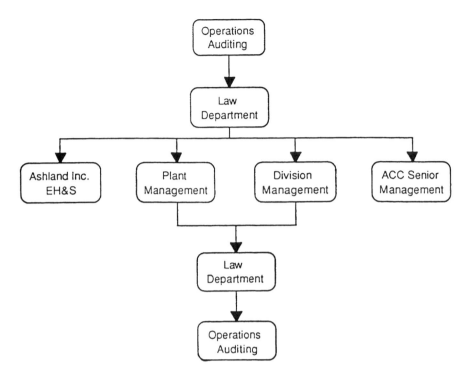

FIGURE 9.5 Ashland Chemical information flow—operations auditing.

9.12 SUMMARY

The audit was very challenging and educational for the Guest Auditor, the Audit Team, and the plant management team. The audit team identified approximately 60 deficiencies in the areas of regulations, company policies and procedures, and management systems that were reported as findings. The plant manager and staff also learned several things about the areas that were audited and asked several questions on how they could improve their overall management systems and operations. The audit team was able to provide guidance to the plant on how to resolve some of the findings. They also provided the plant with names of individuals in the EH&S Department who could give additional assistance. The Audit Team members also benefited from the process as they learned more about the distribution business, especially the unique white room and food grade storage operations. This information exchange between the Operations Auditing Group and the plant is all part of the designed continuous improvement goals of the Operations Auditing Program. This sharing of information and experience helps ACC sustain its business and operating facilities in a corporate and regulatory climate that continues to become more stringent. The Guest Auditor, in particular, benefited from the auditing process. She developed a greater understanding of the distribution business because this was her first trip to a distribution site. She was also able to share some of her manufacturing experience with the plant about a leaking pump that had frustrated them for several months. After sending the pump out to a local mechanical shop and spending over $2000 on several pump rebuilds, the pump continued to leak. The Guest Auditor noted that an incorrect replacement seal had been used and helped to identify the proper seal that would provide for leak-free pump operation.

The Guest Auditor was in Dublin for a corporate training program in August 1995. She met briefly with the Team Leader and learned that all findings for the Walnut Station audit had been resolved. The Team Leader also mentioned that both OSHA and the local waste authority had visited the Walnut Station facility to perform inspections during the first part of 1995. The plant successfully passed each inspection.

Attachment 1
PREAUDIT QUESTIONNAIRE
ENVIRONMENTAL MANAGEMENT

Location: _____
Division/Group: _____
Date Completed: _____
Completed By: _____

This preaudit questionnaire is intended to elicit background information from the operating facility pertaining to its environmental health and safety management activities. This background information will assist auditors in planning and conducting facility audits; thus, accurate and timely completion is requested.

AIR   YES   NO   N/A

1. Is the facility required to register or obtain a permit for air emissions sources?

    a. If yes, which sources?   ___   ___   ___
    _____
    _____

    b. If yes, which agency or agencies?
    _____
    _____

    c. If yes, which air pollutants are present in the facility's air emissions (e.g., beryllium, asbestos, volatile organic compounds)?
    _____
    _____

    d. List current air permit number, if applicable:
    _____

2. Have facility emissions resulted in complaints from the general public owing to:

    a. Odors?   ___   ___   ___
    b. Fugitive dusts?   ___   ___   ___
    c. Other?   ___   ___   ___

3. Does the facility utilize air pollution control equipment?

    If yes, what types: Cyclones, fabric filters, ESP's, scrubbers, flares, absorbers, other: _____
    _____

4. Does the facility dispense fuel to motor vehicles?   ___   ___   ___

TRAINING

                                                                                                                        YES   NO   N/A

1. Are employees trained in any of the following areas:

    a. Maintenance of pollution control equipment?
    b. Oil or hazardous material spill response?
    c. Hazardous waste management?
    d. Notification of significant environmental events?

PCB

1. Are PCB or PCB-contaminated oils in use or stored in the plant:

    a. Transformers?
    b. Capacitors?
    c. Electromagnets?
    d. Hydraulic systems?
    e. Lighting ballasts?
    f. Other?

2. Are there any PCB items in storage for disposal?

3. Does the facility regularly inspect PCB items in storage or in use?

SPILL CONTROL

1. Does the facility store oil in volumes greater than 42,000 gallons in underground storage tanks, 1,320 gallons in above ground storage tanks, or 660 gallons in a single container above ground?

    a. If yes, does the facility have an oil spill control plan?
    b. If yes, does the facility's oil spill plan include provisions pertaining to hazardous substances or hazardous wastes?

2. Is secondary containment provided for:

    a. Oil storage tanks?
    b. Hazardous substance or hazardous waste tanks?

3. Does the facility have any underground tanks in or out of service?

4. Does the facility have a procedure for reporting releases of hazardous substances?

WASTE	YES	NO	N/A

1. Does the facility produce any wastes classified as:

    a. Listed hazardous wastes?
    b. Ignitable?
    c. Corrosive?
    d. Reactive?
    e. Toxic?
    f. Other? _____

2. Does the facility treat, store, or dispose of hazardous wastes on-site?

    If yes, please specify: _____

3. Does the facility accept wastes from other plants for treatment, storage, or disposal?

4. Does the facility engage in the transportation of hazardous wastes?

5. Is the facility required to monitor:

    a. Groundwater?
    b. Leachate?

6. Does the facility have programs in place for regular inspection of:

    a. Waste storage areas?
    b. Waste disposal areas?

7. Does the facility utilize other locations for the treatment, storage, or disposal of hazardous wastes?

    If yes, please specify: _____

8. Describe the type of permits and classification (i.e., transporter, TSD, generator) of the facility:
    _____
    _____

9. Is the facility an Environmental Services hub?

WATER/WASTEWATER YES NO N/A

1. Does the facility discharge any of the following into surface water:

    a. Non-contact cooling water?
    b. Manufacturing/storage area stormwater runoff?
    c. Undeveloped area stormwater runoff?
    d. Dredge and fill solids drainage water?
    e. Wastewater treatment plant effluent?
    f. Process wastewater?
    g. Other? _____

2. Does the facility discharge into a Publicly-Owned Treatment Works Works (POTW) any of the following:

    a. Process wastewater?
    b. Domestic (sanitary) wastewater?
    c. Wastewater treatment plant effluent?
    d. Other? _____

3. Is the facility required to have a permit to discharge to surface waters/POTW's?

4. Is the facility subject to any effluent limitations?

    If yes, please specify: _____

5. Does the facility use an on-site wastewater treatment system prior to effluent discharge?

6. Does the facility conduct any effluent monitoring?
   If yes, are monitoring samples analyzed by:

    a. Plant personnel?
    b. Contractor?

7. Does any portion of the facility's drinking water supply come from on-site wells or surface water sources?

    a. If yes, does the facility monitor on-site drinking water sources?
    b. If yes, what are the monitoring locations?
    _____
    _____

8. Does the facility have an NPDES and/or a POTW permit?
   If yes, please list permit numbers.
   _____

SAFETY MANAGEMENT

EMPLOYEE SAFETY YES NO N/A

1. Does the facility have a safety manual or safety guidelines? ___ ___ ___

    If yes, please specify: _____
    _____

2. Are there programs/practices in place for the following:

    a. Confined space entry? ___ ___ ___
    b. Line breaking? ___ ___ ___
    c. Corrosive handling? ___ ___ ___
    d. Storage and handling of compressed gases? ___ ___ ___
    e. Hazardous materials handling and storage? ___ ___ ___
    f. Equipment guarding? ___ ___ ___
    g. Grounding of electric circuits? ___ ___ ___
    h. Inspection of pressure vessels? ___ ___ ___
    i. Boiler and furnace inspections? ___ ___ ___
    j. Hot work? ___ ___ ___
    k. Lockout/tagout? ___ ___ ___
    l. Forklift truck inspection and maintenance? ___ ___ ___
    m. Hoists, slings, and cranes? ___ ___ ___
    n. Good housekeeping? ___ ___ ___
    o. Fall protection? ___ ___ ___
    p. Start-up/shutdown of equipment? ___ ___ ___

LOSS PREVENTION

1. Does the facility have an emergency action plan, fire prevention plan, or emergency response plan? ___ ___ ___

    If yes, please specify: _____
    _____

2. Are records maintained on work-related injury/injuries by the following methods:

    a. OSHA recordkeeping? ___ ___ ___
    b. Ashland injury/illness reports? ___ ___ ___
    c. Motor vehicle accident reporting? ___ ___ ___

3. Does the facility have a fire brigade?

    If yes:

    a. Is there formal brigade training? ___ ___ ___
    b. Are there fire drills? ___ ___ ___

LOSS PREVENTION (Continued)   YES   NO   N/A

4. Are there programs in place for the regular inspection of:

    a. Fire detection systems?
    b. Sprinkler control valves?
    c. Water flow on sprinkler system?
    d. Fire pumps?
    e. Fixed fire extinguishing systems?
    f. Fire extinguishers?
    g. Alarm systems?

PROCESS SAFETY

1. Does the facility have processes with chemicals which meet any of the following criteria?

    a. OSHA Appendix A material > TQ?
    b. < 10,000 lb of flammables (other than storage and transfer)?

    Please list covered processes:
    _____
    _____

2. Has the facility performed PHAs?

    a. Total number performed: _____
    b. Total scheduled: _____

3. Does the facility have documentation of Process Safety Information (PSI) compiled prior to conducting reviews?

4. Does the facility have a written plan of action regarding Employee Participation which allows access to PSI?

5. Does the facility have current operating procedures for OSHA covered processes?

6. Does the facility have a Preventive Maintenance (PM) Program?

7. Does the facility maintain Equipment Files?

8. Does the facility have training documentation for employees working on OSHA covered processes?

9. Does the facility have a procedure for contractor evaluation and for communicating PSI to contractors?

PROCESS SAFETY (Continued)

	YES	NO	N/A

10. Are there any processes which have required a pre-startup safety review in the past 24 months?   ____ ____ ____

    Please list processes:
    _____
    _____

11. Are there any processes which required Management of Change (MOC) procedures in the past 24 months?   ____ ____ ____

    Please list processes:
    _____
    _____

12. Were there any incidents in the past 24 months which resulted in or had the potential for a catastrophe release of a covered chemical in the workplace?   ____ ____ ____

    Please list:
    _____
    _____

13. Does the facility have an Emergency Action plan that includes all required elements found at 29 CF 1910.38?   ____ ____ ____

HEALTH MANAGEMENT

INDUSTRIAL HYGIENE   YES   NO   N/A

1. Do facility associates handle chemicals?

2. Does the facility have any potential worker exposures to chemical agents?

   If yes, please list examples of sources of potential exposure:
   _____
   _____

3. Has exposure monitoring been conducted?

4. Is there a written exposure monitoring plan in place that specifies priorities and sources on which to conduct monitoring?

5. Have employees been informed of exposures monitoring results?

6. Are ventilation systems used to control chemical exposures? If yes, do ventilation systems have:

   a. Routine maintenance programs?
   b. Regular air flow rate tests?
   c. Regular filter changes?

7. Were ventilation systems designed or approved by a qualified engineer or industrial hygienist?

8. Are any other engineering controls used for control of hazards (e.g., noise enclosures, controlled atmosphere rooms)?

9. Are any administrative controls used for hazard control (e.g., limiting exposure time or operational procedures to minimize exposure)?

10. Do you have any potential worker exposure to physical agents such as noise, radiation, hcat,ctc.?

11. Does the facility have any automatic monitoring or automatic alarm systems to detect chemicals or physical agents?

12. Are any of the following types of personal protective devices in use:

    a. Respirators?
    b. Work clothing?
    c. Barrier creams?

13. Is there a respirator program at the facility?

14. Is there a hearing conservation program which includes everyone exposed to noise in excess of 85 dBA?

MISCELLANEOUS   YES   NO   N/A

1. How many associates are located at the facility?

   _____   Operations
   _____   Sales
   _____   Office
   _____   Technical Service
   _____   Lab
   _____   Other (Explain): _____
   _____

2. What are the normal operating hours?

3. Indicate the number of the following at your facility?

   _____   Storage tanks
   _____   Blend tanks
   _____   Reactors
   _____   Scrubbers
   _____   Load racks
   _____   Size of warehouse(s)
   _____   Trucks
   _____   Trailers
   _____   Drum lines
   _____   Remediation systems

4. Are there any off-site locations (warehouses, processing units) which fall under direct control of plant management? _____ _____ _____

5. Are associates expected to administer first aid and/or CPR? _____ _____ _____

   If yes, have associates received bloodborne pathogen training? _____ _____ _____

6. Does facility import or export materials? _____ _____ _____
   If yes, please list materials: _____
   _____

7. Does the facility manufacture, store, or handle food grade products? _____ _____ _____

8. Are pesticides (insecticides, fungicides, herbicides):

   a. Manufactured, stored or handled at the facility? _____ _____ _____
   b. Sprayed or dispensed for pest or weed control by employees or contractors? _____ _____ _____

TRAINING YES NO N/A

1. Are there training programs related to hazardous materials? ____ ____ ____

2. Are there training programs for:
    a. Use of respirators? ____ ____ ____
    b. Other personal protective equipment? ____ ____ ____
    c. General plant safety? ____ ____ ____
    d. Food grade? ____ ____ ____
    e. Confined space entry? ____ ____ ____
    f. Lockout/tagout? ____ ____ ____
    g. Forklift? ____ ____ ____
    h. Incident investigation? ____ ____ ____
    i. OSHA recordkeeping? ____ ____ ____
    j. Back and muscle strain? ____ ____ ____
    k. RCRA? ____ ____ ____
    l. HAZWOPER? ____ ____ ____
    m. Contractors? ____ ____ ____
    n. First aid and CPR? ____ ____ ____
    o. Bloodborne pathogens? ____ ____ ____
    p. Incipient fire fighting? ____ ____ ____
    q. Emergency action plan? ____ ____ ____
    r. Hot work? ____ ____ ____

Attachment 2: Operations Auditing Report/Response Tracking

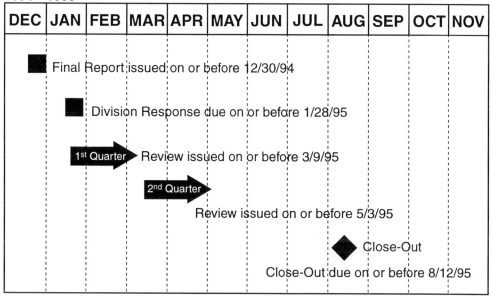

Attachment 3

Form 103

PRIVILEGED AND CONFIDENTIAL

COMPLIANCE REVIEW RESPONSE FORM ATTACHED TO
AND FORMING PART OF AUDIT PROGRAM UNDER GENERAL COUNSEL,
ASHLAND CHEMICAL COMPANY

Division: *Distribution*   Location: *Murphy Station*

Audit Date: *12/5/94*

A. Finding: Corrective Action Needed? Yes _X_ No ___

The plant does not have an air permit to install/operate the six new 10,000 gallon storage tanks.

Citation: *PA Air Regs. Section 3.07*

B. Response: (attach additional information if necessary)

We will work with our EH&S Engineer to prepare and submit permit applications to the PADER.

C. Operations Auditing Response Review: Is the Response Acceptable?
Yes _X_ Go to Section D
No ___ Explanation provided below

Auditor _GRG_ Date _1-15-95_

Revised Response:

Facility Mgr.: _____ Mfg.Mgr./Reg.Svc.Dir: _____ Div. V.P.: _____ Date: _____
(Please return to Operations Auditing when this section is completed.)

Acceptable: Yes ___ No ___   Auditor ___ Date ___

D. Quarter #1 Corrective Activity (Provide explanation of activity undertaken and progress made toward completion. Include date of completion or anticipated date of completion.)

A permit application was prepared and submitted on 1/28/94. We are awaiting agency response.

Facility Mgr.: __JG__ Mfg.Mgr./Reg.Svc.Dir: __REH__ Div. V.P.: __LB__ Date: _2-16-95_
(Please return to Operations Auditing when this section is completed.)

E. Corrective Activity Review/Comment
   1. Is the corrective activity acceptable? Yes __X__ No _____ EH&S Reviewer _GRG_
                                                                              Date _3-5-95_
   2. Is the corrective activity acceptable? Yes _____ No _____ Oper.Aud.Reviewer _____
                                                                              Date _____

F. Quarter #2 Corrective Activity (Provide explanation of activity undertaken and progress made toward completion. Include date of completion or anticipated date of completion.)

Permits to install and operate were received from the PADER on 4/30. Copies are attached.

Facility Mgr.: __JG__ Mfg.Mgr./Reg.Svc.Dir: __REH__ Div. V.P.: LB Date: _5-10-95_
(Please return to Operations Auditing when this section is completed.)

G. Corrective Activity Review/Comment
   1. Is the corrective activity acceptable? Yes_____ No _____ EH&S Reviewer _____
                                                                              Date _____
   2. Is the corrective activity acceptable? Yes X No _____ Oper.Aud.Reviewer _GRG_
                                                                              Date _5-20-95_

Finding resolved

GRG 5/20/95

10 Maxxam Group Inc.'s Takeover of the Pacific Lumber Company Case Study Number 2

Nigel C. Strafford, Raymond F. Gorman, Timothy C. Krehbiel, and Orie L. Loucks

CONTENTS

Part A. History and Takeover 213
10.1   Introduction 213
10.2   Maxxam Group Inc.'s History 215
10.3   The Takeover 216
10.4   Redwood Natural History 216
Part B. Impacts of the Ownership Change, 1986–1995 218
10.5   Introduction 218
10.6   Management by Maxxam Group Inc. 218
10.7   Impacts 219
10.8   References 225
Maxxam Case Technical Appendix: Forest Management Science 225
10.9   Tree Growth 225
10.10 Two Harvesting Approaches 228
10.11 An Old-Growth Watershed under Clear-Cut Felling 228
10.12 An Old-Growth Watershed under Selection Cutting 230
10.13 The Politics of Forest Management 230

PART A. HISTORY AND TAKEOVER

10.1 INTRODUCTION

On September 30, 1985 Maxxam Group Inc.'s Charles Hurwitz offered to purchase Pacific Lumber Company (PALCO) for $36 per share, an $8 premium over the $28 share price two weeks earlier. Two weeks later PALCO's CEO, Gene Elam, accepted $40 per share, ending PALCO's 125 years as a family run company. In 1985, PALCO owned 195,000 acres (305 square miles) of redwood and Douglas fir timberland in Humboldt County, Northern California, three sawmills, a building in San Francisco, and a welding equipment manufacturing company, PALCO Industries. All told, PALCO employed 3000 employees, and had recently lowered its cost structure by upgrading equipment and negotiating a labor contract with a lid on wages and benefits. 1985 revenues of $260 million included $85 million in forest products and $162 million in welding equipment.

The Pacific Lumber Company

PALCO showed the following financial data before the takeover, in $ million:

Sales (% profit margin)	1982	1983	1984	1985
Forest products:	$63 (25%)	$74 (40%)	$80 (38%)	$85 (42%)
Welding equipment:	127 (9%)	129 (9%)	162 (17%)	170 (17%)
Other:	46 (9%)	37 (2%)	38 (7%)	10 (7%)
Total sales:	236 (13%)	240 (17%)	28 (22%)	265 (25%)
Net profit:	23	28	42	38
Net worth:	209	218	159	170
Working capital:	83	86	74	95
Long-term debt:	16	9	46[a]	41
Shares outstanding (mil):	24.1	24.2	21.7	21.8
Dividends per share:	0.90	0.85	1.15	N/A
Beta:			1.10	1.10

[a] Long-term debt increased because PALCO initiated a stock repurchase program.
Source: Pacific Lumber Annual Report.

The *Value Line Investment Survey* reported on November 1, 1985, that PALCO's "Growth prospects remain bright through the end of the decade. PALCO accounts for only 11 percent of total redwood production, but 33 percent of the highly profitable upper grade redwood output. The percentage is likely to increase over time since PALCO, unlike its competitors, has maintained these timberlands on a continuous yield basis. Thus, while we expect industry capacity to decline as it has in recent years, PALCO should be able to maintain current production levels, making it a major beneficiary of this trend." Old-growth redwood is the source of upper-grade redwood lumber. Meanwhile, Maxxam arranged the merger and publicly announced that it planned to sell the welding business and to double the lumber output.

In 1984 PALCO had upgraded its product mix by manufacturing more upper-grade items from its old-growth harvest: At 35 percent of production, the upper grade accounted for 50 percent of lumber revenues.* Demand for these products is less cyclical than for common-grade products. In May 1985, the news from *Value Line Investment Survey* was attractive: It noted that PALCO held $18 million in proceeds from the sale of a welding business segment, and that it generally operated its business at a constant rate year round, without regard to changing demand. *Value Line* felt that PALCO shares had good three- to five-year potential and that PALCO's forest products business, in general, and its upper-grade redwood line, in particular, were likely to perform well over the next few years.†

There is evidence that PALCO traditionally was not operated to maximize profit in the short run. Insiders have reported there was a company rule to consider the social, economic, and environmental effects of business transactions. PALCO had only recently tackled costs when it responded to the 1981 recession by initiating efficiency improvements in its mills, and negotiating wage and benefit lids for labor. It planned to shave management staff in 1986, but still offered rich labor benefits, including partial college tuition for dependents, subsidized rent housing, job openings for dependents, and strong job security. When demand went slack during recessions, PALCO would cut back to a four-day work week rather than lay off employees. It typically operated at a constant rate, allowing inventory and earnings to fluctuate with demand. The pension plan was overfunded by $60 million.

PALCO declared that its forest management policy since 1930 was to operate on a continuous yield basis.‡ The *Value Line Investment Survey* noted PALCO's 33 percent market share in

---

Value Line Investment Survey, February 1, 1985.

†*Value Line Investment Survey,* May 3, 1985.

‡Pacific Lumber Company Annual Report, 1984. Continuous yield forest management is defined in Reference 1.

premium-grade redwood and PALCO's singular success in maintaining its old-growth timberlands on a continuous yield basis. Maxxam's 1985 aerial timber survey showed 45 percent more old-growth than was present in PALCO's 1956 survey.[2] By conserving its stands of old-growth and employing only selective cutting at a steady annual rate, PALCO operated to supply the premium-grade redwood market in perpetuity. A complementary goal was to minimize public opposition to its stewardship, an expensive lesson PALCO learned in 1925 when it battled the Save the Redwoods League. The Sierra Club revered PALCO for this prudent forest management.[2] However, because of the takeover, it will never be known whether PALCO's management would have continued to preserve the old-growth character of its ancient forests.

Selection cutting is a harvesting method in which specific, individual trees are cut, leaving the land fully forested except for a road network. With only selection cutting, PALCO might not have been maximizing the productivity of its younger growth and, hence, the revenue potential in the secondary market for common-grade redwood. Redwoods grow faster after a clear-cut. Furthermore, redwood grows most rapidly (up to 6 percent annually) between the ages of 30 and 70. If parcels exist over age 70, their growth rate would be slowing, and they would be considered for clear falling again, according to simple models of maximum sustainable yield. Yet PALCO neither clear-cut its redwoods nor liquidated its old-growth to replace it with faster growing young trees. PALCO invested in mill improvements before the takeover that improved the upper-grade lumber yield from old-growth trees. The evidence suggests that PALCO was positioning itself to be an upper-grade redwood monopolist, conserving its old-growth redwood as other producers cleared their stands to maximize short-term revenues and their timberlands' production of common-grade lumber.

PALCO's adherence to selection cutting methods ameliorated the environmental impact of the harvest by maintaining some of the original ecosystem's structure and function. Recent research demonstrated that as forest succession proceeds from a clear-cut, significant changes in the forest's ecological structure take place.[3] A selectively cut old-growth forest retains more of its characteristic variety of structure and species than a young forest clear-cut every 70 to 100 years just after its peak timber growth rate. Still, some soil erosion from logging roads seems inevitable, resulting locally in stream siltation. To compensate for the impact on salmon reproduction, PALCO voluntarily built two salmon hatcheries on its property, demonstrating its consideration for the environment and effects of its operations on the local community.

10.2 MAXXAM GROUP INC.'S HISTORY

The Maxxam Company's mode of operation had been to take over under-valued companies and to maximize their cash flow to Maxxam. The driving force behind Maxxam is Charles Hurwitz, a former stockbroker from Texas. "He is the most tenacious human being I've ever met. He doesn't let go," says Barry Mintz, President of Federated Development Co., a Hurwitz company.[2] His business activities gained momentum in 1978 as he took over the cash-rich chain saw manufacturer McCullough Oil, sold assets, and bought Simplicity Pattern. Through his holding company, MCO Holdings, he directed Simplicity Pattern to invest $68 million in real estate before selling Simplicity's pattern business in 1984. Then he acquired MAXXUS and reincorporated Simplicity Pattern in Delaware, renaming it Maxxam Group Inc. Maxxam had cash to spend in 1985 with $175 million in equity, $130 million in cash, and only $45 million in liabilities. By June 1, Maxxam had bought 8.7 percent of UNC Resources, 8.7 percent of Amsted Industries and 13.3 percent of Informatics General. Maxxam had declared publicly that its interest was "to acquire non-real estate operating businesses."* Legally, it needed either to consummate an acquisition before year end or to reregister as an investment company and be subject to additional regulations.

---

* *The Value Line Investment Survey,* #1801, Maxxam Group Inc., March 15, 1985.

10.3 THE TAKEOVER

Maxxam frequently worked with the investment firm Drexel, Burnham, and Lambert. There, Mr. Hurwitz had access to Michael Milken's innovative junk bond financing system. A Drexel analyst, Bob Quirk, reviewed PALCO's assets for Mr. Hurwitz,[4] and during the summer of 1985, Mr. Hurwitz commissioned an aerial survey of PALCO's timberlands that showed more good news: The forests contained significantly more timber than PALCO reported.* Mr. Hurwitz described PALCO as a "sleepy" and "undermanaged" lumber company with hidden assets.[2] By August 1985, Mr. Hurwitz secretly had bought 4.6 percent of the shares, the same amount as held by PALCO insiders. (Later he would be sued for parking stock as a result). In the two weeks before the September 30 initial offer, word leaked to Wall Street of Mr. Hurwitz' intentions, and insider trading drove the stock price from $28 to $33. The total Maxxam paid for 100 percent of the shares of PALCO was $840 million, with miscellaneous merger costs accounting for another $60 million.† Best estimates put the amount financed at $670 million, with $450 million of that provided by sales of junk bonds through Drexel, Burnham, and Lambert.[4] Initially, PALCO's board of directors fought the takeover. They retained two top law firms and sued Mr. Hurwitz, calling him, "a notorious takeover artist…who knowingly engaged in a pattern of racketeering activity consisting of multiple acts of securities fraud."[5] Mr. Hurwitz denied the charges and countersued to overturn PALCO's anti-takeover provisions. The board hired Salomon Brothers, an investment banking firm, to analyze Mr. Hurwitz' offer and to shop PALCO for better offers. Over 100 potential buyers were approached, but no significantly better offer materialized despite Salomon Brothers' own estimate of the timberlands' value of $60 to $77 per share. The battle ended when Mr. Hurwitz met privately with the PALCO CEO who thereafter encouraged the board to accept $40 per share. Apparently, Mr. Hurwitz threatened to sue the board members individually for failure to perform their duty on behalf of the shareholders.[5] "We were scared so we voted to accept," said Suzanne Murphy Beaver, a director and widow of a member of the founder's family.

Five shareholders filed separate class action suits against PALCO to reverse the sale. The suits challenged the board's financing arrangements, estimate of PALCO's assets, protection of minority shareholders' appraisal rights, and consideration of the merger's social impact on the employees, suppliers, and the local geographical area. The suits also alleged that the directors breached their fiduciary duties to shareholders and abused their control of PALCO to further their own interests.‡ Maxxam paid $52 million to settle the combined suits in 1995 after ten years of negotiations.

10.4 REDWOOD NATURAL HISTORY

The biological history of the redwoods goes back 100 million years.[6] Present day descendants live only in a remote valley in China (Dawn Redwood, *Metasequoia*) and in California (Coast Redwood, *Sequoia sempervirens* and Giant Sequoia, *Sequoiadendron giganteum*). PALCO's land contains only the Coast Redwood species. The author John Steinbeck wrote, "The [Coast] redwoods, once seen, leave a mark or create a vision that stays with you always. No one has successfully painted or photographed a redwood tree. The feeling they produce is not transferable. From them comes silence and awe. It's not only their unbelievable stature, nor the color which seems to shift and vary under your eyes, no, they are not like any trees we know, they are ambassadors from another time." Coast Redwoods indeed have stature: They are the tallest living things on Earth. The tallest is 368 ft. They

---

* The actual results of this survey are unavailable, but a Maxxam survey in 1986 showed 30 percent more timber volume with the old-growth share, 45 percent higher than PALCO's data from its last aerial survey in 1956. A Salomon Brothers aerial survey performed in October 1985, for the Board of Directors of PALCO while fighting the takeover, returned a timber valuation of $60 to $77 per share.

† Annual Report, Maxxam Group, Inc., 1987.

‡ Maxxam Group, Inc., Annual Report, 1988.

also anchor an impressive ecosystem: The greatest accumulation of plant mass ever recorded on Earth was in a coast redwood stand in Humboldt Redwoods State Park.[7]

Coast Redwoods first were felled in California in the 1820s. Hence, applying the conservative age definition for old-growth Northwest forest (trees older than 200 years[8]), a clear-cut old-growth redwood forest cannot have been restored until 2020 (2820 if it should contain millennium-old trees). Thus, the hypothesis that clear-cutting virgin redwood ecosystems does not cause irreparable harm is not fully supported by data. The rate of redwood cutting increased steadily from 1820 to the period 1905 to 1929 when it averaged 500 million board ft per year. During the Depression years, the cut dropped to a low of 135 million board ft per year. From 1947 through 1958, cutting rose rapidly to a peak of over 1 billion board ft each year. By 1960, many smaller mills were forced to close owing to the lack of redwood timber available to them. The 1988 redwood harvest was 749 million board ft, 500 million of which was from trees older than 100 years.[9] The quantities of redwood actually shipped from 1973 to 1994 are shown in Table 10.1.

A 1985 USDA Forest Service inventory of commercially available standing redwood revealed 23,566 million board ft. Trees 29 inches and larger in diameter held 9,992 million board ft.[10] Older

TABLE 10.1
Annual Timber Shipments—California Redwood Association[a]

Year	Uppers[b]	Commons[c]	Total
1973	*	*	558.00
1974	*	*	550.59
1975	*	*	673.75
1976	*	*	868.71
1977	*	*	856.95
1978	*	*	699.83
1979	*	*	629.83
1980	*	*	592.20
1981	*	*	608.73
1982	*	*	611.73
1983	154.46	379.51	533.97
1984	167.38	383.50	550.88
1985	138.73	395.35	534.08
1986	148.00	425.70	573.70
1987	153.80	451.80	605.60
1988	202.81**	672.88**	875.70
1989	204.37	774.52	978.89
1990	177.20	693.68	870.88
1991	151.44	691.06	842.50
1992	135.37	622.80	758.17
1993	126.49	552.51	679.00
1994	97.66	563.74	661.40

Note: All figures are in millions of board feet.

* Data not available.

** Projected subtotals (Only the total was recorded for 1988).

[a] The California Redwood Association currently includes 10 redwood producers. From 1973–1994 this number varied from 6–10. The Pacific Lumber Company was a member in all 22 years reported.

[b] Uppers redwood lumber is clear (knot-free) and is obtained from old-growth trees. Old-growth forests produce both uppers and commons, while young-growth forests produce only commons.

[c] Commons redwood lumber contains knots.

second-growth exceeds 29 inches, so it is uncertain what percent of the original old-growth volume remains. Originally, the Coast Redwoods' range covered 1,740,000 acres. Today, 255,000 acres are reserves protecting 157,000 acres of actual redwood forest, 89,000 acres of which is considered outstanding old-growth. The 255,000 acres of reserved land represents 14 percent of the redwood's range. Timber companies own another 52 percent and private owners hold the remaining 34 percent. Hence the 89,000 acres of protected old-growth is 5 percent of the redwood's range. PALCO owned 33,000 acres of old-growth redwood forest in 1985.

Documented efforts to save groves of redwoods began in 1852. A coordination of individual efforts established the Save the Redwoods League in 1918, and one of the League's first actions was to recommend to Congress the immediate creation of a Redwood National Park. On May 23, 1920, the U.S. House of Representatives passed a Redwood Resolution to commission a report on the feasibility of a Redwood National Park. Forty-eight years later, President Johnson signed an act to establish the Park. It preserved 30,000 acres, or 1.7 percent of the redwoods' original range. On March 27, 1978, President Carter signed the Redwood National Park Expansion Act to add essential watershed lands, bringing the Park's size to 78,000 acres. But this protection is not permanent. The Republican Party's 1995 Contract with America advocated privatizing federal land including some National Parks. Representative John Doolittle (R-CA) said, "We should vastly shrink the size of the Redwood National Park, transfer some to the county, and sell the rest of it."[11] Fortunately for Americans like John Steinbeck, who appreciate the intrinsic value of an ancient redwood ecosystem, there are other redwood preserves unlikely to be threatened by politics. The Save the Redwoods League in its first 75 years contributed $43 million to help protect 250,000 acres of Coast Redwood, portions of which have been designated World Heritage Sites and Biosphere Reserves.

PART B.   IMPACTS OF THE OWNERSHIP CHANGE, 1986–1995

10.5   INTRODUCTION

The takeover was held up by lawsuits until February 1986, but Maxxam already had begun to change the way Pacific Lumber was run. All the relationships changed, both explicit and implicit, between the company and other systems in its environment: suppliers, distributors, employees, communities, governmental institutions, and ecosystems. The driving force was the significant increase in lumber sales required by Maxxam's demand for cash flow. Employment in the forest products operation increased from 870 at the takeover to about 1400 when the first lay-off, owing to a log shortage, was announced in 1995.*

10.6   MANAGEMENT BY MAXXAM GROUP INC.

Charles Hurwitz told his new employees early in 1986, "There's a little story about the Golden Rule, 'Those who have the gold, rule.' " First, Mr. Hurwitz split off non-timber related assets into a separate holding company and offered them for sale (the welding business sold in August 1986 for

Maxxam's Forest Products Operations Produced the Following Financial Results

Item ($ million)	1986	1987	1988	1989
Sales	$119.8	$150.8	$160.8	$183.6
Depletion	25.7	50.4	76.9	101.4
Timberland Assets	520.4	495.9	470.1	445.9

*The accuracy of the second figure is uncertain because a second newspaper article (see Reference 14) reported 1500 employees.

$320 million, and the San Francisco office building sold for $32 million). The proceeds paid down the takeover debt. Second, the excess $60 million in the pension fund also was used to pay down the takeover debt, and the remaining $38 million was used to buy annuities* to meet the pension liability. Third, the timber harvest plan was changed. Maxxam reported in its 1986 Annual Report that Pacific Lumber was expected to generate additional cash flow, through increased production and enhanced marketing efforts, and that management was increasing substantially the timber harvest (including the harvesting of old-growth trees), the volume of sales of unprocessed logs, and lumber production. In the first year production increased 50 percent, from 137 million board feet of redwood and Douglas Fir lumber in 1985 to 205 million board feet in 1986, enough lumber to make a million hot tubs.[†]

Lumber production figures understate the current harvest, because sawlogs can be stockpiled or sold directly. *Value Line* reported on August 29, 1986, that "Maxxam has already significantly stepped up the harvest of Pacific Lumber's valuable redwood crop....Since the supply of redwoods has generally been declining anyway, the market should have little problem absorbing the extra lumber at a reasonable price." To implement the new harvest plan, 200 employees were added, bringing the total to 1070 by year end, 1986.

In contrast, PALCO's peak sales year, 1979, produced $86.9 million in forest products sales. Table 10.2 shows how the value of redwood sawlogs at the stump changed after 1978; 1979 was a peak year not surpassed again until 1992.[‡] To the extent that lumber market prices follow stumpage prices, PALCO's peak sales in 1979 reflect market price behavior: The volume PALCO sold was not the sole cause of the strong sales results. By the same reasoning, Maxxam's sales figures indicate that most of the growth was from sales volume increases. Redwood stumpage values increased 34 percent for old growth and 55 percent for young growth from 1985 through 1989, while Maxxam's sales increased 116 percent (see Table 10.3). Old-growth stumpage value increased sharply after 1991 as the housing recession ended.

10.7 IMPACTS

Seeing the old growth liquidated, workers began to question how long their jobs would last. The new workers and the overtime opportunities created boom conditions in the communities, but with expectations of a future bust. Even Wall Street was wondering, "But to deplete Pacific's redwood too quickly could hurt the company's long-term prospects."[§] Some employees organized in 1986 to initiate an employee buy-out to return the company to its former ways, but the effort died.

Expectations of an uncertain future came true late in 1994 when the remaining old-growth was unavailable owing to environmental lawsuits, and Maxxam offered the company and timberlands, excluding the 3000-acre old-growth Headwaters forest, for sale at a price of $1 billion. No buyers came forward. On April 12, 1995, a restraining order to protect salmon spawning beds blocked the company from clearing a 198 acre old-growth grove in the Yager Creek watershed. The next day, Maxxam, short of sawlogs, announced that it would temporarily close one of its mills and lay off 155 of its 1400 employees. Chief executive John Campbell warned, "If the judicial and regulatory

---

*Junk bond-based GIC's were bought from First Executive Life which went bankrupt one year later. First Executive Life had strong ties to Drexel, Burnham, and Lambert and had been a major buyer of the junk bonds that Maxxam issued for the PALCO takeover. Maxxam, subsequently, was sued by the Department of Labor under ERISA laws for failure of its fiduciary duty to manage the pension fund prudently.

[†]A 9 ft diameter, 4 ft high tub with a 1 ft wide internal bench, all made from 1 in thick lumber (a board ft is a $1'' \times 1'' \times 1''$ unit).

[‡]The 1979 peak year was the year after the Redwood National Park Expansion Act. Speculation concerning the diminished supply of commercially available redwood drove up prices which are reflected after a six-month lag in the state's timber tax stumpage values. See Table 10.3. Mid-1979 old-growth redwood values jumped 119 percent and young-growth values jumped 75 percent over mid-1978 values.

§ *Value Line Investment Survey,* May 30, 1986

TABLE 10.2
California Timber Harvest and Price[a]

	Douglas Fir		Redwood		Total	
Year	Board Feet in Millions	$ per 1000 Board Feet[b]	Board Feet in Millions	$ per 1000 Board Feet	Board Feet in Millions	$ Per 1000 Board Feet
1978	1009.9	147.26	806.9	279.64	4343.7	157.02
1979	949.7	162.03	647.4	373.63	3913.7	189.77
1980	810.2	167.12	528.9	337.54	3091.0	183.03
1981	695.6	170.22	535.7	292.69	2671.1	184.55
1982	536.6	113.87	487.5	231.84	2319.1	127.69
1983	803.6	96.31	592.4	217.26	3357.8	119.28
1984	823.2	91.88	676.9	204.04	3510.0	121.08
1985	940.1	73.76	663.4	210.27	3785.6	104.76
1986	1015.8	70.25	767.9	218.20	4099.3	110.21
1987	1151.1	83.33	796.5	217.81	4500.2	126.29
1988	1193.0	107.07	779.2	224.99	4625.8	144.66
1989	1107.5	155.44	740.2	261.28	4364.5	174.76
1990	912.7	228.55	748.2	319.99	3997.9	222.73
1991	716.2	208.00	518.8	350.76	3172.2	208.62
1992	653.8	307.06	533.3	467.41	2958.7	304.99
1993	635.7	444.41	514.5	649.06	2871.2	441.40

[a] Compiled from annual reports of the Timber Tax Division, Property Taxes Department, California State Board of Equalization.
[b] Value of timber immediately before cutting as calculated by the Timber Tax Division, Property Taxes Department, California State Board of Equalization.

TABLE 10.3
Timber Harvest Values[a]

Year	Douglas Fir Old-Growth[b]	Douglas Fir Young-Growth[c]	Redwood Old-Growth[d]	Redwood Young-Growth[e]	Year	Douglas Fir Old-Growth[b]	Douglas Fir Young-Growth[c]	Redwood Old-Growth[d]	Redwood Young-Growth[e]
1978a	215	150	240	180	1987a	140	80	375	185
b	230	165	420	310	b	150	95	385	185
1979a	230	165	420	310	1988a	150	95	385	185
b	250	170	525	315	b	210	150	385	200
1980a	300	185	525	280	1989a	250	170	405	205
b	270	165	445	220	b	375	250	450	280
1981a	295	180	435	220	1990a	400	290	495	310
b	275	150	415	220	b	400	290	515	330
1982a	260	140	400	200	1991a	400	270	550	350
b	195	95	340	155	b	400	225	575	350
1983a	175	80	325	145	1992a	450	265	600	375
b	200	105	345	210	b	600	375	800	550
1984a	175	90	335	190	1993a	550	300	800	565
b	165	85	345	190	b	800	575	900	700
1985a	150	80	335	180	1994a	800	550	900	650
b	140	75	335	180	b	800	500	900	600
1986a	135	75	335	195					
b	135	75	375	185					

Notes: a. Values for January 1–June 30 of each year in-dollars per 1000 board ft. b. Values for July 1–December 31 of each year in-dollars per 1000 board ft.

[a] Values compiled from the *Harvest Values Schedules* adopted and published by the California State Board of Equalization. The harvest schedule is used by owners of timber for preparation of the timber tax harvest report and tax return. The values are derived from a mixture of stump age and log sales to mills. This is then back forecasted to estimate a value for standing trees. A rough rule of thumb is $200 per thousand board feet to log, bark, skid, load, and haul the logs to a mill. Thus, adding $200 to the tax value gives a rough estimate of the mill price. Values are for Timber Area 1 which is comprised of Humboldt and Del Norte counties. All Pacific Lumber Company holdings are located in Area 1.

[b] Price for highest quality old-growth Douglas fir (highest-level of three possible quality levels: lower levels valued at lower rate). Old-growth refers to stands over 150 years old. In 1978–1979, the average net volume per log had to be over 500 board feet to be in the highest category. From 1980–1994, the average net volume per log had to be over 430 board feet to be in highest category.

[c] Price for young-growth Douglas fir.

[d] Price for highest quality old-growth Redwood (highest level of three possible quality levels: lower level valued at a lower rate). In 1978–1979, the average net volume per log had to be over 850 board feet to be in the highest category. From 1980–1994, the average net volume per log had to be over 1000 board feet to be in the highest category.

[e] Price for young-growth Redwood.

bottlenecks continue, it is possible that additional layoffs and curtailments may be required before we can restore a normal log supply."[12] If a "normal log supply" refers to the previous decade of old-growth clear-cuts, and if further lay offs could break the bottlenecks and restore the "normal log supply," then full employment nevertheless would be temporary given the scarcity of remaining old growth.

Original PALCO crews accustomed to selectively cutting trees to conserve the character of the old-growth forest may not have adjusted well to the new corporate culture. One employee of 13 years said in 1986, "We went into an area where probably no man had set foot in a century or more. It was a wonderland in there, and everything was felled. We left a moonscape."[13]

The impact of clear-cutting on the forest ecosystem's health can be significant. Clear-cutting removes vegetative cover from large forested areas, destroying the existing web of life and increasing the rate at which rain erodes the land. Sunlight, penetrating to the once-shaded soil, warms it and alters its ecology. Maxxam's logging in the Yager Creek Basin demonstrated the impact of clear-cutting. PALCO began logging the 135 square mile basin in 1948. Six years later, 80 percent of the ancient forest remained. In April 1995, less than 10 percent remained. California Department of Fish and Game surveys of Yager Creek and its tributaries in 1961, 1964, and 1972 reported abundant salmon and steelhead trout. 1995 surveys found no salmon and a greatly reduced population of steelhead trout.[14] Erosion had silted over the gravel spawning beds. When trees along the streams are removed, the lack of shade allowed the water to warm, lowering oxygen concentrations, and the stream's ability to support cold-water aquatic life. A local fisheries biologist noted that Pacific Lumber had logged so much of the canopy on the lower main stem of Yager Creek that 96 percent of the section no longer had shade. "(Yager Creek) was running dirt-brown. It was awful looking. I can't believe this kind of habitat degradation is good for the fish, and increased logging will only exacerbate the situation," said Ken Hoffman, a U.S. Fish and Wildlife Service biologist. An environmental specialist with the California Department of Fish and Game, observing the reduction in fish stock, declared, "You can't remove hundreds of trees without having an impact—and we've had so many impacts (in the Yager Basin) in such a short period of time."[14]

The resilience, or capacity of the redwood ecosystem to rebound in the Yager Creek basin also is affected by the intensity of the logging. PALCO's selective logging had not impacted significantly its salmon and steelhead trout populations as late as the 1972 survey, indicating that the ecosystem could sustain the selection harvest. In contrast, Maxxam's clear-cuts compromised the structure and function of the upland forest ecosystem to the point that it could no longer sustain the salmon reproductive stages. Whether the ecosystem has lost resiliency and will never recover to its original condition is unknown, but Maxxam was sued in 1995 to block further logging in the basin. "The whole redwood forest system has to be protected if there's any chance of resuscitating the coho (salmon) as well as chinook salmon, steelhead, and other species such as bald eagles and marbled murrelets. You can't just preserve a few little islands of redwoods and expect that to work," said Doug Thron, a plaintiff in the lawsuit.[14]

Reforestation on the clear-cut sites also is difficult. Once the complex structure of an old-growth forest is cleared and soils warm and erode, site conditions may no longer be suitable for a shade-tolerant species such as redwood. Many industry reforestation programs plant only tree species valuable in the marketplace. Routinely herbicides are used to kill early successional plants that could crowd out tree seedlings. The resulting monocultures are more susceptible to disease and provide less habitat diversity than mixed forests. Tree farming forces the short-term production of young trees, but it is not clear how long the land can repeat production of the crop at the same rate. Alternatively, selection cutting that preserves all age classes and the productivity of the forest may assure ultimate sustainability. The forest should continue to survive and produce valuable resources perpetually.

The California Department of Forestry is required to apply California environmental law to proposed timber harvests and to reject those that would cause excessive environmental damage.

Maxxam's forest management practices raised public scrutiny. Eight lawsuits within three years of the takeover were filed by environmental interest groups against Maxxam and the California Department of Forestry to halt old-growth clear-cutting plans.* The lawsuits delayed the harvests, and in some cases produced modifications and mitigating practices, but the most significant effect was a legal review of the Department of Forestry. Apparently, it had been ignoring recommendations of the California Department of Fish and Game and rubber-stamping harvest plans without complying with the California Environmental Quality Act and the California Forest Practices Act. A public agency was on trial for failing to act on the public's behalf.

Greater cooperation between California's Departments of Forestry and Fish and Game resulted, but the former retained the final word on logging issues. By 1995, less than 5,000 acres of the 33,000 acres of old-growth redwoods alive at the time of the takeover remained on Maxxam's property. The remaining stands were prized for ecological reasons and attempts to log them were costly. Environmentalists sued to block harvest plans, Maxxam defended, and the Department of Forestry worked to mitigate the environmental damage. Regarding the Yager Creek basin, Tom Osipowich, Deputy Chief of Forest Practices for the Department of Forestry, said in 1995 that logging is "at an acceptable level at this time. What you see from the air is not necessarily representative of the impact from the ground. The trees are gone, no doubt, but the most notorious producers of sediment are the roads and landings, many of which date back to the 1930s. We're constantly working to mitigate any problems that arise."[14]

The Endangered Species Act requires state and federal measures to protect endangered species. One species that nests in Maxxam's old-growth is the shorebird, the marbled murrelet. Negotiations between Maxxam, the California Department of Forestry, and the U.S. Fish and Wildlife Service to protect this species suspended logging in the prized 440-acre Owl Creek Grove for several years. Over Thanksgiving 1992, Maxxam walked away from the table, and for five days logged at Owl Creek before an injunction stopped them.[14] On February 27, 1995, federal Judge Louis Bechtle permanently protected the remaining 137 acres of the Owl Creek Grove. It was the first time a U.S. court used the federal Endangered Species Act to permanently prevent a lumber company from logging on privately owned timberland. The judge found "clear evidence" that Pacific Lumber's procedures for locating birds "either were designed to fail" or were administered "with the intent to grossly understate" the bird's presence. Maxxam responded, "The result illustrates the serious depredations being inflicted...by sweeping interpretations of the Endangered Species Act." The plaintiff, the Environmental Protection Information Center (EPIC) added, "It is the most sweeping indictment of Pacific Lumber's methodology and lack of sincerity that one could hope to find."[15] EPIC was awarded $1.1 million in the lawsuit, and Judge Louis Bechtle said that if state and federal officials had "fulfilled" their duties to protect the bird, the lawsuit could have been avoided."

Maxxam's defense in the Owl Creek lawsuit was the Sweet Home standard, a federal appeals court ruling in a concurrent case, Babbitt vs. Sweet Home, that interpreted the Endangered Species Act (ESA) as protecting only individuals of the species and not its habitat on private land. Maxxam had argued that as long as logging the grove did not kill any murrelets, they were conforming to the ESA. Judge Bechtle's ruling against Maxxam set a precedent that the ESA, in making illegal any "harm" to an endangered species, also protected habitat critical to an endangered species. Meanwhile, Babbitt vs. Sweet Home went to the U.S. Supreme Court which decided the case on June 29, 1995, in favor of habitat protection "even if it is on private property."

The largest remnant of PALCO's old growth in 1995 was the 3000 acre Headwaters forest, a unique high elevation (mostly over 5000 feet) watershed populated with giant redwoods. Maxxam has been consistently blocked from logging it directly. Finding a loophole, nevertheless, in 1990 it clear-cut a mile and a half swath into the middle of the Headwaters Forest and called it "our wildlife-biologist study trail."[16] Congressman Dan Hamburg of Northern California sponsored the

---

*Maxxam Group, Inc., Annual Report, 1989.

Headwaters Forest Act in 1993 to purchase 44,000 acres from Maxxam, including the Headwaters Forest and several scattered ancient groves; the remaining acreage was secondary growth or recent clear-cut. The Act passed the House of Representatives on September 21, 1994, and advanced to the Senate which adjourned for the year before considering it. Mr. Hurwitz only wanted to sell 4500 acres: The Headwaters Forest and a 1500-acre logged-over buffer zone. He requested $600 million and stated, "If the government is not prepared to agree to acquire the property on a timely basis at fair market value, it should not impair our ability to manage the property according to its intended and permitted use. After all, this is America."[17] When California originally zoned the region, it recognized the forest's value only as timberland. John DeWitt, Director of the Save-the-Redwoods League, estimated the Headwaters' timber value at $100 million."[2]

Since Mr. Hurwitz, as Chairman and a Director, played a "leading role" in the collapse of the United Savings Association of Texas,[2] some argue that he ought to donate the Headwaters Forest to U.S. taxpayers to offset the $1.6 billion bailout. Commenting on the likelihood of this debt-for-nature swap, FDIC's Andrew Potterfield said doubtfully, "Very complex finances and very sophisticated people are involved."[2]

The first quarter of 1995 found Maxxam unwilling to await resolution of the Headwaters forest conflict. Maxxam notified the California Department of Forestry on March 2 that it wanted to proceed on two fronts to log the Headwaters forest. First, it proposed a conventional road building and logging plan that would be subject to public review. Second, it served notice that it planned to use a California Forest Practices Act provision that allows the harvest of dead, diseased, or dying trees, up to 10 percent of the forest's timber volume. The provision waives any public review. Diseased and dead trees are a natural component of an old-growth forest. Left to rot, they can be seen as "waste" from an economic perspective. Whether to harvest them in an ancient forest with heavy equipment is a matter of contemporary debate. Congressman Charles Taylor (R-NC) proposed a law, which the U.S. House of Representatives passed on March 15, 1995, that mandated salvage logging in public forests, suspending applicable environmental laws, waiving public review, and justifying the public expense* with industry claims that salvage logging improves forest health. Simultaneously, Senator Slade Gorton (R-WA) and Senator Mark Hatfield (R-OR) introduced similar legislation attached as a rider to a Senate Appropriations bill. After a House-Senate conference combined the two bills, the Appropriations Rescission Bill including the Emergency Two-Year Salvage Timber Sale Program rider was sent to President Clinton for signing. President Clinton vetoed it on June 9, 1995, and said, "There's another thing in this bill which I really object to, which would basically direct us to make timber sales to large companies, subsidized by the taxpayers…that will essentially throw out all of our environmental laws and the protections that we have that surround such timber sales."[†]

Some fire-damaged forests are well served by salvage logging, and others benefit from cutting in combination with prescribed burning to help them maintain their original biological diversity and character. But there is no consensus on the definition of "forest health" and no procedures for ensuring that the harvests lead only to "acceptable" damage. The Denver Post editorialized on June 9, 1995, "The President was wise to veto the rescissions bill because the measure had become festooned with special-interest riders on its trip through Congress. The most notorious example is the so-called 'Emergency Two-Year Salvage Timber Sale Program.' It defines salvage timber so broadly that virtually any tree would qualify." Nevertheless, the reworked Rescissions Act that won President Clinton's signature on July 27, 1995, included the timber salvage rider modified to mandate the sales for one and a half years instead of two.

California's Department of Forestry did approve Maxxam's request to salvage log the Headwaters forest subject to the constraint that Maxxam wait until the murrelet's nesting season con-

---

*The Taylor amendment estimates the public cost of the program, including logging road construction into currently roadless areas, at $400 million.

[†]Quoted in a Western Ancient Forest Campaign release dated May 17, 1995 from President Clinton.

cludes. Subsequent legal efforts to block the plan were based on potential harm to the marbled murrelet. Representatives Ron Dellums (D-CA) and Nancy Pelosi (D-CA) asked Interior Secretary Bruce Babbitt to review Maxxam's documentation that the salvage logging would not harm wildlife. Senator Tom Hayden (D-CA) wrote to President Clinton, "Before it is too late, we believe every effort must be made to obtain public ownership of these ancient redwoods."[18] Federal and state legislation authorizing public acquisition has since been passed, but appropriation of the necessary funding has faced difficulty at both state and federal levels.

10.8 REFERENCES

1. Greber, B. and Johnson, K. N., What's all this debate about overcutting?, *J. For.,* 89(11), 25, 1984.
2. Parrish, M., Western environmentalists' enemy no. 1, *Los Angeles Times,* August 19, 1990.
3. Booth, D. E., Valuing nature, the decline and preservation of old-growth forests, 1994. Franklin, J. F., et al., U.S. Department of Agriculture, Forest Service, General Technical Report, PNW-118, 1981.
4. Henriques, D., The redwood raider, *Barron's,* September 28, 1987.
5. Schultz, E., A raider's ruckus in the Redwoods, *Fortune,* April 24, 1989.
6. Chaney, R., Redwoods of the past, Save-the-Redwoods League, 1990.
7. California Park Service, *North Coast Redwoods State Parks,* Sacramento, 1994.
8. Franklin, J., Cromack, K., Denison, W., McKee, A., Maser, C., Sedell, J., Swanson, F., and Juday, G., Ecological characteristics of old-growth Douglas fir forests, U.S. Department of Agriculture, Forest Service, General Technical Report PNW-118, 1981.
9. USDA, California's forest products industry, U.S. Department of Agriculture, Publication A13.80, PNW-RB181, 1988.
10. USDA, U.S. Department of Agriculture, Forest Service, Publication A13.80, PNW-131, 1985.
11. NRDC, The raid on public lands, *Amicus J.,*17(2), 11, 1995.
12. Pelline, J., Embattled Pac Lumber trims jobs, *San Francisco Chronicle,* April 14, 1995.
13. Walters, M., California's chain-saw massacre, *Reader's Digest,* November 1989, 144.
14. Martin, G., New battle over old redwoods: logging dispute heats up in Humboldt's Yager Creek Basin, *San Francisco Chronicle,* April 28, 1995.
15. NW forests: court rules to protect birds on private land, *Greenwire*, February 28, 1995.
16. Skow, J., Redwoods, the last stand, *Time,* June 6, 1994, 58.
17. Gannon, J., "Banker, spare that tree," *Common Cause Mag.*, 20(1), p. 8, 1994.
18. Lucas, G., Humboldt forest's ailing redwoods, firs to be cut, *San Francisco Chronicle*, March 16, 1995.

MAXXAM CASE TECHNICAL APPENDIX: FOREST MANAGEMENT SCIENCE

Forest management for sustainable timber harvests is a field with established principles going back 100 years. These principles require understanding the growth pattern of trees and of whole forests, and outcomes from the two major options for harvesting, clear-cutting often followed by tree planting, or selection harvest, usually with natural regeneration. The following is a general summary of these relationships.

10.9 TREE GROWTH

The volume of wood in a tree can be calculated as a cone: The formula for V, volume, is $AL/3$, where V is the wood in cubic feet, A is the cross-sectional area at the base of the tree in square feet, and L is the length of the tree trunk in feet.

The annual growth of a tree is a new layer added onto the outside of the tree. A cross section reveals the layer as a ring, but the new layer actually covers the entire tree, its branches, trunk and roots. Since loggers do not harvest the branches, the volume of wood in the trunk can be measured as a cone using trunk diameter (measured at breast height), and tree height. The tree's *mean annual volume increment* (MAI) is its long-term growth, expressed as the average wood volume gained, in cubic feet, over the life of the tree (Figure 10.1). Such growth also can be expressed as a percent increase annually in total volume. Both can be plotted in relation to the tree's age, typically showing a sustained positive slope.

However, trees, typically, do not grow alone, they grow in stands. If a stand of trees is very dense, the trees crowd each other at their widest points (in the canopy) and diameter growth of each tree slows as they compete for light, water, and nutrients. Individuals either adapt to the crowded conditions, for example, by self-pruning lower branches and growing tall but slender, or they die. However, total volume growth of the stand is not limited very much until later years. If the mean annual volume increment for the entire stand of trees is plotted against time, an effect from maturation and crowding reverses the long-term positive growth trend (Figure 10.2).

Alternatively, tree *mortality* in the stand can be plotted against time (Figure 10.3). Initially, mortality on a volume basis is nearly zero. It then drifts up but levels off in a long optimal retention phase where net growth is the greatest. This phase represents the stand's maximum capacity for real growth. Thereafter the stand approaches old-growth status (200+ year old trees), and mortality increases until it equals the annual growth rate. As the stand ages, MAI, expressed as a percent of the preexisting volume, decreases, and as mortality increases, *net* growth is 0 in old-growth forests. When net growth (growth minus mortality) is zero, the stand is at a dynamic equilibrium in total mass, but net annual return to an owner is zero.

Ancient forests in moist sites, undisturbed by crown fires, logging or significant pest or disease outbreaks, can persist in this steady-state for hundreds of years. Examples include old-growth redwood groves and the rain forests of Washington's Olympic Peninsula.

The goal of forest management for timber is to plan an extraction system that maximizes net growth (and return to the owner) in the retention phase. Harvesting trees whose growth would be slowing from old age, or crowding from more vigorous trees, keeps the mortality low in the retention phase, maximizing net growth. By removing some of the wood that otherwise would eventually fall to the forest floor (and decay), the stand is maintained in its most productive condition.

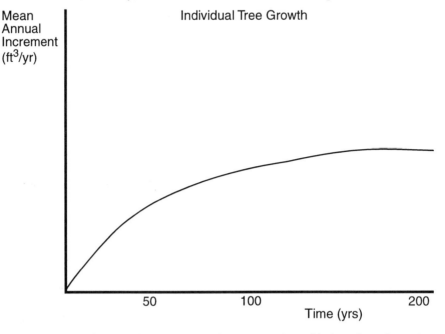

FIGURE 10.1 The growth of trees begins slowly (when expressed as cubic feet of wood growth annually), then increases more rapidly each year for a period, and as the tree ages (and the surface area of the branches and bole increases to a maximum), the rate of annual increase levels off.

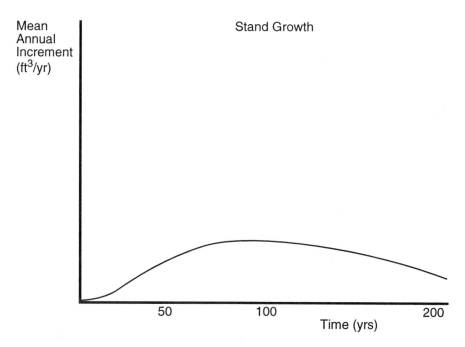

FIGURE 10.2 For a forest (or stand) made up of perhaps 1000 individual trees, growth in the early years is similar to growth of one tree. With time, many of the small trees die annually, bringing the mean annual increment of wood volume into lower values than are seen during the period of peak growth. At this lower rate of growth retention, the forest is said to be over-mature, a characteristic of old-growth.

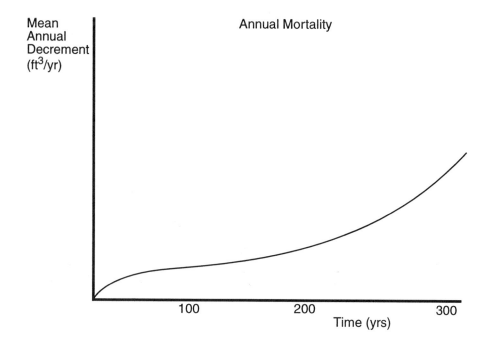

FIGURE 10.3 Mortality of small trees begins early in the life of a forest, but because small trees have little volume, the losses are not large until the trees are quite old and the volume of mortality from each tree is large.

10.10 TWO HARVESTING APPROACHES

There are two general harvest approaches: selection cutting and clear-cut felling. Selection cutting removes the largest trees at intervals of 10 years or so. It necessitates returning to the stand at these intervals, requiring good roads so that the harvesting equipment can access every part of the stand.

Clear-cutting removes all the trees at one time when the mean annual increment begins to taper off, and saves money on long-term road maintenance, as the roads are abandoned after one use. With replanting, the clear-cut stand starts growing anew, but the new crop of trees will be all the same age. Hence clear-cut harvesting is sometimes called even-aged management. It anticipates a cycle, called a rotation, over periods ranging from 50 to 200 years (depending on the species, climate, and soils), after which the clear-cut would be repeated. In mountainous areas, clear-cutting risks significant erosion and loss of nutrients for the next rotation, as well as loss of habitat for species such as the northern spotted owl.

10.11 AN OLD-GROWTH WATERSHED UNDER CLEAR-CUT FELLING

A watershed is an area of land drained by a single watercourse from the ridgeline to the outlet. Any rain falling on land within the watershed will eventually drain out through that stream or river unless it filters into an aquifer. Measurement of ecosystem health and dynamics is done in watershed units to simplify the expression of inputs, outputs, and storages. The trees within a watershed often are cataloged according to an age-class distribution, such as shown in Figure 10.4. When the individual trees are grouped into age classes, the volume of wood in each age class can be calculated. Typical volumes for a western old-growth watershed are shown in Figure 10.4.

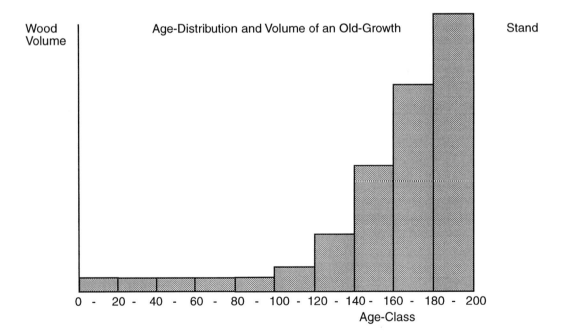

FIGURE 10.4 In an old-growth watershed, such as Headwater's Forest in the California redwood region, most of the trees present are 200 years or more in age (some may be 1000 years old). Only a few trees are even modestly younger than this oldest age class and the wood volume in the youngest age classes is negligible.

Note that most of the wood here is in the oldest trees. This watershed would not be at a peak in productivity, since most of the space is taken by the old trees whose growth is slowed. Thus, this watershed is judged as an opportunity for harvest and conversion for a younger system. A large volume of valuable timber could be removed and converted to other forms of capital that would offer greater return (although other values would be lost). One goal might be to create an all-aged distribution as depicted in Figure 10.5, thereby maximizing net growth, and defining areas for harvest each year in a sustainable pattern of rotations.

The all-aged distribution depicted in Figure 10.5 could be accomplished during 200 years of patch clear-cutting. The watershed's tree stands would be mapped into patches, and a group of patches could be clear-cut during each 20-year period. The watershed could yield a sustainable supply of timber for present and future generations, although a large portion of the nutrients are removed with the clear-cutting, and erosion may follow. Economically, the watershed as a whole should provide steady and reliable economic rent to the present and future landowners.

The most serious problem arises, however, in determining how quickly the landowner should transform an old-growth watershed like the Headwaters Forest in the Maxxam case into an all-aged, timber-producing forest. Many other landowners have clear-cut the old-growth rapidly, producing an excess of young even-aged stands that will not produce timber for perhaps 50 to 100 years. Once all the old-growth is gone, there will be a long period of low timber supply, causing devastation to local communities. In a sense Maxxam is cutting the next generation's redwood supply now, in the interest of creating a younger forest, and for financial reasons. The alternative of clear-cutting only patches of old-growth every 20 years for 200 years, would leave slow-growing older forests for a long time.

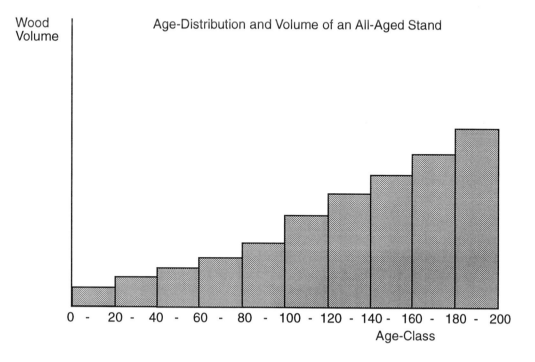

FIGURE 10.5 In a truly all-aged forest, as can be developed with "selection logging," a sample of tree diameters from hundreds of acres will show a similar number of trees from the youngest age classes to the oldest. Because of the larger volume in each of the older trees, however, wood volume is distributed in the older age classes almost linearly in relation to the tree age, as shown here.

10.12 AN OLD-GROWTH WATERSHED UNDER SELECTION CUTTING

Actually, PALCO harvested mostly by selection cutting. Essentially, their practice in other areas was to have the entire watershed partially logged over in 5- or 10-year intervals, with only specific trees being cut from the mostly older age classes. The method mimics natural mortality, although the trees are taken before their time. Conceptually, selection cutting differs from clear-cutting mainly as a matter of scale: Selection cutting is very small patch clear-cutting, because only the area around a felled tree is opened up in the operation. The method can transform an old-growth watershed into an all-aged forest as depicted in Figure 10.5, but it is achieved in micropatches, spread over 200 years of short-interval logging. The method is more expensive than clear-cutting, because there are few opportunities for economies of scale, and the felled trees must be threaded out of the forest very carefully so as not to damage other trees. On the other hand, more of the characteristic variety of plants and animals are preserved in the forest, erosion is avoided, and nutrients are conserved. The watershed's function and stream fisheries are less disrupted, increasing the likelihood that both land and water will remain healthy and productive.

Redwood forests are well-suited to selection cutting. They are relatively open forests making tree removal easier. Also, many of the cut redwoods sprout new trees from the stump, facilitating regeneration. The young redwoods need a lot of light, however, so more than one tree in a given small patch may need to be taken to facilitate replacement. (Other tree species, however, often need a cleared forest in order to regenerate. Douglas fir, for example, grows better after larger patch clear-cutting than after selection cutting.) Thus, the question of whether to clear-cut or to do selection cutting depends on the tree species present, as well as on topography, watershed ecology, the biodiversity present, the costs of road building, and the financial returns expected from the land.

10.13 THE POLITICS OF FOREST MANAGEMENT

The USDA Forest Service announced in 1993 a new commitment to achieving sustainable forest management in the Pacific Northwest by the year 2000. Rather than focusing forest management solely on timber production, the Forest Service will begin to consider other values and uses of forests on the National Forest lands, such as wildlife habitat, recreation, flood control, climate moderation, carbon sequestration, existence value, bequest value, and biodiversity maintenance. By the year 2000, only a very small percentage of the original old-growth forest in the Pacific Northwest will remain. Thus, the liquidation of older trees will be nearly complete, and most of our forests will have to be managed sustainably as all-aged, timber producing forests as they slowly recover from over-aggressive clear-cutting. Whether patch clear-cutting or selection cutting will be used in the future will depend on the other values for which the forest is managed.

Since clear-cutting was extensive over the last fifty years, a timber shortage is likely until the second-growth forests mature. This expected shortage represents a mismanagement of our nation's forests in the past, a departure from the sustainable forest management that was intended throughout the 20th century. To some extent, the large capital investment and employment in the forest products industry today represents a recent past, when large volumes of old-growth were liquidated. Downsizing is proceeding as the supply of old-growth disappears. Mills configured to saw only large diameter trees will be modified or closed. There will be other controversies such as import fees levied on cheap timber from Canada needed to meet U.S. shortages. The Northwest debate over old-growth preservation and the spotted owl may unintentionally blame endangered species regulation for the general downsizing, by framing the issue as "owls versus jobs." The central question, however, is the scale of economic rent vs. natural capital conversion that each generation expects from forested land.

11 Walnut Acres, Organic Farms Case Study Number 3

Jan Willem Bol, Allison R. Leavitt, and David W. Rosenthal

CONTENTS

11.1	Introduction	231
11.2	The Beginnings	232
11.3	The Company	232
11.4	Operations	235
	11.4.1 Organic Farming Practices at WA	235
	11.4.2 Food Processing	237
	11.4.3 Order Processing	238
11.5	The Whole Foods Industry	238
11.6	The Whole Foods Consumer	240
11.7	Marketing at Walnut Acres	241
11.8	The Future	242
11.9	Appendix I: Sample Page from Catalog	244
11.10	Appendix II: Organic Certification at Walnut Acres	245
11.11	Appendix III: Laws on Organic Farming	245
11.12	References	247

11.1 INTRODUCTION

In the early winter of 1993, Walnut Acres (WA) faced a number of significant opportunities that, given WA's unique nature and history, did not seem to have easy solutions. Throughout its history, WA had always been a company that had held to its own unique values throughout all aspects of its operations. The challenges facing the company, however, raised some questions regarding WA's strategic direction. It was important to management that growth could be maintained without making unacceptable compromises to the nature of the company, its owners, and its customers.

Foremost on management's mind was an unexpected sales decline. Whereas sales had always increased steadily, WA experienced its first sales decline in 1993. Also, market conditions had changed significantly and, in order to sustain its growth objective, WA had to reconsider its tried-and-proven marketing strategy. Finally, WA faced a number of ethical issues that could influence the future of the company.

A number of alternatives existed to deal with these challenges ranging from entering additional distribution channels to opening a health spa and buying another organic farm. Management was aware of the potential long-term consequences of each of these alternatives, yet it was felt that a decision needed to be made rather quickly. Whatever WA decided to do, it was clear that balancing economic, operational, and ethical issues was at the heart of defining the sustainability of WA's unique operations.

11.2 THE BEGINNINGS

In the summer of 1938, Paul Keene, a young mathematics teacher at Drew University (a Methodist seminary) in New Jersey, was sent by his church as a missionary to India. Although Keene was dedicated to his calling of teaching math in local schools, he felt strongly drawn to India's independence movement. The church, disapproving of Keene's political involvement, asked him to go back to the United States and to resume his teaching at Drew. Just prior to his departure and somewhat uncertain about his future, Keene had a chance to ask Mohandas K. Gandhi, whom he had met during his travels in India, what he should do with his life. Sitting in Gandhi's clay hut, the wise man told Keene, "Ah, my friend, when you return to your home in America, you must give away everything you have. Don't own anything. Then you will be free to talk and to act. Doors will open for you." Although the young Keene could not envision ever completely giving up all his worldly possessions, it later became evident how much Gandhi had been an inspiration for the rest of his life.

In 1940, Keene, now married to the daughter of a British missionary, returned to the United States and enrolled at The School of Living in Suffolk, NY, a pioneer in the teaching of organic farming methods. Keene eventually was hired as farm manager and instructor at Kimberton Farms, near Philadelphia. The agricultural methods taught by this farming school were very much like the scientific organic farming principles professed by Sir Albert Howard, who had been teaching farmers in India at the time that Keene was there.

In 1946, Keene and his wife purchased a 186 acre farm, called Walnut Acres for $2500 in Penns Creek, Pennsylvania, a small community in the heart of the Susquehanna Valley. The Keenes worked hard, often without mechanical tools, and lived modestly. Throughout even the toughest years, the Keenes stayed true to their principles. "Always ask nature what is the very best way from her point of view," Keene said. "Our guiding principle has always been not to get the most out of nature by planting certain grains year after year in the same field. In doing so, you ultimately drain the soil of its nutrients, and that's greedy and self-defeating. It's not how much we can get out of nature, but how we can build up nature and make it last generation after generation," Keene explained. Following the principles of organic farming that they had taught at the farm school in Pennsylvania, the Keenes split up each tillable acre into fields separated by grass strips to ward off erosion and rotated crops every year. In addition, every three or four years, they gave each field a rest by not planting on it. After almost fifteen years their methods actually resulted in the insect population diminishing in number and variety.

Soon after having moved in, the Keenes used the remains of a crop from an apple orchard behind the house to make apple butter (which they called Apple Essence.) The apple butter made its way to New York City, where it received a favorable review in the *New York Herald Tribune*. The successful sale of Apple Essence was the earliest sign that there was a need for wholesome products. Looking back over almost fifty years of business success, Keene commented, "We never really knew where each commercial effort would take us, but somehow we always believed that the Universe would take care of us. And, it did!" In 1993, WA was the oldest and one of the largest organic growers and processors in the country.

Since 1946, the Keenes had bought two additional farms. On the original farm they grew most of WA's vegetables and some of the grains. On the 277 acre "hill top" farm, two miles from the main farm, they grew most of the other grains, and on the 126 acre Beavertown farm, twelve miles from the main farm, they raised beef and chickens. Of the total 589 acres, about 350 were tillable, which, in turn, were subdivided into 130 fields. WA's farming operations were considered very large for an efficient organic farm.

11.3 THE COMPANY

When Keene retired in 1987, his son-in-law, Bob Anderson, who had been president of WA, took over the day-to-day management of the company. Anderson, who with his wife, Ruth, lived on the farm, had been a philosophy major in college.

Anderson reported directly to the Board of Directors, which consisted entirely of employees. Four of the seven members were chosen by the founders (Keene, Anderson, and his wife) and the remaining three were chosen by the employees. Members of the Board of Directors could only serve for a maximum of two consecutive years, and then they had to be off for the same amount of time. The rotation of members was instituted in order for more employees to have a chance to gain an understanding of the complexity of the business.

The inclusion of the three employee members on the Board of Directors had often highlighted differences in perception and priorities. The employees tended to be more risk averse, and management often found itself having to "sell" an idea. For example, real opposition was encountered when, in 1992, management proposed the remodeling of the organic foods store that was located on the farm. Since total company sales had been running behind plan, no raises were given that year, and the employees felt that capital expenditures were not as important as continued raises.

The company was owned entirely by its 110 employees and 4 founders. Each employee received 5 (common) shares for each year of employment up to a maximum of 100 shares. In addition to the common shares, founder shares (of which there were 120) were owned by Paul Keene and his family. (Remembering Gandhi's advice, Paul Keene had insisted that the only other part of the company the family would own was the actual farm and the land, which WA rented from them.) The value of a founder share was $1, whereas the value of a common share was determined at the end of each fiscal year. In 1993, the book value of a common share had gone up by 1.4 percent, whereas in previous years owners had seen at least double that increase. Profits from operations were also distributed amongst the owners at that time.

Sales tended to be very seasonal; its strongest quarter was from September to December, with over 70 percent of that quarter's sales falling between late November and early December. Distribution expenses included all costs that were related to getting a packet out of the door, including hourly wages for packers, material costs, and even the salaries of the store clerks. Selling expenses included catalog and marketing expenses. "We don't do a very good job in allocating expenses," Anderson said. "For example, wages for order entry selling clerks fall under general administrative expenses, which makes it hard to accurately calculate the cost of goods." In late 1991, he started a revision of WA's accounting methods. Table 11.1 gives the 1993 Income Statement.

The transition in management from Keene to Anderson had not been easy. For example, when it was decided that the store would stay open on Sundays, Keene, a strong advocate of family time, opposed. However, with Keene concentrating more and more on the writing of essays for the catalog, Anderson was able to put into place a structure that facilitated loose job descriptions that were typical of WA's corporate culture.* Reporting to Anderson was his wife, Ruth, who was Vice President in charge of catalog production, office support, and customer service. Paul Shaw, Assistant General Manager, and Bob Row, Plant Manager, also reported directly to Anderson. Shaw, a former camp director with a degree in geography, was in charge of wholesale operations as well as the organic certification process. He joined WA in 1988 as the store manager and was largely responsible for the successful introduction of many new (non-WA) products in the store. In 1992, with the help of many employees, he also remodeled the store, doubled the amount of shelf space, and added a book section and a small restaurant, causing store sales to double by the end of 1992. Row was in charge of the entire production process, including the farming operations. The planning of the yearly crop planting and rotation, however, was done by a larger management team, including Paul Keene.

Since the successful introduction of Apple Essence in 1946, sales grew steadily to $13.3 million in 1992. Averaging 15 percent annual growth in the early eighties, growth slowed down to approximately 10 percent in the late eighties. In 1993, with an objective of 10 percent growth, sales suddenly declined to $12.9 million. With more than 90 percent of sales coming from its catalogs, and the remaining 10 percent from the store, Anderson felt that he needed to understand the reasons behind the sudden decrease in sales.

---

*Under Keene there had been no formal organization structure or job descriptions.

TABLE 11.1
Statement of Income
for the Years Ending March 31, 1993 and 1992

	1993	% of Net Sales	1992	% of Net Sales
Sales				
Mail order	12,714,812		13,223,900	
Store	1,099,182		1,091,682	
Claims	10,554		17,036	
Total Sales	13,824,548		14,332,618	
Less:				
Postage, freight	611,238		713,974	
Sales returns	96,498		70,572	
Credit card charges	192,368		205,032	
Net Sales	12,924,444	100.00	13,343,040	100.00
Cost of Sales				
Materials	4,814,355	37.50	5,087,701	38.13
Farming production	489,836	3.79	468,341	3.51
Manufacturing production	2,113,147	16.35	2,141,558	16.05
Distribution	987,428	7.64	861,960	6.46
Inventory change	−153,801	−1.19	44,032	0.33
Total cost of sales	8,250,965	63.84	8,603,592	64.48
Gross profit	4,673,479	36.16	4,739,448	35.52
Expenses				
Selling	2,206,203	17.07	2,246,968	16.84
General and administrative	2,387,145	18.47	2,425,765	18.18
Total expenses	4,593,347	35.54	4,672,733	35.02
Income from operations	80,132	0.62	66,715	0.50
Other income (expense)				
Interest: expense	−41,358	−0.32	−53,372	−0.40
Interest: income	6,462	0.05	9,340	0.07
Discounts earned	18,094	0.14	22,683	0.17
Miscellaneous income	9,047	0.07	34,692	0.25
Gain on sale of assets	0	0.00	4,003	0.03
Income before income taxes	72,377	0.56	84,061	0.63
Provision for income taxes	14,217	0.11	18,680	0.14
Net Income	58,160	0.45	65,381	0.49

After an initial review of the data, Anderson attributed the unexpected decline in sales to a number of factors. A stagnant economy was believed to have had a negative effect on consumers' purchasing behavior of organic foods. There were also signs that consumers had become more price sensitive and had switched to less expensive alternatives, and, in 1993, the size of the catalog had been reduced slightly (3 percent fewer pages). But, by far the greatest challenge that Anderson had identified came from a change in the competitive structure in the Whole Foods industry. In the past WA competed mostly with other organic foods mail order companies, and, to a lesser extent, with organic or natural food stores. Recently, however, large retail chains had recognized the growing need for natural foods and opened entire Whole Foods sections inside their stores.

Finally, the increase in postal rates also contributed to the decrease in earnings. In each of the last three years, rates went up by 8 percent, and because they were based on a sliding scale, WA was hit particularly hard because its packages were mostly at the lower end of the weight scale. This translated into a 14 percent increase in postal costs for each of those years.

An important aspect of WA's corporate culture was its financial commitment to a number of causes. Although the company did not contribute any particular yearly amount, it was strongly committed to its role in the community. In 1960, WA established the WA Foundation, Inc., and borrowed $200,000 to build a community center in a nearby Appalachian village. Each fall, WA explained to its mail order customers what some of the projects were that the community center was developing. On the order form, customers could indicate any contributions they wished to make to these projects. In 1992, for example, the WA Foundation raised $50,000 from its customers. In the same year, the annual cost of running the community center was $80,000. The administrative cost of the fund raising effort was absorbed by WA.

Another project that received significant funds from WA was located in India, where WA contributed toward the construction of a number of buildings that housed orphans. And, in 1993, large quantities of food were given to the victims of the floods in Iowa. Similarly, excess inventory was often donated to organizations such as the Salvation Army.

11.4 OPERATIONS

WA was unique in that it grew, processed, and marketed most of its products. Although it competed with a number of organizations in each of these areas, there was no other company in the United States that had similar operations. At one time WA even distributed its products by truck, but they had gotten out of that side of the business because of the high cost of operating a truck fleet. Similarly, a decision was made to cease the raising of chickens by 1994. The labor requirements needed to run a 2500 hen chicken farm turned out to be prohibitive and nearby farmers were able to supply WA with organic eggs.

There were three sides to WA's operations: the farm, the factory, and order processing.

11.4.1 ORGANIC FARMING PRACTICES AT WA

Organic farming at WA revolved around two major organic farming concepts. The first concept was the exclusion of inorganic fertilizers and chemical pest and weed controls. The second was building soil structure and improving fertility through sustainable agricultural methods such as crop rotation, composted fertilizer, strip and contour farming, and hand and mechanical weeding. Walnut Acres' entire farm had been certified organic, based on these farming techniques, except for livestock production. The National Organic Standards Board and the USDA had not yet established livestock standards.

At WA crop rotations were used to maintain the fertility of the soil and control pests. Grains were grown on a four-year rotation of corn-oats, wheat-grass. Clover, alfalfa, timothy, and brome grass made up the grass seed mixture. Vegetables were grown on a five-year rotation with grass in the fifth year. Walnut Acres established rotation schedules predominately from the previous year's crop. No crop could be grown in consecutive years on the same field. However, alterations in the schedule could be made because of particular crops needed in the food production operations.

Field rotations were also an important part of cattle production. Walnut Acres' herd size and number of pastures was sufficient to graze the herd on an individual pasture for a week and then move the herd to the next pasture. The rotation allowed the pastures to be free of grazing for approximately three week periods, preventing over grazing and subsequent impoverishment of the pastures.

Composted fertilizer provided two key services in WA's operations. On the farm, composted fertilizers were used to build the soil. But in food processing, it was a way of disposing of production residuals. Composted fertilizers and sheet composting (current year's grass crop was chopped and the hay left to decompose on the field—it was not harvested) were the major methods used to add nutrients to the soil. No raw manure was added to the soil, and no chemical additives were used on the compost to speed up activation. Composted fertilizers were added to the soil prior to planting heavy feeder crops. For grains, compost was applied during cultivation to fields being planted in corn. On vegetable fields, the composted fertilizer was added the year tomatoes were grown. Compost was applied the year it was composted and plowed under the same day. Immediate plowing prevented nutrient runoff.

Walnut Acres added approved organic products to the soil when necessary. Soil tests were taken every year to determine soil pH, humus content, nitrogen levels, and percent organic matter. Also, annual tests were conducted to determine if heavy metals were present in the soils. Soil tests helped identify the type and quantity of soil building additives needed for that year's crop. Organic inputs would include such products as fish fertilizer 5-2-2, seaweed, and dolomite limestone.

Strip and contour farming provided an organically approved method to control soil erosion. Almost all farms, conventional and organic, in this part of Pennsylvania, used strip farming because of the rolling topography. Many years ago, the Agriculture Extension Agent assisted WA with the subdivision of their fields in strips. Each strip was approximately 100 ft wide. The three farms were divided into 130 strip fields, including the cattle pastures. The fields were set and remained the same year after year; only the crops were rotated. Contour farming was used to prevent erosion on extremely steep land. The crop rows were planted parallel to the slope, breaking the flow of water across the surface, minimizing erosion.

Organic weed and pest control were often accomplished through mechanical means. Some weeds such as red root were controlled through crop rotations, but also mechanical cultivation and hand weeding were used. Other weeds like Jerusalem artichoke required fallow till when the outbreak was acute. Insect control in the grain bins was accomplished by a mechanical system of stirrers and fans added to the bins. Every inch of the bin was stirred once a month. The stirring dislodged the weevil eggs and the fans blew them out of the bins. This method had been very effective for weevil control, however, it was relatively expensive. Live traps were the major form of rodent control.

Water for irrigation was pumped from a pond located on the farm or from Penns Creek. Penns Creek was considered a prime trout stream. Trout streams are indicators of clean water that are free of excessive contamination, because trout do not survive in heavily contaminated waters. Penns Creek's sewage disposal facility was located downstream of WA and no industrial facilities were located upstream.

Required for organic certification were 30 ft buffers from all crop fields adjacent to other non-certified properties. The intent of the buffer was to prevent contamination of the organically grown crops by spray from conventional farming practices. The major concern was with blowing sprays, although surface or ground water contamination also could be a consideration. Any equipment used in the application of inorganic materials was cleaned in an appropriate manner, preventing contamination of organic processes or products.

When organic plants were grown in the same facility as inorganic plants, they were grown far enough away from conventionally grown plants to prevent commingling. For example, seedlings purchased from third-party growers and grown in conventional nurseries were kept in a location segregated from inorganic plants and the equipment used for both was thoroughly cleaned prior to use on organic products.

Accounting practices were designed to meet the audit trail requirements of a third-party certifier. Farm production audits ensured that the company did not sell more produce than it raised. This practice prevented a company from buying inorganic produce and attempting to sell it as organic. Walnut Acres kept field history records on each of the 130 fields for both auditing and management purposes. The certification process required a field history from the three previous years and the current year being certified. Information on the farm that was required as part of the application included a map of the farm with fields numbered and legal descriptions of the properties. Specific information on each field was necessary, including major items such as acreage, ownership, type and variety of crop grown each year, yield estimates per acre, storage location for current year's yield, and a list of all inputs used on the fields.

11.4.2 FOOD PROCESSING

When produce was harvested, it entered the "factory" where it was processed. Grain, for example, was milled into flour and bagged. The flour and other products were then refrigerated. Unlike most millers, grain itself was stored in one of the 42 separate refrigeration bins. Peas and other vegetables, used for making soups, were canned. In response to the growth in sales, WA expanded its buildings to allow for growth of its operations. Although some of it had been automated, the many phases of its production process were housed in separate parts of the buildings which made difficult integration. Other parts of its production process simply could not be automated because they required manual labor.

WA's food processing facility included a mill, a cannery, a bakery, peanut butter processing, rebagging, and in-house label printing. Each of these processing units functioned at very small scales compared to traditional industries. For example, when the storage supply of a particular item, such as Mediterranean Bean Soup, was low, food processing scheduled that item for production. WA has kept their products fresh by minimizing the amount of product on their shelves, maintaining a just-in-time inventory. Also, when processing a particular product required special equipment, the production line was refitted with special equipment and production continued on the same line the next day.

Organic certification of the food processing facility paralleled the production process. A third-party also certified the food processing operation. From the first third-party certification, came the recommendation for WA to implement a product auditing system. The audit system provided the means to trace organic ingredients in each product to the farm where they were grown.

Tracking products from production to retail was crucial in maintaining the integrity of certified organic products in the marketplace. Food materials shipped to WA were recorded on a receiving report, including the supplier's name, quantity of item received, and supplier's lot number. The material was assigned a WA lot number, and a removable lot number sticker in the form of a bar code was placed on each container and on the receiving report. The new WA lot number consisted of a WA item number and the date received. Materials were then added to the computer inventory. When production requisitioned food materials, using a Materials Requisition Sheet, those materials were deducted from the inventory.

At each production area, a Daily Production Report was completed that detailed the quantity of product produced, the date, and the lot number sticker from each ingredient in the product. Any product or ingredient that was lost through spillage, for example, was listed on a Damage or Waste Report. Date codes were placed on each product and recorded in the computer inventory. The same basic procedure also was used on food materials grown and processed at WA. The audit system allowed each organically grown ingredient in WA's processed food items to be traced to the grower.

11.4.3 ORDER PROCESSING

WA's order processing department was staffed by mostly full-time telephone operators who took a customer's phone order and produced an order form. With this form in hand, a second employee collected the ordered products in much the same way as a consumer takes a cart through the aisles in a grocery store. Once the products were gathered, a third employee prepared a package for shipment. By 1993, over 1200 orders per day were filled.

Over the years, WA had received a number of suggestions from customers to change some of its packaging material. For example, bubble wrap, which was used to protect the heavy glass bottles in which WA packaged some of its products, was manufactured by Sealed Air, Inc., and was made out of plastic. In addition, the manufacturing of this wrapping material required the use of CFC's in its blowing agent. In response to WA's pressure, Sealed Air was one of the first companies in the packaging industry to totally phase out CFC's in its production process, and started to make the wrapping material from over 50 percent recycled plastic.

11.5 THE WHOLE FOODS INDUSTRY

Whole Foods was one of the fastest growing segments in the food industry (see Table 11.2.) Including both health and natural foods, such as low-fat foods, vegetarian dishes, organic foods, and natural remedies, total 1993* industry sales increased, average store sales were higher, profits were up, and the number of retail outlets increased. Driven by a growing desire for healthier and more natural products, an increasing number of consumers were shopping at health food stores.[†]

According to a *Whole Foods* [‡] Retailer Survey, total industry sales reached $4.84 billion in 1993, up 7.2 percent from 1992. The 8,450 stores in the United States[§] had an average gross sales per store of $573,693, up 6.7 percent from 1992. An increase in store sales was reported by 67 percent of the

TABLE 11.2
The Whole Foods Marketplace: A Three-Year Perspective[a]

	1993	1992	1991
Total Sales	$4.84 billion	$4.51 billion	$4.03 billion
Ave. Store Sales	$573,693	$537,918	$480,400
Number of Stores	8,450	8,300	8,390
Store Size	2,326 ft^2	2,425 ft^2	2,301 ft^2
Retail Selling Space	1,681 ft^2	1,805 ft^2	1,627 ft^2
Sales/Square Foot	$247	$222	$209
Daily Customer Traffic	158	160	117
Avg. Transaction	18.74	16.07	16.55
Avg. Store Profit	39,588	32,563	31,760
Retail Inventory	67,260	65,230	57,948
Inventory Turns	8.5	8.2	8.3

[a]*Whole Foods*, December 1993.

*September 1992 to September 1993.
†The recent FDA approval of the growth hormone drug, BST, to stimulate livestock milk production, had been yet another contributor to general consumer fears regarding the effects of certain products on health.
‡*Whole Foods*, December 1993.
§Including independent health food stores, health food chain stores, and natural food co-ops, but excluding mass market outlets.

surveyed stores. Although the average store's retail selling space had decreased slightly, the average transaction in 1993 had gone up to $18.74. Average store profits were just below $40,000, up 22 percent from 1992, while average gross profit margins were around 32 percent.

Whole foods were broken down into packaged foods (15 percent of total industry sales), vitamins/supplements (41 percent), bulk foods (4 percent), herbs (14 percent), cosmetics/body care (6 percent), refrigerated foods (4 percent), non-foods (4 percent) and other products (12 percent). Almost 70 percent of the products were ordered through wholesalers and distributors, 25 percent came directly from manufacturers, and 6 percent from other sources. On average, stores ordered from 13 different wholesalers and distributors.

The distribution of Whole Foods was done through a wide variety of retail outlets. The *Whole Foods* Retail Survey grouped these stores on the basis of their size. Level 1 stores were small stores with up to 1,000 ft^2 of selling space, whereas Level 6 stores had more than 5,000 ft^2 of selling space. Sales levels and profit margins varied significantly across different levels, not just because of the difference in store size, but also because smaller stores tended to focus on one or two categories, such as dietary supplements, while larger stores typically carried most of the Whole Foods items. Level 1 stores represented 25 percent of all industry stores and had average sales of $208,000, while spending close to $3,000 on advertising. There were approximately 2,000 Level 2 stores, with average store sales of $273,000, that spent $4,000 on advertising. The remaining categories had the following characteristics:

Level	Ft^2 Selling Space	No. of stores	Average Sales	Advertising
3	1,501–2,000	1,183	$322,000	$7,000
4	2,001–3,000	1,600	596,000	7,900
5	3,001–5,000	1,000	937,000	11,600
6	> 5,000	600	2,800,000	25,000

In 1993, approximately 36 percent of the stores indicated that they had participated in supplier co-op advertising. The most often used type of advertisement was newspaper ads, which accounted for 31 percent of ad dollars spent. Yellow pages represented 18 percent of average ad expenditures, followed by flyers/posters (11 percent), direct mail (10 percent), newsletters (10 percent), and radio ads (8 percent).

The average retail outlet sold a variety of products. For example, in 1993 a typical store carried thirteen different brands of dietary supplements, and sold an average of $235,000 worth of supplements, with a gross profit margin of 48 percent. Packaged foods accounted for over $87,000 in sales for the average store with a gross profit margin of 33 percent. Herbs sold about $83,000 in a typical store and had a similar gross profit margin as supplements. Cosmetics/body care products represented 6 percent (or $34,000) of an average store's sales and had a gross profit margin of 40 percent. Bulk foods, refrigerated foods, and frozen foods accounted for $25,000, $20,700, and $13,840, respectively, of an average store's sales. Gross profit margins for these three categories were around 35 percent. Finally, other products, such as produce, juice bar/restaurant, accounted for 4 percent of a typical store's sales.

Sales in 1993 also varied by geographical region. In the midwest, the average store sales was the lowest at just over $400,000. Stores in the Rocky Mountains/Plains, the East and the South sold around $520,000, $540,000, and $549,000, respectively. A typical store in the West sold on average $709,000 worth of Whole Foods.

According to the 1993 *Whole Foods* Retail Survey, over 30 percent of retailers had remodeled their stores and nearly 16 percent of retailers planned to increase the number of stores they owned in 1994. Other methods to increase sales that were mentioned by retailers were in-store promotions, discounting, and in-store sampling. Overall, 78 percent of the retailers surveyed expected to see a sales increase in 1994.

11.6 THE WHOLE FOODS CONSUMER

The Whole Foods consumer was different from the general consumer. Over the years, WA's management observed its customers in the store and concluded that its products were appealing to a rather up-scale, well-educated clientele that was deeply concerned about and committed to a healthy life style. Some customers traveled as many as 50 miles just to visit the store, and on Saturdays and Sundays out-of-state license plates (especially from New York, Washington, DC, and New Jersey) were often seen in the parking lot. Some of their regular customers even planned their vacations in the valley just to be able to visit the farm and the store. An initial marketing research study done by students of a local university had further shown that quite a few of WA's customers were employed in the health care industry, while a few others, who were on food stamps, bought their staples, such as beans and rice, exclusively from WA in order to get more nutrition for their dollar.

WA's customers also appeared to be very active politically. Frequently, WA would ask their customers to lobby for a change in a law that was preventing WA from applying its principles to its business practices. For example, when WA started to make peanut butter in the sixties, they intended to use unblanched peanuts (skin left on) without adding any oil or sugar. However, because the USDA's Standard of Identity specified that any product that was called "peanut butter" had to use blanched peanuts with added oil and sugar, WA was forced to call its peanut butter "imitation peanut butter." In response to WA's request, customers flooded the Food and Drug Administration with letters that were, in turn, largely responsible for a change in the standards of identity for peanut butter. In 1993, WA's peanut butter was its number one retail item, using a quarter of a million pounds of peanuts.

Because not much industry data on the Whole Foods consumer were available, WA's management commissioned a marketing research study. The basic objectives of the study were to get a better understanding of who WA's customers were and to identify why they purchased organic products. Additionally, the survey aimed at identifying WA's competitors. In early 1993, an independent consultant sent out 2,000 questionnaires. The four-page, twenty-five-question survey was sent to four categories of WA's customers who were classified on the basis of the number of times they had ordered and the amount of their total purchases. Overall, a 28 percent response rate was obtained.

Although, in retrospect, a few of the questions in the survey had been worded somewhat ambiguously, the data confirmed that WA's typical customer was affluent and well educated. Overall, customers were satisfied with the contents of the catalog and did not purchase many products from other organic food mail order companies. The vast majority of the respondents also reported that they were consuming the same or more organic foods over time.

In addition, insights were obtained about areas of customer concern, some surprising to WA management. For example, Anderson was not sure what to make of the fact that almost half of the customers reported being dissatisfied with one or more of their purchases, and that many were questioning the value of WA's products with regard to its prices. In addition, it appeared that when customers experienced difficulty with their orders they would not return the products since WA's returns were only around 9/10th of 1 percent (in the mail order industry returns typically ran around 8 percent). Anderson was afraid that WA's customer philosophy of total satisfaction of each and every customer no matter what the problem was, did not seem to work well. Finally, respondents had also noted the presence of organic foods in their local retail stores.

11.7 MARKETING AT WALNUT ACRES

Historically, marketing at WA had gone through several phases. Whereas in the early days WA sold only products it made, more recently it greatly expanded its product line in order to meet rapidly changing consumer needs. In 1993, WA offered over three hundred different food items ranging from cookies and granolas to fruit cakes, chicken soup, salad dressings, and mayonnaise. In addition to its regular products, numerous gift items were featured in the catalog. These were handwrapped combinations of some of their more popular products, and proved to be especially successful during the Christmas season. All these gift items were made to order or in small batches because of the perishability of some of the products. Consequently, WA guaranteed that a customer would receive the gift item two or three days after it had been put together. Although the process of producing these gift items was rather labor intensive and, therefore, did not offer many opportunities for economies of scale, the freshness of its products made WA unique. Given that some products went straight from the land to the on-premise cannery, WA could guarantee its freshness, something that was important to its customers. "There isn't much reason for us to be in business if we cannot maintain our commitment to high quality; we [will] never compromise on [that]," Anderson said.

In the early 1990s, WA added various skin care products, such as body lotions and soaps, all of which were made in its own, small laboratory. Given that no preservatives were used they believed these products to be the purest on the market. WA designed, developed, and produced all of its own products, their promotions, labels, packaging, and pictures.

New product development was done primarily by a consultant hired by WA. This part-time job was recently added in order to give the new product development process some structure. The new product development process started with management analyzing its sales figures, competitive catalogs, and other marketing data in order to obtain insight into the latest consumer trends. For example, early in 1992, a trend toward vegetarian foods was identified. Using one of their more successful products, beef chili, WA asked the new product consultant to develop a vegetarian chili. The consultant then developed a number of recipes, which were given to a taste panel within the company. After making some changes to one of the recipes, the new chili was put in the store where customers could sample it free of charge. Reactions were then obtained through a self-administered questionnaire. Early in 1993, plans were also being considered to include WA's mail order customers as part of the taste panel. By the end of 1992, WA had successfully introduced a black bean chili and a southwestern chili. Their new product introductions did not go unnoticed. In 1992, WA was recognized in the organic food industry as the number one innovator with twenty-eight new products. WA's closest competitor had launched ten new products. Only two of WA's twenty-eight new products had been unsuccessful.

A comparison of WA's prices to those of its competitors often led consumers to perceive significant differences. For example, a can of Health Valley soup, one of WA's main competitors in that segment, might cost $1.79, whereas a can of similar WA soup might cost $3.25. Actual prices, however, were often less than those perceived by consumers. If a consumer would add water to WA's soup to make it comparable in concentration, the consumer could observe that prices were in fact similar.

In part, driven by its labor intensive farming and production processes, WA always maintained that little room existed to reduce prices if the same quality and freshness of its products were to be guaranteed. It went to great length to "tell its story" to its customers by explaining its labor intensive process in the catalogs. The ability to communicate directly to its customers was believed to have been critical in consumer acceptance of its prices.

By the early 1990s, WA's products were found in approximately three hundred stores, most of which were in the eastern part of the United States. Yet, only 10 percent of its sales came from its retail and wholesale efforts. The vast majority of its sales came from its mail order business. On average, WA sent out nine 50 to 60 page catalogs per year (three runs of three versions) to 40,000 of its core customers (those buying at least semi-annually). Using its own customer mailing list and those obtained from other sources, WA often obtained a 25 percent response rate. Up until the late 1980s, an independent marketing consultancy in LaCross, Wisconsin, produced the catalog. However, given the distance and recent increase in cost of outside catalog production, WA moved the entire operation in-house in the early 1990s.

Each catalog showed many four-color pictures and detailed descriptions of most of its products (see Appendix I). On the inside cover, next to a picture of Paul Keene and the Andersons, a short history of WA introduced new customers to the company. The remaining pages listed the products with the order numbers and prices. Frequently, an insert highlighted a new product, such as an energy efficient space heater or a top-of-the-line juicer. Over the years, the catalogs were greatly expanded and, in 1993, included such products as full-spectrum light bulbs, reading lamps that shielded the user from x-rays that were emitted by its fluorescent tubes, complete air filtering systems, and water purification systems. Paul Keene's essays, often describing periods of his life, also were printed in the catalog.

Finally, WA treasured some of its annual events that had become a tradition in Penns Creek. Each year, the Walnut Acres Country Fair attracted several thousands of people who visited the farm and the store. And, throughout the entire year, visitors of the farm could take special, self-guided walking tours that showed them the fields where the organic produce was being grown. WA had found that customers who visited the farm tended to become loyal customers in the future.

11.8 THE FUTURE

In order to address the strategic issues that faced WA, management identified a number of options. The choice of one or more of these alternatives was driven by the Board's desire to reach $20 million in sales by 1998. Some of the alternatives, such as a health resort where people could be educated about healthy lifestyles, were rejected because of the amount of capital needed and the lack of potential. Another option, such as franchising its store operations, was also put on hold because of its capital intensity.

The sale of WA to a large food company, although not ruled out, was considered difficult given that the purchaser would have to convince WA's owners* that the nature of its corporate culture would not be changed. "We will only do business with people who have philosophical and technological competencies that are compatible to ours," Keene said. A number of WA's competitors, however, had been successfully sold to large companies and had been able to keep their management philosophy in place. For example, Cascadian Farms was now owned by Welches, and Knudsens was recently purchased by Smuckers. Given the unattractiveness of this option to WA's owners, no estimates had ever been generated of WA's market value.

Since opening its own retailing outlets would require a $250,000 investment per store, WA's management felt it needed to carefully consider the option of marketing its products directly to wholesalers and retailers. Given that large retailers were opening up Whole Foods sections in their stores, WA's management believed that this distribution channel probably represented the greatest potential. Yet, in the past WA had always discouraged this side of its business and it would be difficult to convince wholesalers and retailers that it was sincere about its attempt to reenter this segment. Besides, it would put severe pressure on WA's financial resources as well as on some of its principles. Slotting allowances, for example, were often required to get retailers to allocate shelf space to a product. These charges could run high and had never been paid for by WA.

---

*As with every decision, the approval of a majority of founder shares was needed as well as a majority of the employee shares.

On the farming side, serious consideration needed to be given to buying additional farms. Although WA had successfully developed a network of organic food suppliers, it had received a number of requests for opening up a "Walnut Acres—East," or "Walnut Acres—West." This would allow them to reduce their distribution costs while further guaranteeing the freshness and quality of products to the customers in that geographical area. In addition, depending upon the location of these farms, it would then be possible to grow organic fruits, which were often hard to come by in the off-season. Prices per acre varied widely across the country, but averaged around $3,000 per acre.

Given the capital requirements of many of these options, WA's management began to look into the possibility of publicly trading its stock. Although it would remove ownership from its control, it could provide much needed funds for expansion. One alternative to this option was to offer shares to its customers only, who, given their commitment to and involvement in the organization, would probably be more than willing to buy shares. This would allow WA to continue its close relationship with its owners. Again, to date, no estimates of capital requirements had ever been generated.

Finally, WA had received numerous inquiries from international companies or individuals for exclusive distribution rights. To date, though, it had no experience in the international market and it would take some time and effort to establish itself to deal with international customers.

"There are so many potential opportunities out there for us," Anderson explained, "that our challenge is to find the one or two that are going to be successful for us. We cannot go into too many directions."

"I hope," Keene added, "that no matter what we decide to do, we can keep our principles. We have seen that this is one of the ways it can work and I hope that we can continue that. In any case," he added, "the next year will be critical for us."

11.9 APPENDIX I SAMPLE PAGE FROM CATALOG

Vegetables, As Fresh & Delicious as Home-Canned!

Golden Sweet Corn... Full of Natural Sweetness

Immensely popular. Golden drops of sun captured in a can! These naturally sweet morsels are canned the very same day they're harvested. You should see the flurry of activity: The clattering of corn pickers in the early morning, followed by the short trip up to our cannery, where the corn is husked and washed. We blanch our corn with steam while it's still on the cob, preserving more of the milk in each kernel—and more of the natural flavor. After blanching, the corn is immediately cooled and canned—all so we can get it to you in tip-top condition. This extra effort really pays off with a flavor and texture that reminds many customers of their own home-canned corn.

■ **Sweet Corn.**
Organic sweet corn, deep well water. 1 lb. tin.
#66102 (12 tins) **$25.25** (2 tins) **$4.29**

■ **Beets.**
Delicious early crop. Diced. An excellent source of vitamin A. Organic beets, deep well water. 1 lb. tin.
#66104 (12 tins) **$26.39**
(2 tins) **$4.49**

■ **Sauerkraut.**
Months of natural fermentation takes a lot of extra effort. But once you taste our old-fashioned sauerkraut, you'll be glad we've taken no shortcuts! Just our organic, hand-picked cabbage and salt. 1 lb. tin.
#66106 (12 tins) **$25.79** (2 tins) **$4.39**

The Beauty, the Magic of Our Farm-Grown Peas

There is something magical about the life cycle of organic peas, our earliest vegetable. Only a brief time after April planting, they seem to come in a flood, to break through the soil and rise up overnight. In June, the harvest is upon us. Oh, what an exciting time that is! Our harvester picks up each row of peas and lifts the peas, pods, vines and all into a truck. From there, the entire stalk is fed into our old peaviner, which separates peas from pods. Within hours of harvesting, the peas are brought to our farm's small cannery. We wash and hand-inspect them, then pack them in our own deep well water. All the freshness, vitality and flavor is sealed in, ready for you to enjoy.

■ **Peas.**
Just organic green peas and deep well water. 1 lb. tin.
#66101 (12 tins) **$25.25** (2 tins) **$4.29**

Organic Symbols
■ Walnut Acres-certified organic.
● Certified organic by others.
♦ Transitional organic, or unsprayed.

See page 3 for a full explanation of our symbols.

Protein & Fiber-Rich Beans, Ready to Enjoy!

Beans...a marvelously versatile food. They're equally at home in casseroles, stews, soups and salads—not to mention the fact that they're rich in fiber, protein and vitamins. And now, we've made it convenient to enjoy this goodness. No overnight soaking, just open and enjoy!

● **Garbanzo Beans.**
Mellow-flavored and richly textured. High in fiber, too. Try them on salads, in soups or as a main dish. Increasingly popular. Organic garbanzo beans, deep well water. 1 lb. tin.
#66107 (12 tins) **$25.79** (2 tins) **$4.39**

● **Kidney Beans.**
Large, tender beans. Naturally rich in fiber, low in fat. Wonderful in chili, beans and rice, soups and much more. Organic kidney beans, deep well water. 1 lb. tin.
#66110 (12 tins) **$25.25** (2 tins) **$4.29**

● **Pinto Beans.**
Your best choice for Mexican dishes. Enjoy these fiber-rich, high-protein beans in so many ways. Organic pinto beans, deep well water. 1 lb. tin.
#66120 (12 tins) **$25.79** (2 tins) **$4.39**

● **Great Northern Beans.**
We've saved you the trouble of soaking and cooking these delicious beans. Just open and heat, or use in your favorite recipes. No salt needed. Rich in dietary fiber. We hand-inspect these beans, selecting only the best. Packed with our deep well water. 1 lb. tin.
#66121 (12 tins) **$25.25** (2 tins) **$4.29**

● **Navy Beans.**
A favorite for baked beans, soups and casseroles. Organic navy beans, deep well water. 1 lb. tin.
#66111 (12 tins) **$25.25**
(2 tins) **$4.29**

■ **North Woods Rice.**
A mellow, nutty blend of pre-cooked organic brown rice and 20% organic wild rice. No fuss or guesswork! Simply open, and it's ready for soups, casseroles or side dishes. Deep well water, organic brown rice, org. wild rice, pressed canola oil. 1 lb. tin.
#66112 (12 tins) **$25.25** (2 tins) **$4.29**

We Accept MasterCard, VISA, or Discover 1-800-433-3998

11.10 APPENDIX II ORGANIC CERTIFICATION AT WALNUT ACRES

Until 1991, WA maintained their own organic certification. They discontinued this practice, in part, because they were involved in establishing standards for the 1990 Farm Bill. Walnut Acres felt strongly that if they were playing a leadership role in the creation of organic standards, they also had to be leaders in the certification process. In order for the organic labeling to have credibility and not appear self-serving, independent third-party certification was needed. Walnut Acres also struggled with requiring their suppliers to meet WA's own certification standards. As a result, it was one of the first major farms to be third-party certified in the country.

Walnut Acres first applied for third-party organic producer, processor, and marketer certification from the Organic Growers and Buyers Association (OGBA). They selected OGBA for several reasons. First, WA liked the leadership of the organization. Second, even though OGBA charged fees for the inspection and certification, WA owned the paperwork. This issue was an important factor. If the paperwork was owned by the certification agency, WA would have to request processing of their certification each time verification of their certification was required. The process could be time consuming, causing delays, and costing additional money. The third reason WA selected OGBA was the cost of certification. Some certification organizations required royalties on every item sold under the certification label. The extra royalty fees could increase WA's certification costs to approximately $18,000 to 20,000 per year. When WA first contracted with outside certifiers, the fees were $2,400 per year.

In the inspection process, OGBA's representative inspected WA's growing, processing and marketing processes as outlined in an application. A field inspection of the property was a major part of the certification process. The certification verified that the methods implemented on the farm and the food processing met the certifier's standards for organic. An individual certifier's standards may vary slightly from other certifiers, as long as they met national standards.

Walnut Acres used products from other organic growers. They accepted organic products that were certified by most recognized, long-standing organizations, including OCIA, Oregon Tilth, Farm Verified Organic, and many state certification programs such as those in Kentucky and Iowa. Many of the non-governmental certifiers were non-profit organizations. Walnut Acres also researched new certification organizations as they became known. Products were only selected for use in WA's products if they were certified by reliable organizations.

Currently, Pennsylvania does not have a state certification program. However, if the fees for certification increased, WA would support state certification. Pennsylvania was a strong candidate for developing a state program, partly because of the large commercial organic farms located there, such as WA and Rodale's farms.

Walnut Acres believed in the organic industry ethically certifying itself as opposed to government regulating the industry. The Organic Foods Production Act of 1990 allowed certifying agents to be private organizations, individuals or public agencies. The certification process would develop better certifiers by using the already established knowledge base within the industry, and certification costs would be more competitive. The role for the government, then, was in licensing or certifying the certifiers.

11.11 APPENDIX III LAWS ON ORGANIC FARMING

In the late 1980s organic growers, including WA, were concerned that the term *organic* would become loosely defined, like the term *natural* did several years earlier. The growers understood what organic meant, but felt that it should be codified to insure a consensus and greater public understanding. They began working directly with Senator Patrick Leahy (D-VT), Chairman of the Senate

Agriculture Committee, on the *Food, Agriculture, Conservation, and Trade Act of 1990* (the 1990 Farm Bill).

Every five years Congress develops the program that focuses the agricultural funds for that period. The Organic Foods Production Act of 1990 (S. 2108) started to build a niche for organic food production within the 1990 Farm Bill. The organic foods industry was growing and Congress recognized the need for a nationally accepted definition of "organically grown" products.

The Organic Foods Production Act created a nationally level playing field for the industry.[1] In the late 1980s as many as 22 states regulated organic food production, but the regulations were inconsistent. Interstate commerce of organic products, as well as individual consumers, were troubled by the different regulatory standards. Marketing organic products by large food chains was difficult owing to tracking product certification, resulting in major food chains refusing shelf space for organic products.

The Organic Foods Production Act of 1990 had three main purposes:

1. To establish national standards governing the marketing of certain agricultural products as organically produced products.
2. To assure consumers that organically produced products meet a consistent standard.
3. To facilitate interstate commerce in fresh and processed food that is organically produced.[2]

The bill required that on or after October 1, 1993, to sell or label a product as organic, it must have met the following requirements:

1. Have been produced and handled without the use of synthetic chemicals, except as otherwise provided in this title.
2. Except as otherwise provided in this title and excluding livestock, not be produced on land to which any prohibited substances, including synthetic chemicals, have been applied during the 3 years immediately preceding the harvest of the agricultural products.
3. Be produced and handled in compliance with an organic plan agreed to by the producer and handler of such product and the certifying agent.[3]

The Organic Foods Production Act of 1990 established a national definition of organic and an organic labeling program, but it also called for a farm plan. The idea behind the farm plan was not just to avoid inorganic chemicals and sprays (residues), but to create an element for building and sustaining the productivity of the soil. Residue testing, although provided as a part of the bill, was a reactive position. The act provided a means to develop a more proactive process by having a goal of sustainable agricultural practices through the farm plan.

Processed foods that bear the organic label are covered under this bill also. Processed foods labeled organic must contain 95 percent organically produced ingredients by weight. The remaining 5 percent can be inorganic, but must be permitted ingredients on the National List.

The Organic Foods Production Act of 1990 was basically a marketing bill, not a food safety bill. There was an important distinction between the two. While the industry wanted to define the terms and concepts of organic products, they did not want the bill to become punitive. Even though the bill implied food safely, it was specifically oriented toward labeling and marketing organic products.

The real issue was providing an organic product alternative. Organic labeling was a consumer issue, providing a level of consumer confidence that what was on the label was, in fact, in the package. A level playing field was created for all organically labeled products. All producers have to meet the same conditions and presumably the same expenses. To that end, the organic growers' interests were to define terms and increase the restrictions, not reduce them.

A National Organic Standards Board was established to make recommendations to the Secretary of Agriculture for the bill's implementation. This process was underway in 1993. Bob Andersen had been involved, testifying before the Board. The recommendations from the National Organic Standards Board had been through the USDA process and ultimately had public comment like any law.

Much of the subcommittee's work had been done through investment of time from individual companies within the organic industry. The industry had been heavily involved in the entire political process, forming a consensus among themselves and presenting a strong front before the National Organic Standards Board and its committees. Even though Bob Andersen had been involved in the process, WA's investment in terms of hours was small, particularly compared to the major commitments of other organic foods companies.

11.12 REFERENCES

1. Legislative History Senate Report No. 101-357, Title XVI-Organic Certification Program, (4943-4960), 1990, 289.
2. TITLE XXI—Organic Certification, Food, Agriculture, Conservation, and Trade Act of 1990, Section 2102, P. L. 101-264.
3. TITLE XXI—Organic Certification, Food, Agriculture, Conservation, and Trade Act of 1990, Section 2105, P. L. 101-264.

12 From the Earth to the Table: Fetzer Vineyards and Its Bonterra Wines Case Study Number 4

O. Homer Erekson and Allison R. Leavitt

CONTENTS

12.1 Introduction 249
12.2 The History of Fetzer 250
12.3 The Grape Vine and Its Growing Conditions 252
    12.3.1 Growing Conditions 252
    12.3.2 Agrichemicals and Soil Ecology 254
12.4 The Conversion Process from Conventional Farming Practices to Organic ... 257
    12.4.1 Conventional Viticulture Practices 257
    12.4.2 Modern Organic Viticulture Practices 260
    12.4.3 Conversion to Organic Practices 260
    12.4.4 Organic Farming at Fetzer 260
12.5 The Wine Industry and Economic Feasibility of Bonterra Wines 262
    12.5.1 The Wine Production System 262
    12.5.2 The Organic Certification Process 263
    12.5.3 Economic Feasibility of Bonterra Wines 264
    12.5.4 Sales and Public Relations 266
12.6 Vision for the Future 268
12.7 References 268

12.1 INTRODUCTION

It was a beautiful sunny day, June 22, 1995, as we walked the five-acre experimental garden at Fetzer's Valley Oaks Food and Wine Center in Hopland, California. The garden had more than 1000 varieties of organically-grown fruits, vegetables, herbs, and edible flowers. Moments later, we were enjoying a delicious lunch featuring meals prepared directly from the garden, along with two wonderful Fetzer wines. As we ate, Pat Voss, the Senior Vice President for Production, and Steve Dorfman, Director of Grower Relations, conveyed the Fetzer philosophy that biological balance and diversity provide the foundation for producing award-winning wines and flavorful foods. They went on to describe how Jim Fetzer, in 1986, began to wonder if the same marvelous flavors that they were able to create in the garden could also be created in their grapes and, thus, in their wines. This led to a decision by Fetzer to invest significantly in the development of wines using organically-grown grapes. Paul Dolan, Fetzer company president, has described how, as a young winemaker, he would stroll through the vineyards sampling grapes. He noted how at one vine the grape flavor would be full of fruit. Then, at a vine just ten feet away, he said, "it would be flat and insipid, and I couldn't figure it out. Now, looking back, I get it. What we had done, is that we had sprayed so much that we had killed the microbiological life in the soil. The reason I believe that now is you can go back into that same vineyard and go from vine to vine to vine, and [the grapes] have great consistency of flavor and depth of fruit, and the wines that we're making from that vineyard now go into some of our best sauvignon blanc. Prior to that, it just went into our simple table wine."[1]

12.2 THE HISTORY OF FETZER

The history of Fetzer Vineyards begins almost 40 years ago when Barney Fetzer purchased land in the North Coast region of California to grow grapes for the home winemaking market. Fetzer's Valley Oaks Ranch is located in Mendocino County (see Figure 12.1). Grape production began in Mendocino County in the 1860s. By 1910, 5800 acres were in production. In 1967, only one winery was operating in the county. The next year, Fetzer was bonded and wine production in the county began to expand. By 1992, 12,364 acres were in production.[2]

In the 1950s, Fetzer was in the lumber business, not wine production. It was not until 1968 that he began a commercial enterprise in wine production. From 1968 until his death in 1981, Barney managed Fetzer Vineyards. After his death, his family remained in the business, and 10 of the 11 siblings took over its management.[3]

The second generation of Fetzer wine producers were successful in the management of the company as it grew from 200,000 cases of wine produced annually in 1981 to 2.2 million cases by 1995.[3] Today, Fetzer's list of wines includes classic varietals such as Johannisberg Riesling, Gamay Beaujolais, and White Zinfandel; Barrel Select wines, Cabernet Sauvignon, Sauvignon Blanc, Zinfandel, Pinot Noir, and Chardonnay; super-premium Reserve wines, Chardonnay, Cabernet Sauvignon, and Zinfandel; and additional popular wines, Sundial Chardonnay, Valley Oaks Cabernet Sauvignon, Eagle Peak Merlot, and Fetzer Gewurztraminer.[3]

As early as 1985, Fetzer Vineyards began experimenting with organic growing practices. They became interested in organic farming, because they saw that organic methods could improve the quality and flavor of vegetables. Fetzer converted 200 acres of vineyards to organic vegetable and fruit farming for use in their Culinary Center.[4] They began to consider that, since organic farming practices improved the vegetables and fruits, might it also improve the quality and flavor of grapes.

Fetzer began to convert the vineyards from conventional growing practices to organic. The North Coast region of California has a climate that makes organic farming functional and economical. It has cold winters, minimal summer rainfall, relatively low humidity, high daytime temperatures, and winds that alleviate the conditions conducive to mildews and molds that are devastating to grape production elsewhere. By the early 1990s, Fetzer Vineyards introduced two new wines made exclusively from organically grown grapes and certified by the California Certified Organic Farmers[5]: the Fetzer Organically Grown Cabernet Sauvignon and the Fetzer Organically Grown Chardonnay. Production of these wines has increased significantly since their introduction into the market. For instance, production of the Organically Grown Chardonnay increased from two thousand cases for the 1991 vintage to four thousand cases for the 1993 vintage. For 1996, Fetzer expected to introduce three new organic varietals: Sangiovese, a popular Italian red grape; Syrah, a full-bodied Rhone red varietal; and Viognier, a floral Rhone white grape.[6]

By the 1990s, the Fetzer family had decided to move away from the wine production business, and in 1992 sold the winery operations to Brown-Forman, a company from Louisville, KY. Brown-Forman purchased the Fetzer and Bel Arbor brand names, the Hopland winery that included 360 acres of vineyards, and the Fetzer Food and Wine Center. The family maintained ownership of the

From the Earth to the Table: Fetzer Vineyards and Its Bonterra Wines 251

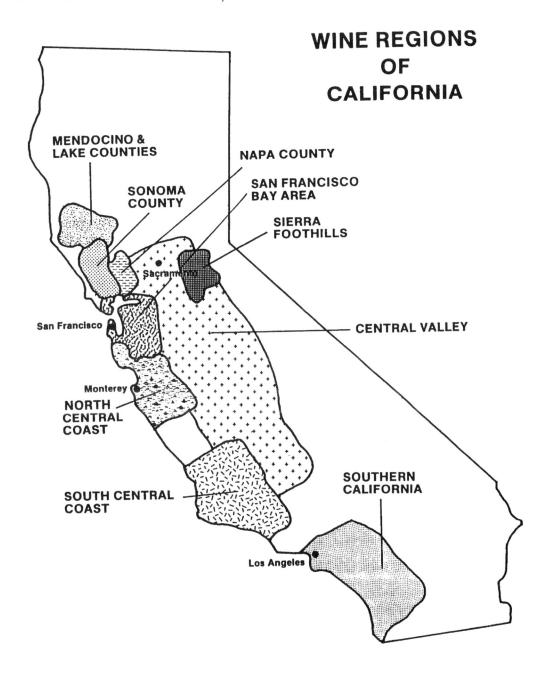

FIGURE 12.1 Wine regions of California. *Source:* Developed by Miami University Applied Technologies.

1100 acre Kohn Properties farm, and today grows grapes for Brown-Forman Fetzer under an exclusive contract.[3,7] The entire 1100 acre Kohn Properties vineyard is certified organic by the CCOF.[7]

At present, Fetzer Vineyards is a national leader in organic grape production practices. Currently its entire 360 acre vineyard is either organically certified or pending certification by the CCOF.[8] Besides their own 360 acres, Fetzer contracts with the Kohn Properties farm and about 200 independent growers to supply grapes from more than 2000 acres of organically certified land.[3,7]

12.3 THE GRAPE VINE AND ITS GROWING CONDITIONS

12.3.1 Growing Conditions

Vineyards in the best wine producing regions are thought to have a harmonious relationship between variety, soil, and climate. The French call this relationship *goût de terroir*. All grape varieties will ferment when crushed, but only a few are of a quality suitable for wine production. Wine quality is measured by its bouquet, flavor, and general balance. These qualities are derived from the special characteristics of each grape variety. The inherent characteristics of the grape variety, such as aroma and flavor, determine the type of wine produced. Within each type of wine, quality is determined by the soils and temperature of the region where it is grown.[9]

All cultivated varieties of grape are believed to have originated from the species *Vitis vinifera* (known as the Old World Grape) in Asia Minor.[9] Cultivated varieties spread to much of the world through European colonization. The Church was a key institution in the spread of both varieties and winemaking in the seventeenth and eighteenth centuries. In California, the first vines were thought to have been planted at the Mission San Francisco Xavier around 1697 by Father Juan Ugarte. However, early settlers found native grape varieties growing in all parts of the United States, from New England to the Pacific Coast. There are approximately 20 distinct species and many natural hybrids, unlike the Old World's single species.[10] In North America, 70 percent of the world's grape species are found.[9]

As is true with all modern monoculture crops, species are selected, bred, and grown to produce the best quality. Varieties of the *Vitis vinifera* species produce 90 percent of the world's grapes. California's grape production also uses mainly varieties of *V. vinifera*. However, these varieties are susceptible to freezing, Pierce's disease, and other pathogens and insect pests that thrive in humid climates, so they are not grown east of the Rocky Mountains. In these areas varieties of a native American species, mainly, *V. labrusca*, are grown.

Grape varieties also have been selected to control some pests, particularly phylloxera (the grape root louse, *Daclylosphaera vitifoliae*). Phylloxera, along with several other deadly diseases such as powdery mildew, downy mildew, and black rot, is native to the Mississippi Valley.[9] All grape vines native to this region are resistant to phylloxera. However, phylloxera was carried to Europe where it destroyed many of the European vineyards in the late nineteenth century. The viticulturalists from Europe tested American species for resistant species and propagated their own vines to the roots of the American species. Then they replanted the damaged vineyards. The rootstock of *V. riparia*, *V. rupestris*, and *V. berlandieri* are all particularly suitable for this purpose.

In the 1970s, a hybrid cross of *V. rupestris* and *V. vinifera* was developed and recommended for planting, because it was particularly suited to California's microclimates and thought to be resistant to phylloxera.[11] Most of the new vineyards were planted with the new hybrid. However, the phylloxera mutated, and the new pest strain feeds on the roots of the *rupestris/vinifera* hybrid. Now, California's vineyards are at risk of being destroyed like the European vineyards were destroyed a century ago. Around the world, today, most vineyards are being planted with rootstocks of hybridized American varieties.

The second element of *goût de terroir*, soil, contributes to the quality of the grapes. Grapes grow on a variety of soils, reflecting varietal preferences. Appropriate soil conditions include gravelly alluvial soils, clay loams, and slopes formed on limestone and chalk subsoils. Soils to be avoided include heavy clays, poorly drained, and shallow soils, as well as those with high amounts of salts and alkali metals. The best soils are usually found on terraces, lower slopes of hills, lower sides of valleys, and on gravelly alluvial fans.

Because grapes grow in many types of soils, it has been suggested that soil structure is more important than soil type. One of the most important aspects of the soil structure is its water holding capacity. Grapes do best with a steady, but moderate, supply of moisture. Soils must be free-draining, well-aerated, and promote deep root growth. The structure of the soil also must have moderate water retention capacity. If soils are well-drained, but still provide adequate moisture, then extreme droughts or excessive rain will have minimal effect on the vines. Because soil structure is so important to growing quality grapes, we will consider soil ecology in more depth later on.

Finally, climate is a key factor in viticulture. Grape species grow in warm temperate regions of the world, generally between 34° north and 49° south latitudes (see Figure 12.2). However, regional and local climate has a major effect on the quality of wine grapes. Regional climatic and soil conditions determine the sugar-acid ratio, total acidity, and tannin content of the grape. For example, because of the climate, the California North Coast varieties do not reach high sugar levels before flavor develops, and they retain high levels of malic acid. This is particularly true for Chardonnay.

The best climate for grapes is one that has even accumulation of heat. Grapes ripen quickly in this type of climate. Optimum temperatures for grapes to ripen are from 68 to 72°F. Once the temperature rises above 73°F (the maximum rate of photosynthesis for the vine), the plant loses excessive moisture through transpiration. To protect against dehydration, leaf photosynthesis, involving the uptake of carbon dioxide and release of water, can be shut down.[12]

California's position on the Pacific coast provides microclimates that are exceptionally well-suited to growing grapes. The winds off the ocean and the mountain valleys create a natural cooling system. California's growing season is long, with few unseasonable frosts. The mild to moderate winters, hot dry summers, and wide ranges of rainfall are optimal for many wine grape varieties. From October through February, 80 percent of California's rain falls, with little or no rain during the ripening period.

V. vinifera is well-suited to the hot dry summers of California. Rainfall at the wrong time in its growth cycle can damage the vine or its fruit. Winter rains are desirable, but insufficient rain during the early summer can be managed through irrigation. Disease and insect pests are more difficult to control if rains come early in the growing season, but vine growth is not affected. Rains and clouds can adversely affect the set of fruit during blooming, and rains may cause fruit rot if they come during ripening or harvest. Because of the hot dry summers, the pest phylloxera does not spread as

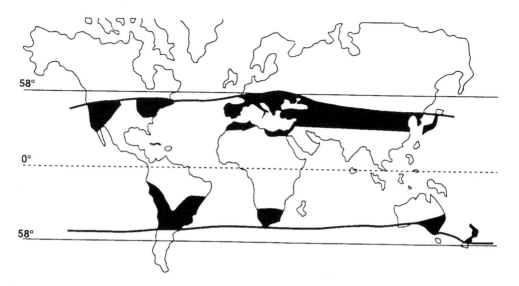

FIGURE 12.2 Climatic limitations of grape-growing. *Source:* Adapted from Figure 1.4 (page 7) in Reference 10.

quickly, the diseases of downy mildew, black rot, and anthracnose are not present. Also, grape and berry moths do not live in this habitat. Basically, the California climate is almost ideal for the implementation of organic viticulture.[2]

12.3.2 AGRICHEMICALS AND SOIL ECOLOGY

Maintaining appropriate soil structure for optimum grape production requires maintaining the health of the natural soil/detrital ecosystem (see Figure 12.3). Soils provide plants with mineral nutrients and water. The major supply of nutrients comes from the recycling of detritus (dead organic matter) and native soil-animal wastes. Water is held in the soil after rains largely by the organic matter. The amount of water that can be stored is called the water-holding capacity of the soil. If the water-holding capacity of the soil is good, then the grape plants can have access to water during the annual summer-drought. Plant roots get oxygen from the soil and expel carbon dioxide into it. The capacity of the soil and roots to participate in these functions is achieved through soil aeration.

Soil biota have been recognized as important elements in the process of maintaining soil aeration and decomposition in terrestrial ecosystems (see Figure 12.4). The structure of soil ecosystems is based on below-ground production of fine roots and the detrital or decomposer food chain. The function of the ecosystem is to maintain organic matter while breaking down detritus and releasing stored nutrients for use by plants and other micro-organisms. Besides nutrient cycling, soil biota improve the structure of the soil (aeration) for good drainage and plant growth. The soil insects and micro-organisms aerate the soil, add essential forms of organic matter, and generally improve the physical properties of the soil by their tunneling, excretions, and body remains.

The food chain consists of a variety of decomposer species, classified from micro- to megafauna (see Figure 12.5). Bacteria make up about 2 to 5 percent of the living matter in the soil. The hyphae of fungi (hair-like cells found in decomposing leaf-litter) also are important decomposers,

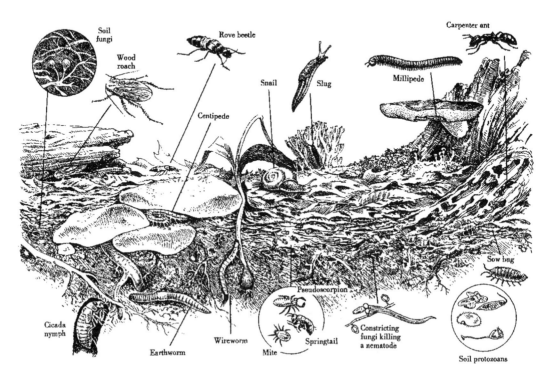

FIGURE 12.3 Soil organisms: reducing detritus to humus and the development of soil structure. See Reference 28.

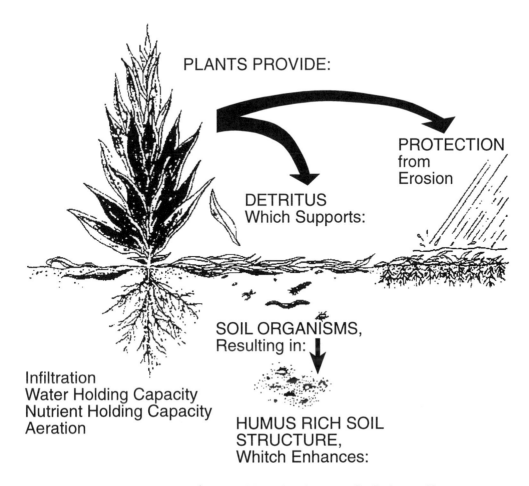

FIGURE 12.4 Relationship between plants and the soil environment. See Reference 29.

breaking down about 5 to 15 percent of the living matter. Nematodes are small worm-like organisms that make up anywhere from 1 to 15 percent of the soil animal activity, while protozoa make up around 2 percent. Mites are predators that feed on nematodes and bacteria and make up 15 percent of the living matter in soil. Collembola are mesafauna that feed on hyphae and make up 15 to 20 percent of the living matter. Soil organic matter, much of it from the fine roots produced annually by the grapevines, is critically important in the soil ecosystem because it is an essential component of soil structure and texture. Organic matter retains water-soluble ions and nutrients and improves the permeability of the soil.

Some agriculture practices have a devastating effect on soil ecosystems. Crops that are harvested and removed extract nutrients from the soil that are not then replaced by normal detritus processing. In addition, tillage of the space between the rows can destroy the organic matter in the soil by exposing it to the sun during the summer. However, one of the greatest effects of modern agriculture on the soil ecosystem comes through the use of pesticides.

Increased use of agrichemicals after World War II has been the major means of dealing with pest control problems, while maintaining crop yields. Monoculture crops are particularly at risk to pest and disease outbreaks. Before the twentieth century, pest control was achieved mainly through manual control, such as hoeing and hand removal of insects. Natural controls included crop rotations and encouraging birds as predators of harmful insects. As early as the turn of the century, chemicals such as arsenic came into use. Other metal-based compounds from copper and zinc also were part of the pest control tool-kit.

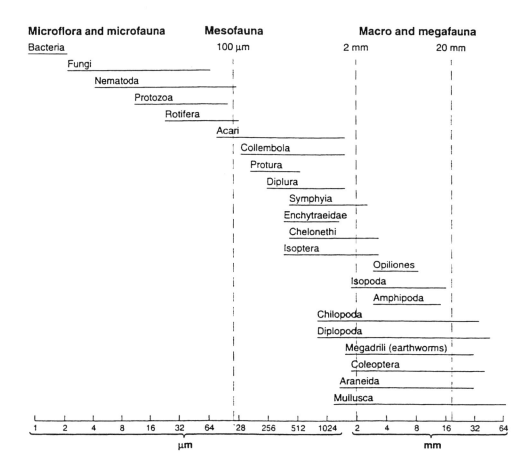

FIGURE 12.5 Size classification of organisms in decomposer food webs by body width. See References 12 and 27.

By mid-century, DDT was a widely utilized agricultural chemical. It was a known nerve poison, but was thought to be benign for humans, affecting only insects. DDT was valued because, with its ability to persist in the environment, farmers had only to apply it once a year. It was not known until later that the persistent characteristics of the chemical would have disastrous effects on all soil insects and, through bioaccumulation, result in toxic concentrations at the top of the food chain. DDT also was registered for control of bats. However, bats are mammals, and if DDT could affect the nervous system of small mammals, then it probably should have been recognized as a risk to humans as well.

Since the banning of DDT, agrichemists have been developing new chemicals that break down in the environment more quickly, but still tend to kill non-target organisms in soil. These agrichemicals also break down quickly and need to be stronger to work as effectively. They also must be applied more often. Finally, the residuals of these compounds, although small in quantity, tend to be mobile in the environment, penetrating more deeply into soils and groundwater.

Modern farming practices are moving toward integrated pest management using biological control, crop rotations, and organic farming as alternatives to chemicals. Integrated pest management may use chemicals when there is no alternative, but generally seeks restoration of the function of soil fauna. It also evaluates the use of chemicals against an economic threshold of risk to the harvest crop. It tries to reach a balance between how much of the crop may be lost from pest damage vs. the cost to control the pest chemically or biologically. Organic farming is different in that these farming practices optimize the use of low-cost natural controls to achieve comparable economic thresholds. They maintain the natural predators of pests and the moisture retention in the soils by mulching and retaining crop residues. Nutrient cycling is maximized through the use of organic matter and retention of organisms in the soil.

12.4 THE CONVERSION PROCESS FROM CONVENTIONAL FARMING PRACTICES TO ORGANIC

12.4.1 CONVENTIONAL VITICULTURE PRACTICES

Vineyards are typically monoculture crops grown as an agroecosystem. In traditional agricultural systems, diversity of plant species is limited to the crop of interest. Row cropping, herbicides and other land treatments are used to maintain the monoculture. The biodiversity of birds and insects, including beneficial and pest species, also is limited. Management practices, such as the use of pesticides and cultivation, tend to eliminate vertebrate, invertebrate, and microbial species, whether harmful to crops or not. Traditional viticulture is no exception.

As with many types of agricultural practices, modern viticulture has relied on technological advances that emerged after World War II: new varieties of crops, agrichemicals, and labor-saving, energy-intensive farm machinery.[13] New technologies in pest control and new crop varieties allowed the same crops to be grown on the same land every year.

Traditional viticulture management in the North Coast has to contend with weather conditions, insect pests, and weed control.[14] Frost damage in the fall is a major issue. Young vines with immature wood growth are particularly at risk. Also, several days of high temperatures during the flowering period may cause poor fruit set. Temperatures of 105°F can cause the fruit to burn, ripen too rapidly, or "raisin."

After heavy winter rains, nitrogen usually is applied early enough to allow spring rains to move the nitrogen into the root zone. Cover crops are recommended on hillside vineyards to prevent erosion, particularly in young fields. In the North Coast region, young vineyards and new plantings are irrigated in the summer months, but mature vineyards are irrigated only during times of drought.

According to the Division of Agricultural Sciences at the University of California (1982), "Many north coast vineyards seldom or ever require treatment for insect and mite pests, but in some years the following pests may require control measures: grape leafhopper, orange tortix, spider mites, cutworms, false chinch bug, grasshoppers, thrips, branch and twig borer, phylloxera, nematodes, powdery mildew, botrytis bunch rot, eutypa dieback, black measles, Pierce's Disease, birds, deer, and weeds."

Weeds are generally pioneer plants that become established in disturbed soils. They are considered a management problem because they compete with the grapevines, particularly young plants, for light, water, and nutrients. Also, undesirable plants may make harvesting the grapes and maintaining the grapevines difficult. Figure 12.6 lists the type of weeds that often plague vineyards along with the herbicides used in conventional viticulture practices to control these weeds.

The control of pests in the conventional viticulture practice of the last 30 years has relied heavily on the use of chemical pesticides. Major pest problems come from insects and mites,* and most control agents have been applied using sprayers. Figure 12.7 shows several different types of

---

* There are three major categories of insecticides and miticides: organochlorines, organophosphates, and N-methyl carbamates. Organochlorines are persistent in the environment, but less toxic to humans. However, they can have long-term effects because they are soluble in fat and are stored there, thus chronic poisoning is possible. Organophosphates are less persistent, but more directly dangerous to humans through effects on the human nervous system. N-methyl carbamates have a similar effect on the human nervous system. Fumigants and fungicides also are used in vineyards. Chlorpicrin, ethylene dibromide, and methyl bromide are fumigants used in the soil to kill unwanted soil insects and pests. Included in the list of fungicides are benomyl, captan, sodium arsenite (inorganic arsenic compound), and sulfur.

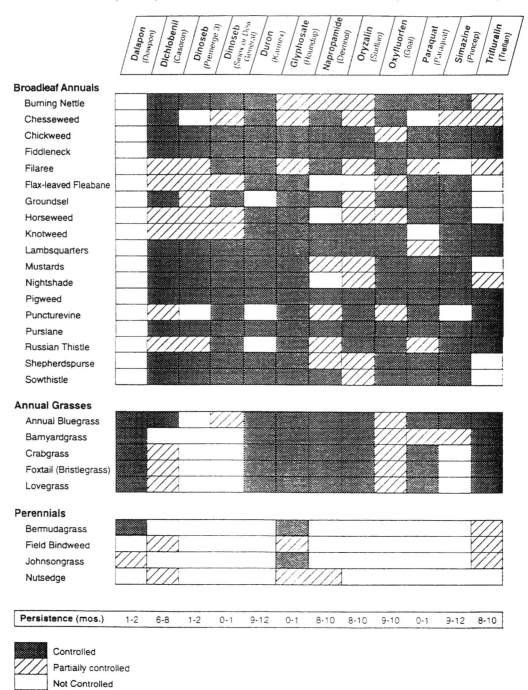

FIGURE 12.6 Relative susceptibility of common weeds to herbicides registered for vineyards. See Reference 14, page 286.

Over-the-vine, high-pressure sprayer.

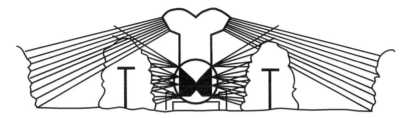
Air-carrier, double-row sprayer covering four half rows.

Air-carrier, single-row sprayer.

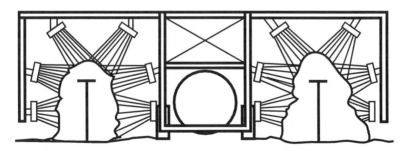
Air-carrier, double-row sprayer covering two complete rows.

FIGURE 12.7 Sprayers for application of chemical pesticides. See Reference 14, pages 291–293.

spraying apparatus. The sprayer applicator is not specific for any one pest, and even though the sprays are directed, perfect targeting onto the vines is not possible. Chemicals are deposited on non-targeted species as well as the targeted ones. People who apply the chemicals are required to wear protective clothing and to follow strict guidelines for application.

12.4.2 Modern Organic Viticulture Practices

With increased emphasis on sustainable agriculture have come practices to produce quality foods, protect the resource base, reduce off-farm inputs, and be profitable,[13,15] many of them developed under the framework of organic viticulture. Sustainable agriculture is seen as a "system-level approach to understanding the complex interactions within agricultural ecologies."[13] It builds on the relationships and interactions of biotic and abiotic resources. Farming practices that emphasize building up soil, diversifying crops, controlling pests naturally, and utilizing local resources that are economically feasible are key components of sustainable agriculture.

A key linchpin in sustainable agriculture and organic viticulture is seeing production as a biological/ecological system. Biological diversity is a key in the design and functioning of the system. Organic viticulture practices are mainly established to promote long-term soil health and crop viability and to eliminate the use of inorganic chemicals in pest management. Long-term soil health and viability are improved or maintained by increasing biodiversity in the soils and the field vegetation. Cover crops increase diversity and improve soil fertility. Also, some cover species fix nitrogen in the soil, improve water penetration, infiltration and retention, promote microbial activity that, in turn, increase decomposition and soil stability, control erosion, provide habitat for predatory insects, and suppress weed growth.[16]

The major strategies for pest management in organic farming are designed around preventing pest outbreaks. The use of synthetic chemicals to control pests is prohibited on certified organic farms, and organic methods take longer and are less effective for short-term, immediate control than conventional methods.[16]

12.4.3 Conversion to Organic Practices

Gradual conversions to organic culture practices are usually recommended. Abrupt conversions can cause reduced yields. Short-term effects also can be a major drawback. Often, integrated pest management practices (with occasional use of chemicals) are recommended as an intermediate step.[15] Converting one field or area at a time to organic viticulture may be beneficial.[13] Reducing pesticides and fertilizers on soils that have not had sufficient organic inputs to rebuild the soils or helpful insect populations, can increase weeds and harmful insects. It takes time for the ecological system to adjust to the new inputs and populations.

12.4.4 Organic Farming at Fetzer

Following Fetzer's decision to produce organically-grown grapes, the Fetzer Bonterra wines were the first to be 100 percent from certified organic vineyards. No pesticides, herbicides, fungicides, or synthetic fertilizers are used in the growing of these grapes. Rather, organic practices at Fetzer, and the contract farms that produce for Fetzer, follow four major programs: soil improvement techniques, disease prevention strategies, insect control, and weed control.[17] Fetzer has three major strategies to improve soil fertility and overall health, including cover-cropping, cultivation, and applications of composted material. Fall cover-cropping after harvest is the major soil improvement practice. Cover crops provide organic matter that improves soil texture, structure, and friability. Also, cover crops provide nutrients for the grapevines, habitat for beneficial foliage, and soil insects. Typically, Fetzer plants an oat-legume cover mix, but rotates every third year with another legume to increase organic matter. The oat-legume mix grows well in the North Coast area and is easy to maintain. Cover crops that flower are particularly beneficial because they attract useful insects. Less tractor traffic minimizes soil compaction and erosion. Perennial cover mixes may become effective for cover crops in the future. These crops are managed through mowing, not discing. The major concern with cover crops is competition with the grapevines for water and nutrients but, to limit competition, drip irrigation systems are installed.

Cultivation is the second practice that Fetzer implements to improve soils. Cultivation increases the organic matter in the soil when the standing cover crops are disced into the soil so that the organic material decomposes deep in the soil, building the soil structure. This practice is called green manure. Cultivation improves the soil structure and fertility more quickly than surface decomposition. Cultivation also aids in the control of weeds. Cover crops can become a weed problem because of the large amount of rain in this part of California.

Fetzer also applies composted material to improve soil fertility and structure. Composted material consists of pomace and stems, the by-products of the grape crushing process. Sometimes, manures are imported and mixed with the pomace and stems. Fetzer applies compost at a rate of 1 to 2 tons per acre on about 100 acres each year.

The soils in Fetzer vineyards are maintained close to neutral or slightly acidic (pH of 6.8 to 7.0). The pH is amended by adding either lime to acidic soils or gypsum to alkaline soils. Fetzer believes that this pH level is most beneficial for nutrient availability, increased bacterial activity, high cation exchange capacity, and good soil texture.

Diseases are a concern in viticultural practices. Fetzer deals mainly with powdery mildew, botrytis, and eutypa. Powdery mildew and botrytis are consistently a problem. Their strategies to keep these two diseases under control are "removal of early season sources of inoculum, enhancement of air circulation, careful timing of irrigation, and a consistent sulfur program." Their specific practices to implement these strategies are "removal of mummies post harvest, careful pruning, cover crop management to reduce humidity, leaf pulling at berry set, cane cutting as necessary, and sulfur application every 7 to 10 days unless rain or heat dictate otherwise."[17] Also, high trellising methods improve circulation. Tom Piper,[18] Fetzer vineyard manager, notes, "We use 14 different types of trellising to get maximum leaf exposure."[6] Recently, Fetzer has reduced the sulfur used in the fields, but they have found no effective organic practices that substitute for the use of sulfur as a means of limiting disease infestation. Eutypa also is difficult to control, but Fetzer tries to minimize pruning cuts (a path of infection), and uses tree sealants to protect the larger cuts. The effectiveness of these procedures is difficult to verify, however.

Fetzer controls insects through natural predators, as much as possible. Piper says, "You can't control every pest, but if your grapes are healthy, they tolerate some. The important thing with pests is to attract the good ones, those that feed on the damaging insects".[6] Fall cover crops and vegetation around the vineyard provide habitat for beneficial predator species. Fetzer is tolerant of some insect damage, if the health of the plants is not jeopardized. Overall, healthy plants can withstand minor attacks, so maintaining healthy plants is a key part of the strategy. More specifically, leaf-pulling is used to remove newly hatched mites and leafhopper nymphs on new foliage. Early spring growth is encouraged so that the leaves mature before the insects are ready to hatch. Insecticidal soaps may be used to control mites, leafhopper nymphs, and thrips.

Weeds are a problem in two areas, under the vines and between rows. Weeds under the vine are particularly hard to control, but Fetzer has implemented a combination of ridging dirt, hand hoeing, and mechanical cultivators. The program of weed control under the vines also must be timed carefully. Cover crops, mowing, and cultivation are used to control weeds between the rows.

Finally, Fetzer considers monitoring to be a key component of their organic farming practices. Soil tests, pH sampling, and leaf petiole analysis are conducted annually. Brix readings on leaf blades are done four times a year. At least once a week during the growing season pest/predator populations, disease symptoms, soil texture and structure, and surrounding vegetation are monitored. Many elements of Fetzer's organic program relies on preventative measures, and monitoring is an essential element of the system.

12.5 THE WINE INDUSTRY AND ECONOMIC FEASIBILITY OF BONTERRA WINES

12.5.1 THE WINE PRODUCTION SYSTEM

Wine is fundamentally an agricultural product. As noted above, the quality of wine is dependent upon the quality of the grapes, soil, and climate. The "taste of the land," the *goût de terroir*, is determined by the interaction of these variables. But the quality and distinctiveness of wine also is dependent upon the taste, personality, and passion of the winemaker.*

The process of producing red wine has been standardized, and is illustrated in Figure 12.8.[†] The grapes are placed into a hopper (1); they then pass through a must pump (2) to be crushed and destemmed in the destemmer (3). At this point (4), the waste must can be hauled away and used as compost or fertilizer. The natural yeast on the grape skins will have grown and begun fermenting the sugar into ethanol and related by-products. Although it is possible to ferment directly in a barrel (5), it is more common to follow the process illustrated, where the juice passes through several stages, including the use of fermentation tanks (7), a presser (8), fining tanks (9), and a filter (10) before being aged in oak barrels.

In this sequence, the wines are modified through the use of legal additives (6) to remove the undesirable flavors and aromas that might arise from wild yeasts and bacteria, and to address the residual bitterness from the stems, skins, and seeds, both at the fermenting stage and before the wines are moved from barrels into bottles. Three compounds (sulfur dioxide, sorbic acid, and benzoic acid) are approved additives in the United States that may be used to prevent the growth of fungi and bac-

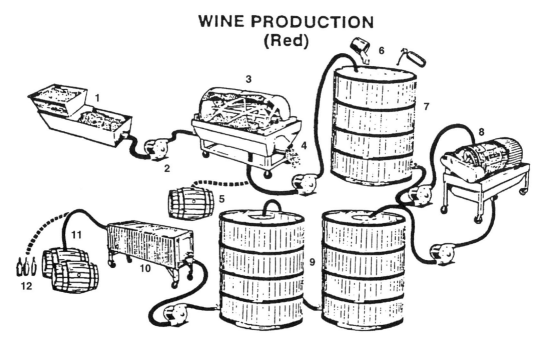

FIGURE 12.8 Production process for red wine. *Source:* Developed by Miami University Applied Technologies.

---

* Sterling has written a provocative story of the life of a vinter in California and the personal struggles and impact that a winemaker has on the quality of wine, see Reference 19.

[†] The process for the production of white wine is similar. Significant departures include a different place in the production process for pressing and aging of white wine (other than chardonnay) in stainless steel barrels rather than in oak barrels.

teria in wine. In particular, sulfur dioxide is one of the most important additives used in winemaking. Sulfur dioxide is either bubbled through the wine as a gas or added as a solid salt or aqueous solution. Sulfur dioxide is produced naturally in small amounts (3 to 12 mg per l) by *Saccharomyces* (the normal wine yeast). However, it is added by the winemaker to control spoilage yeast (such as *Brettanomyces*). Wines that have not been stabilized by sulfur dioxide to control spoilage have distinctive off-odors and disagreeable flavors.[20]

12.5.2 THE ORGANIC CERTIFICATION PROCESS

Regulations that govern organic production processes are tied to the California organic certification processes. Only if a farm wants to label and market its produce as organic is it required to follow organic production regulations. All organic commodity growers must be registered with the state.

The 1990 Farm Bill passed regulations that defined organic production nationally. However, there is no national certification process. California farmers must meet the requirements of the California Organic Foods Act of 1990. The Act has requirements similar to the national 1990 Farm Bill and specifies that food labeled or advertised as *organic* may not contain "any prohibited material residue as a result of spray drift or any other contamination beyond the control of the producer . . . unless the amount of residue does not exceed 5 percent of the federal Environmental Protection Agency tolerance level." Both require organic practices to build soil fertility and overall health, and to minimize the use of external farm inputs, such as fertilizers and pest control. California requires that a farm adhere to the state regulations at least twelve months and maintain detailed and comprehensive records for three years after a crop has been sold before they can become registered as an organic grower.[5]

In California, the need for third party certification is filled by the California Certified Organic Farmers (CCOF) organization. CCOF was organized in 1973 to define uniform organic farming standards and to establish a certification system. It is a non-profit, grass-roots, democratically operated voluntary organization of organic farmers and organic food processors. As the largest statewide organization of organic producers in the nation, its purpose is to promote and support a healthy, ecologically accountable, and permanent agricultural system in California and elsewhere; to develop standards and certification programs for organic farming and processing of organic foods; and to provide verification of adherence to those standards for distributors, retailers, and consumers. They adhere to the national and state standards, as well as having specific standards of their own. As shown in Table 12.1, the certified California wine grape acreage has increased from 630 acres in 1990 to 5450 acres in 1994. Interestingly, of the 72 wineries certified by CCOF, 55 are located in the Mendocino and Lake Counties, Sonoma County, and Napa County wine regions.[21]

With respect to certification standards for wine, the addition of sulfur dioxide to ensure stability and prevent oxidation is a central issue. The National Organics Standards Board (NOSB),

TABLE 12.1
CCOF Certified Wine Grape Acreage, 1990–1994

Year	Total Acreage
1990	630
1991	1808
1992	6059
1993	6165
1994	5450

Source: California Certified Organic Farmers, *Certified Organic Membership Directory and Product Index,* May 1995.

established by the 1990 Organic Food Production Act, allows zero sulfites to be added to wine if it is to be labeled as an "organic wine." The concern is that some persons are sensitive to moderate levels of sulfur dioxide, where bound forms of sulfur dioxide may hydrolyze in the small intestines (where the pH is higher), putting a severe load on the sulfite elimination system. At levels less than 100 mg per l, only asthmatics and steriod-deficient persons may be adversely affected. In fact, at legal levels for wine (350 mg per l), sulfur dioxide has been shown to be harmless for healthy people.[20]

Additionally, the California CCOF standards allow wines to be classified as "organic" only if the wines show 10 parts per million of sulfites or less. Since the Bonterra Chardonnay wines display approximately 65 parts per million and the Bonterra Cabernet Sauvignon wines display approximately 85 parts per million, Fetzer cannot label their wines as "organic," but rather specifies that these wines use only "organically grown grapes." Interestingly, European Community Standards, which limit the addition of sulfur dioxide to 90 parts per million for red wines and 100 parts per million for white wines, would allow Fetzer to label the wines as "organic." John Dickerson, of Hidden Cellars Vineyard, has said, "Since the Europeans allow the addition of sulfites, we'll have another situation where our regulatory requirements are not in compliance with theirs, and that will make everything ever more complicated."[22]

12.5.3 Economic Feasibility of Bonterra Wines

The first economic challenge in producing profitable wines is to grow the grapes cost-effectively. Whether growing grapes conventionally or organically, there are numerous approaches that involve choices such as differential trellis design and various mixes of other agricultural inputs. However, there are high labor and machinery costs needed to properly maintain organic vineyards. Fetzer President Paul Dolan notes, "In the first two to three years of farming organically, our costs increased $250 per acre because of our initial equipment investment."[23]

Table 12.2 shows a comparison of the total operating costs per acre to establish conventional and organic wine grape vineyards in the North Coast region of California, taken from two 1993 reports from the University of California at Davis.[16,24,*] The studies provide reasonably comparative data for 1992, although there are slight differences in the assumptions made in each study. The most significant differences are that the conventional grape study used a 35 acre vineyard with a yield of six tons per acre, while the organic grape study used a 37 acre vineyard with 4.75 ton per acre yield.[†] However, both studies are within reasonable limits, as yields of organically grown and conventionally produced wine grapes are similar in the North Coast, ranging from approximately three to nine tons per acre.[24] The yields are dependent upon factors such as grape variety, vineyard design, and yearly growing conditions. The data on operating costs are presented on a per acre basis to make for more meaningful comparison.

As suggested previously, organic wine production reflects higher labor costs and interest on operating capital. But of even more interest are the differences in the mix of agricultural inputs. The conventional grape production requires $196.73 per acre of expenditure on herbicides, fungicides, fertilizers, and pesticides. Alternatively, organic production uses only $12.51 of fungicide, with no herbicides, fertilizers, or pesticides. Instead, the organic grape production involves $197.22 per acre of soil amendments (mined limestone, sulfate of potash, and compost), custom spread lime and compost, insecticidal soap, and botanical pyrethrin. Other costs for organic grape production are the reg-

---

* These data do not include overhead costs (e.g., property taxes, insurance, or depreciation). While there are differences among these figures for conventional and organic grape production, the principal differences in total costs of production arise in the total operating costs data.

† The most significant cost difference due to variable assumptions about yield is the cost of harvesting, which is assumed to be contracted out with at a constant rate of $120 per ton per acre.

TABLE 12.2
Operating Costs to Produce Wine Grapes

	Conventional Grape Production	Organic Grape Production
Water	$27.29	$41.70
Labor	1017.33	1100.50
Fuel/labor	27.96	28.20
Machinery repairs	45.67	41.54
Interest on operating capital	48.06	69.55
Herbicides	58.92	
Fungicides	56.91	12.51
Fertilizers	47.40	
Pesticides	33.50	
Harvest costs	720.00	570.00
Soil amendments		160.30
Custom spread lime compost		20.00
Insecticidal soap/botanical		16.92
Certification assessments		38.45
Total operating costs per acre	2083.08	2099.62

Sources: See References 16 and 24.

istration, membership, and inspection fees associated with certification by the California Certified Organic Farmers and the State of California.

These data show that total operating costs per acre in 1992 to establish and operate an average North Coast conventional vineyard to be $2083.08, while the total operating costs per acre for an organic vineyard are $2099.62. These costs are obviously quite similar. In fact, Paul Dolan believes, "Natural farming in the long run is less expensive than farming with chemicals,"[6] and "after five years, cost savings were 10 percent."[23] Along with the growth of organic production in the wine industry has come the development of new technology that was not available ten years earlier, including seeders to plant cover crops, bucket loaders to lift compost into a spreader, new weeding equipment, and narrow-gauge tractors. Additional savings have come from recycling, where Fetzer has cut its landfill dump fees by 60 percent in two years.[25] In addition, there is general agreement that organically grown grapes are less susceptible to phylloxera. Moreover, with reduction of the use of chemicals, the work environment for vineyard laborers has improved significantly.

An important dimension of the economic viability of alternative grape agriculture is having sufficient certified acres of organically grown grapes. Fetzer obtains grapes for its wines from several sources. The Fetzer (Brown-Forman) company provides 5 percent from its own vineyards. The Fetzer family (under the name Kohn properties) provides approximately 8 percent of the grapes Fetzer uses. All of this acreage is being grown organic, although, in 1995, only half have met the necessary three-year certification standards. Another 5 percent of the organically-produced grapes come from certified organic growers. But over 80 percent of the grapes used in Fetzer wines come from conventional non-Fetzer growers. The non-Fetzer growers are often quite small operators with limited ability to devote to the bureaucratic paperwork needed to reach organic certification. To facilitate their commitment to organic grape production, Fetzer established Club Bonterra and provides advice on organic grape production and support in filing certification materials. Currently, there are 30 members of Club Bonterra who have made a commitment to producing only organic grapes. At the present, there is more than sufficient acreage of organically-produced chardonnay grapes to meet the market demand for the Bonterra chardonnay.

12.5.4 SALES AND PUBLIC RELATIONS

Fetzer has experienced success in the marketplace with its various wines, those conventionally produced and organically produced. As shown in Figure 12.9, it experienced steady growth in sales from 1990 to 1994. However, within the wine industry, organically produced wines are definitely not regarded as mainstream. There are relatively few major wineries who produce organically-based wines, most of which are regarded as being on the fringe of the wine industry. In fact, Fetzer Vice President and Director of Brand Development, Mark Fedorchak would like "to see another major competitor like Beringer enter the organically-produced wine market to provide greater credibility for what we are doing."

Nonetheless, Fetzer has moved ahead vigorously in defining a market niche and establishing legitimacy for these wines. It notes that the natural foods and organic foods markets have experienced dramatic growth over the past decade and a half (see Figure 12.10). In marketing the Bonterra wines, Fetzer utilizes a variety of public relations techniques. These include programs that emphasize Fetzer's commitment to environmental causes. For instance, in 1992 Fetzer sponsored a "Celebration of the Organic Harvest" at the Valley Oaks Center in Hopland and raised $25,000 to support the Organic Farming Research Foundation.[4]

Of course, of particular significance is the consumers' demand for the product. Wineries emphasize various characteristics of their wines in attempting to establish the value of their wines. Figure 12.11 provides a marketing value matrix that compares the points of emphasis for Fetzer's mass-produced and popular Sundial Chardonnay and the Bonterra Mendocino County Chardonnay (produced with organically-grown grapes). To differentiate its wines made exclusively from certified organically grown grapes, Fetzer assigned a new proprietary name, *Bonterra*, which is derived from Latin and French roots, meaning "good earth." Moreover, Fetzer adopted a distinctive Burgundy-style bottle made from recycled glass to differentiate its Bonterra wines from other Fetzer wines,

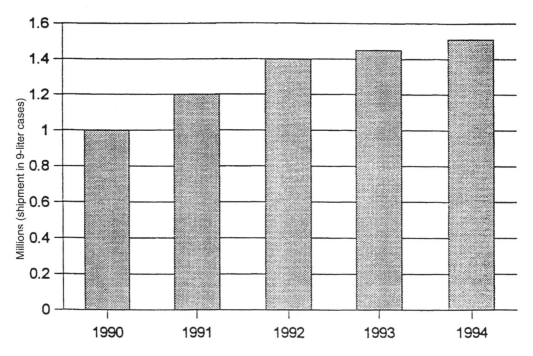

FIGURE 12.9 Fetzer Vineyard sales.

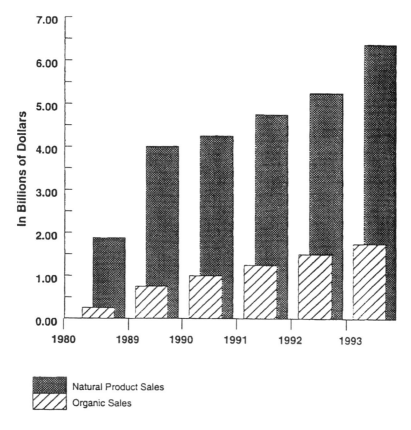

FIGURE 12.10 Market for natural foods. See Reference 26.

eliminated the foil caps, and created new distinctive labels made from "treeless" paper called Kenaf (from bamboo-like grass). Fetzer believes that the wine style of the Bonterra wines is distinctive. In contrast to the light taste of the Sundial Chardonnay, the Bonterra Chardonnay is described as having a "focus on the pear and tropical notes of the organically-grown fruit, making the wine quite versatile with different foods and cuisine." With respect to wine features, where the Sundial Chardonnay is marketed for crisp, clean character coming from stainless steel cooperage, the Bonterra Chardonnay is described as having "a bright, appealing character and is less heavy and oaky than many barrel-aged Chardonnays."[4,] * Obviously, the emphasis on the Bonterra wines being produced from organically-grown grapes is a point of marketing differentiation as well.

As Fetzer further tries to establish product differentiation and justification for a higher price for the Bonterra Chardonnay, it emphasizes other characteristics in the value matrix that it would ignore for the Sundial Chardonnay. The first vintages of the Bonterra wines have been priced in the $9 to 12 range, which is generally regarded as appropriate for mid-level quality wines. However, Fetzer must contend with consumer perceptions that center largely around its well-known Sundial Chardonnay that typically retails for under $8. A useful point of emphasis is the ratings of wine in industry publications. The Bonterra wines have received broad acceptance as high quality wines by the *Wine Spectator*, a leading reviewer of wines, which rated the first Bonterra vintages in the mid to upper 80s (out of 100). Finally, linking a wine to a particular region or district (a.k.a. appellation) increases the value of the wine because consumers know the wine comes from a select geographic area of consistent quality. Thus, the Bonterra wines carry the Mendocino County appelation on the label.

---

*Bonterra wines are aged in American oak barrels, exclusively hand-crafted at Fetzer's own cooperage.

Characteristic	Sundial Chardonnay	Bonterra Mendocino County Chardonnay
Appellation (Region or District)		X
Ratings		X
Wine features (eg., barrel aging, organically-grown grapes)	X	X
Taste	X	X
Packaging	X	X

FIGURE 12.11 Chardonnay marketing value matrix.

12.6 VISION FOR THE FUTURE

As we were finishing our time at Fetzer, Pat Voss, Senior Vice President for Production, reminded us of Fetzer's vision and commitment to the future. She noted that their goal was, by the year 2000, to produce all of their wines from organically grown grapes. This is a tall order, but it was clear that Fetzer had adopted this goal as a driving objective for the company. We left thinking about the various challenges Fetzer would face in achieving its vision, but confident in its commitment to it.

12.7 REFERENCES

1. National Public Radio, California's vineyards discovering organic farming, *Morning Edition,* transcript #1206-13, November 1, 1993.
2. Halliday, J., *Wine Atlas of California,* Viking Penguin, New York, 1993.
3. Fetzer Vineyards, Live right. Eat right. Pick the right grapes™, brochure, 1995.
4. Turner, J., Organics move to mainstream, *The Wine News,* April/May 1993.
5. California Certified Organic Farmers (CCOF), *Certification Handbook,* Santa Cruz, 1995.
6. Gugino, S., Vineyard going big time with organic wines, *Star-Ledger, Newark (NJ),* January 17, 1996.
7. Fetzer Vineyards, Fetzer Vineyards organic farming fact sheet, brochure, 1995.
8. Fetzer Vineyards, Fetzer Vineyards goes organic, brochure, 1995.
9. Winkler, A. J., Cook, J. A., Kliewer, W. M., and Lider, L. A., *General Viticulture,* 2nd ed., University of California Press, Berkeley, 1974.
10. Pongrácz, D. D., *Practical Viticulture,* David Philip, Capetown, 1978.
11. Lewin, R., California's lousy vintage, *New Sci.,* 17, 27, 1993.
12. Freckman, D. W., Life in the soil, soil biodiversity: it's importance to ecosystem process, Report of a workshop held at the National History Museum, London, England, Natural Resource Ecology Laboratory, Fort Collins, CO, 1994.
13. Reganold, J. P., Papendick, R. I., and Parr, J. F., Sustainable agriculture, *Sci. Am.,* 262(6), 112, 1990.
14. Division of Agricultural Sciences, University of California (DASUC), *Grape Pest Management,* The Regents of the University of California, 1982.
15. Ingels, C., Sustainable agriculture and grape production, *Am. J. Enol. Vitic.,* 43(3), 296, 1992.
16. Smith, R., Klonsky, K., Livingston, P., and Tourte, L., *Sample Costs to Establish a Vineyard and Produce Wine Grapes in Sonoma County—1992,* University of California Cooperative Extension, 1993.
17. Fetzer Vineyards, Fetzer Vineyards organic viticulture program, brochure, 1994.

18. Piper, T., Soil fertility report—two Fetzer Vineyard locations, 1994.
19. Sterling, J., *Cultivated Life*, Villard Books, New York, 1993.
20. Ough, C. S., Chemicals used in making wine, *C&EN*, January 5, 19, 1987.
21. California Certified Organic Farmers (CCOF), *Certified Organic Membership Directory and Product Index*, Santa Cruz, 1995.
22. *Wine Business Insider*, Sulfite ban in organic wines upheld by board: one last chance for vintners to object this fall, 5(5), Feb. 7, 1995.
23. Just "say no" to vineyard chemicals, *Wines Vines*, September, 1995.
24. Klonsky, K., Tourte, L., and Ingels, C., *Sample Costs to Produce Organic Wine Grapes in the North Coast with Resident Vegetation*, University of California Cooperative Extension, 1993.
25. Hamlin, S., Organic wine goes mainstream, *House Beautiful*, July, 1993.
26. Fetzer Vineyards, Bonterra Strategic Positioning, 1994.
27. Swift, M. J., Heal, O. W., and Anderson, J. M., *Decomposition in Terrestrial Ecosystems*, Blackwell Scientific, Oxford, 1979.
28. Smith, R. L., *Ecology and Field Biology*, 2nd ed., 1974, 540, as cited in Nebel, B. J., *Environmental Science: The Way the World Works*, 3rd. ed., Prentice-Hall, Englewood Cliffs, NJ, 1990, 166.
29. Nebel, B. J., *Environmental Science: The Way the World Works*, 3rd ed., Prentice-Hall, Englewood Cliffs, NJ, 1990, p. 169.

13 General Electric Aircraft Engines: Pollution Control Investment Analysis Case Study Number 5

Raymond F. Gorman and H. Gregory Hume

CONTENTS

13.1 Introduction 271
13.2 The General Electric Company 272
13.3 Brief Summary of Other Major Divisions 273
      13.3.1 Appliances 273
      13.3.2 Broadcasting 273
      13.3.3 Industrial 273
      13.3.4 Materials 273
      13.3.5 Power Systems 274
      13.3.6 Technical Products and Services 274
      13.3.7 GE Capital Services—Financing 274
      13.3.8 GE Capital Services—Specialty Insurance 274
      13.3.9 GE Aircraft Engines (GEAE) 274
13.4 The Evendale Plant 275
13.5 Chemical Use at GE Evendale 275
13.6 Regulatory Context 276
13.7 Environmental Cost Containment 276
      13.7.1 The Case of Chromic Acid 277
13.8 New Directions? 277

13.1 INTRODUCTION

In the early 1970s, a fire ignited in a paint booth used to apply an aluminum coating to jet engine parts at the General Electric Aircraft Engine manufacturing plant in Evendale, OH. Investigation of the fire led to the conclusion that a build-up of powdered aluminum metal created conditions favorable to spontaneous ignition. To prevent a recurrence of the incident, a procedure was implemented calling for daily cleaning of the booth using water as a cleaning solvent. Unfortunately, during a subsequent routine cleaning, an alkaline soap was apparently accidentally introduced into the cleaning water. The soap reacted with the aluminum creating a build-up of hydrogen gas. The hydrogen ignited, causing a violent explosion that killed two workers.

    The aluminum coating was essential to prevent oxidation of ferrous alloy surfaces exposed to the corrosive conditions found inside jet engines—no other coating process met the engineering requirements. Unable to find a substitute for the basic process, the company modified its procedures to require the daily cleaning of paint booth surfaces with an ultra-dilute chromic acid solution to assure employee safety and avoid explosive conditions. This solution was not without its own problems, however, since chromic acid is itself highly toxic. The hexavalent or +6 form of chromium found in the acid, when absorbed into the body at high levels, can severely irritate the gastrointestinal tract, leading to circulatory shock and renal damage. In compliance with regulatory requirements, wastes from cleanup of the paint booth were pumped into a tank truck for disposal as hazardous waste.

In the two decades following the explosion, the waste-related cost of this process continued to escalate. From low disposal costs for hazardous waste, the cost had jumped to more than $200 per barrel by 1990 as hazardous waste facility operators began to comply with more stringent regulations designed to assure the integrity of hazardous waste sites by isolating toxic materials from the environment. Labor costs rose significantly over the same period and the chemical cost index increased as well. This scenario was often the case for businesses engaged in state-of-the-art machining and metallurgical processes. The operating management of GE Aircraft Engines (GEAE) was concerned with meeting regulatory limits while addressing the escalating costs of all forms of regulatory compliance. In addition, proactive steps to minimize the risk of environmental damage took on added importance considering the potential for significant costs from the cleanup of Superfund sites and the threat of civil or criminal prosecution. The toxic chemical reporting requirements of the 1987 Emergency Planning and Community Right-to-Know Act (EPCRA) increased public scrutiny on corporate operations.

In the late 1980s, regulatory compliance responsibility within GEAE was distributed throughout the business. Each of the 17 aircraft engine locations was solely responsible for its own regulatory compliance matters since many regulatory standards varied on a state by state basis. The cost of preparing permit applications was increasing each year. Something needed to be done, but what?

13.2   THE GENERAL ELECTRIC COMPANY

General Electric had grown to be one of the largest corporations in the world. In 1993, GE had assets of $251.5 billion; consolidated revenue was $60.6 billion; net earnings were $4.3 billion; and employment numbered 220,000. A highly diverse company, GE's divisions included Aircraft Engines, Appliances, Capital Services, Industrial and Power Systems, Lighting, Medical Systems, NBC, Plastics, Electrical Distribution and Control, Information Services, Motors, and Transportation Systems.

Because of its world-wide customer base, many GE businesses were required to comply with a wide array of international environmental laws in addition to domestic legislation. These laws influenced not only how GE products perform in term of noise and emissions, but also the specifications for materials used in service maintenance and overhaul, as well as the processes used to manufacture and assemble these products. Although there was optimism that regulatory requirements would eventually be clarified and rationalized, the company was highly concerned about the impending transition period. CEO Jack Welch commented:

> "In the end, there's going to be a global standard for the environment, and anyone who cuts corners today will wind up with enormous liabilities down the road."

Dr. Welch spoke from experience. In 1993, GE spent $80 million on remediation actions to clean up prior environmental releases as required by state and federal law, and this level of spending was expected to continue for the next few years. Although there remained uncertainties about the regulatory requirements and remediation technology related to individual disposal sites, GE believed that its existing level of liability recognition and programmed expenditures were sufficient to meet environmental liabilities through the end of the century and more than likely through the year 2015.

Therefore, the company considered itself to be in compliance with current laws regarding recognition of environmental liabilities; and did not believe that these expenditures would have a material effect on its earnings, liquidity, or competitive position.

GE's New York transformer plants in Hudson Falls and Fort Edward offered a striking example of the penalty for the release of toxic effluent. Beginning in the mid-1940s, the plants manufactured transformers containing polychlorinated biphenyls or PCBs, valued for their insulating and heat resistance properties. PCBs are practically indestructible to boot, useful in a product but not in the environment. Effluent containing PCBs was discharged into the Hudson River under permits from the state of New York according to the company.

In 1977, as scientific studies began to reveal the toxicity of PCBs and their ability to bioaccumulate through the food chain, the company ceased the practice of discharging PCBs in wastewater. In 1978, Congress banned their manufacture. In 1993, GE agreed to pay up to $7 million to fishermen who lost income from a state ban on striped bass fishing in the Hudson River since 1976 and in marine waters off Long Island between 1986 and 1990 due to PCB contamination. Cleanup of contaminated sediments is also proceeding and it is expected that total costs to GE will exceed the profit made in the transformer business over the course of fifty years.

In addition to pollution remediation, GE had capital expenditures of $140 million for environmental projects in 1993; a $30 million increase over 1992. These capital expenditures included pollution control devices such as wastewater treatment, groundwater monitoring devices, air strippers, and incinerators. The company expected to spend similar amounts over the next few years.

13.3 BRIEF SUMMARY OF OTHER MAJOR DIVISIONS

13.3.1 APPLIANCES

GE Appliances produces and sells major appliances such as refrigerators, freezers, electric and gas ranges, dishwashers, clothes washers and dryers, microwave ovens, and air conditioners. Revenue from operations in 1993 amounted to $5.6 billion, compared to $5.3 billion in 1992. Most of its sales are in North American retail outlets, and to building contractors and distributors. It also sells in global markets under various GE and private label brands.

13.3.2 BROADCASTING

GE's primary broadcasting operation was represented by NBC. With over 200 affiliated stations, the broadcasting division's 1993 revenues were $3.1 billion, down from $3.4 billion in 1992. NBC produces television programs, operates six VHS television stations and is involved with investment and programming activities in cable television.

13.3.3 INDUSTRIAL

The industrial segment of GE produces and sells lighting products, electrical distribution and control equipment, transportation systems products, and electric motors. With 1993 revenues of $7.4 billion (up from $6.9 billion in 1992), its markets are extremely varied. Products are sold to commercial and industrial end users, equipment manufacturers, retail outlets, and transit authorities.

13.3.4 MATERIALS

This division makes and distributes high performance plastics used in automobiles and in housings for computers and other business equipment, ABS resins, silicones, superabrasives, and laminates. Its 1993, revenues were $5.0 billion, up from $4.9 billion in 1992.

13.3.5 POWER SYSTEMS

The products manufactured by this division are mainly for the generation and distribution of electricity. Power systems also include power delivery and control products such as transformers, meter relays, capacitors, and arresters for the utility; nuclear reactors; and fuel and support services for GE's installed boiling water reactors. From 1992 to 1993, revenues grew from $6.4 billion to $6.7 billion.

13.3.6 TECHNICAL PRODUCTS AND SERVICES

This division produces medical systems such as magnetic resonance and CT scanners, x-ray, nuclear imaging, ultrasound and other diagnostic equipment. This segment of operations also includes a full range of computer based information and data interchange services for use in-house as well as for external commercial and industrial customers. Revenues in 1993 were $4.2 billion, down from $4.7 billion in 1991.

13.3.7 GE CAPITAL SERVICES—FINANCING

Among the operations of this division are private label and bank credit card loans, loans and financial leases for major capital assets including portfolios of commercial and transportation equipment, asset management services, and operating leases for middle market customers. Revenues in 1993 were $12.4 billion, compared to $10.5 billion in 1992. The assets of this division were $106.9 billion. This is over twice the assets of all of the previously mentioned divisions and represents about 40 percent of all GE assets. Only a modest portion of the items financed by GE capital are manufactured by GE.

13.3.8 GE CAPITAL SERVICES—SPECIALTY INSURANCE

This division writes multiple-line property and casualty reinsurance, and certain directly written specialty insurance in addition to financial guaranty insurance, private mortgage insurance, and creditor insurance covering international customer loan repayments. Revenues increased from $3.9 billion in 1992 to $4.9 billion in 1993.

13.3.9 GE AIRCRAFT ENGINES (GEAE)

This division produces jet engines and replacement parts and offers repair services for commercial aircraft ranging from wide body planes to executive and commuter craft. In addition, it manufactures and sells a wide variety of engines for military planes and helicopters. Its marine and industrial product line includes jet powered marine propulsion systems and power generation equipment. More than half of its sales are to customers outside of the United States. Due to cutbacks in defense spending, its 1993 revenues were $6.6 billion, down from $7.4 billion in 1992.

GEAE is a success by a key measure of CEO Welch—it dominates its market for high technology jet engines. In the large engine market, only Pratt Whitney in the United States and Rolls Royce in Great Britain are competitors. GE has a 50 percent market share. A joint venture with the French company, SNECMA, has helped GEAE to win a large share of the European Airbus business.

GEAE's roots lay in the defense business and this structured its business perspective through the 1980s—even though the business was successful in the competition for commercial orders. Department of Defense procurement procedures have traditionally had a strong tendency toward hierarchical top down requirements. Military planners conceived the aircraft and its mission and devised performance and test specifications accordingly. Engine and airframe manufacturers were not encouraged to question these specifications—just meet or exceed them.

Moving into the 1990s, GEAE management had a number of concerns. Defense Department spending was down and expected to go lower with a depressing effect on sales. Deregulation of the airlines meant increasing competition and pressure to reduce prices. Employee productivity at many of its satellite locations was higher than at the Evendale plant for a variety of reasons. The GEAE work force—swollen by the defense build-up of the 1980s and an entrenched bureaucracy—was overdue for a major layoff.

13.4 THE EVENDALE PLANT

The GE Evendale plant is located about ten miles north of Cincinnati along Interstate 75 (I-75) and is the headquarters for the Aircraft Engines Group. The Evendale plant is the largest of the Group's 17 plants, test centers, and field service shops that design, develop, and manufacture a wide range of military and commercial aircraft engines as well as marine and industrial gas turbines and gas generators.

The Evendale plant is one of the largest industrial employers in the Cincinnati area with employment ranging from 8,000 to 20,000. The plant property extends for more than one mile along I-75 encompassing about 400 acres. The plant site has 10 major buildings, 58 smaller ones, 72 research and testing laboratories, and more than 60 component and engine test cells. The total space under roof is over 6.5 million ft^2.

The primary purpose of the plant has not changed since the 1950s, (i.e., the design, assembly, and testing of jet engines). The majority of engine parts are manufactured outside the plant and delivered ready for assembly. The balance arrive as rough castings or forgings that require extensive processing.

13.5 CHEMICAL USE AT GE EVENDALE

It is during this extensive processing—primarily the machining, coating, and cleaning operations—that most of the chemicals are used. Using conventional technology, parts are machined to the desired contour by turning, drilling, milling, or broaching, coolant is constantly sprayed on the cutting area to act as a lubricant and carry away the heat and chips generated by the process. For precision complex shaping of exotic alloys, four types of chipless machining processes are also utilized at the Evendale plant: chemical machining, electrochemical machining, electrodischarge machining, and laser machining. After machining, each part is chemically cleaned to remove all oils, dusts, oxides, and other foreign materials that might adhere to the surface. In the 1980s, the primary methods of chemical cleaning processes were:

1. Alkaline cleaning—alkaline agents such as caustic soda are mixed with some type of soap. The mixture emulsifies oils and greases and removes them from the parts.
2. Vapor degreasing—non-flammable solvents, primarily 1,1,1 trichloroethane (also known as TCA or methylchloroform), are heated to boiling in a tank to produce vapors. Parts are then suspended in the vapors. As the vapors condense on the parts, oils and greases are removed.
3. Pickling—parts are dipped in a dilute acid solution, generally 10 percent sulfuric acid or muriatic acid heated to about 165°F, to remove oxides. After pickling, parts are thoroughly rinsed and dipped in a slightly alkaline bath to prevent rusting.
4. Ultrasonic cleaning—parts are suspended in a liquid bath that contains an ultrasonic transducer operating at a frequency that produces cavitation in the liquid. As the partial vacuum in the liquid collapses, pitting on the surface of the part is produced and gross oils and greases are removed.

13.6 REGULATORY CONTEXT

In 1987, the Emergency Planning and Community Right-To-Know Act (EPCRA) went into effect and immediately exerted a significant influence on the manufacture and use of industrial chemicals. Beginning that year, local authorities were required to establish emergency procedures for notification and evacuation of the community in the event of a hazardous chemical release. The act required companies using 366 extremely hazardous substances in quantities above specified threshold amounts to report such use to local authorities, maintain material data safety sheets (MSDS), and immediately report any spills. Annual toxic chemical release quantities were also to be reported and made public. For many companies, the effect of the act was to focus management attention on the toxicity of its process chemicals for the first time.

By the late 1980s, federal environmental statutes began to focus more on pollution prevention and source reduction actions as a more effective means of avoiding pollution than "end of pipe" treatment. In 1990 for example, the Clean Air Act was amended to initiate a national research program to achieve the prevention and control of air pollution as a means of promoting the health, welfare, and productive capacity of the population.

Also in 1990, Congress passed the Pollution Prevention Act, noting that:

> "There are significant opportunities for industry to reduce or prevent pollution at the source through cost-effective changes in production, operation, and raw materials use. Such change offers industry substantial savings in reduced raw material, pollution control, and liability costs as well as help protect the environment and reduce risks to worker health and safety."

The initiators of the act further observed that opportunities for source reduction are often not realized because existing regulation is focused on treatment and disposal rather than the fundamentally more desirable source reduction.

The Pollution Prevention Act directed the Environmental Protection Agency to establish a source reduction program to collect and disseminate information on pollution prevention as well as provide financial assistance to the states. In February 1991, EPA announced the 33/50 goal for a voluntary reduction in the release of targeted chemicals from 1988 levels—by 33 percent by the end of 1992 and by 50 percent at the end of 1995.

13.7 ENVIRONMENTAL COST CONTAINMENT

In 1990, senior vice president Brian Rowe, then head of GEAE, acted to improve control of regulatory risk and mounting environmental costs. A new staff level organization responsible for overall GEAE environmental health and safety was created and Herb Coulter was recruited from GE's plastics business to manage it. Coulter was experienced in the chemical business and was involved in the founding of the chemical industry's Responsible Care® initiative aimed specifically at pollution prevention activities. Coulter, in turn, hired Mark Singleton, a Ph.D. process engineer, to develop a program to promote source reduction activities throughout GEAE.

Singleton's first task was to collect all available data describing the various categories of wastes generated by the 17 GEAE locations by volume and cost to the business. Waste data were reasonably easy to obtain because of the reporting requirements contained in EPA regulations. However, assigning total costs to waste streams proved difficult because the accounting systems employed in the business tended to obscure certain related cost elements.

The next step was to conduct a Pareto analysis of waste streams by volume and cost at each GEAE site in order to establish priorities for action. Pareto analysis was based on the observation that, for most distributions, a small number of high value items constituted a majority of the total value and thus comprised those items with the highest potential for change and greatest need for management. At Evendale, chromic acid from spray booth cleaning was at the top of the hazardous waste list, contributing 55 percent of the total hazardous waste generated at the Evendale site and was selected as one of targets for improvement.

13.7.1 THE CASE OF CHROMIC ACID

Singleton met with shop employees and engineers responsible for the aluminum coating area and invited them to participate on a continuous improvement team. He had expected the hazardous waste stream to be principally associated with paint overspray and was surprised that most of the waste was generated in the daily cleaning of the paint booth.

The improvement process began by questioning some very basic assumptions about the coating and the coating booth cleanup procedures. The first question asked was: "Do engine parts really have to be coated this way?" The answer was "yes" because no alternative coating process was viable. Next, "Is it necessary to clean the paint booth?" The risk of spontaneous ignition caused by powdered aluminum accumulation made the answer obvious.

The next question was: "Can chromic acid be avoided in the cleaning process?" Here the answer was "no" because chromic acid appeared to be the only practical method to reduce the reactivity of the powdered aluminum while avoiding the risk of generating hydrogen gas. Even if pure water were used to clean the booth, hexavalent chromium leached from the overspray would contaminate the solution, creating a hazardous waste. Since the cleaning process could not be modified, the next question asked was: "How often should the booth be cleaned?" This turned out to be the critical question.

The minimum overspray thickness necessary to create the potential for a spontaneous ignition condition was known. Test metal coupons were placed in the spray booth to determine how fast the overspray actually accumulated. Analysis of the coupons indicated that an explosive condition would not be reached for 150 days. In theory, the frequency of cleanup could be decreased from once per day to once every five months before reaching a critical buildup thickness. These data were presented to insurers and on-site fire safety professionals who—after requiring additional confirmatory studies—gave GE Evendale permission to reduce cleaning frequency to once every two weeks coupled with continuous monitoring of buildup thickness.

This single action reduced the volume of chromic acid waste by approximately 75 percent and in doing so reduced the total amount of hazardous waste from the Evendale plant by 40 percent. The cost of the evaluation process was about $10,000 for test coupons and analysis.

These costs were dwarfed by annual savings. By simply changing the frequency of cleaning, the GE Evendale plant saved:

- $100,000 per year in waste disposal costs.
- $30,000 per year in hourly labor costs.
- $10,000 per year in salaried labor.

In addition, GE saved a modest amount in the quantity of chromic acid and water purchased each year.

This story was repeated many times for other waste streams and processes. Throughout the 17 GEAE locations, as processes creating waste and pollution were analyzed, better and less polluting ways to produce engine parts were identified. The use of 1,1,1 trichloroethane and CFC113 as degreasing solvents was eliminated entirely, yielding a reduction in ozone depleting solvent releases to the air of 2.2 million pounds per year.

13.8 NEW DIRECTIONS?

At the end of 1994, as Singleton reflected on the success of GEAE focus teams in reducing waste and pollution, he was surprised at how effectively GEAE had been able to implement process changes. An important measure of that success was the $3 in annual savings realized for each $1 of applied program cost, and this was an important factor in a high profile for the program. In the process, potentially higher indirect or "hidden" costs also were eliminated.

Singleton was somewhat concerned, however, about whether this high level of focus on pollution prevention would be maintained. After all, hadn't most of the "low hanging fruit" been picked? Perhaps the business had eliminated just about all of the effluent it could, given its current technology. Although many cost effective improvements were still to be identified and changes made, the major sources of air, water, and hazardous waste pollution from manufacturing operations had already been addressed.

Several key issues had recently surfaced, however, and promised to make the next year a busy one:

1. The U.S. Military, facing increasing public scrutiny with respect to its environmental record, had developed pollution prevention policies that translated into new material-specific contract requirements for its suppliers.
2. Important commercial airline customers based in northern Europe (where regulatory requirements can be considerably more demanding than the United States) were demanding that GEAE reduce and eliminate certain materials from its product and processes specified in engine overhaul and maintenance.
3. Domestic airlines, also affected by EPCRA reporting requirements, were anxious to know if GEAE was moving away from hazardous materials in its overhaul protocols. GEAE also was aware that improper disposal of materials by their customers could conceivably create a problem for GEAE.
4. Although manufacturing was reducing the discharge of hazardous materials, product design engineering was continuing, by technical necessity, to specify much the same materials and processes.
5. The cost of EPCRA reporting for a given location was quite high but did not vary much if the reportable amount of a toxic substance was slightly higher than the threshold quantity or many multiples of it.

14 The David J. Joseph Company: Closing the Loop in the Automobile Industry Case Study Number 6

Jan Willem Bol and Allison R. Leavitt

CONTENTS

14.1	Introduction	279
14.2	The David J. Joseph Company	279
14.3	The Scrap Business	281
14.4	Automotive Industry and Vehicle Scrappage	283
14.5	Designing for the Environment	284
14.6	Environmental Issues	287
14.7	Conclusion	289

14.1 INTRODUCTION

In the early 1990s, the David J. Joseph Company (DJJ) had to consider a wide range of new strategic directions necessitated by significant market changes in the automobile, steel, and scrap industries. A common thread that ran through many of the changes facing DJJ was driven by environmental issues, such as limited availability of landfill space and the need to design "green cars." Although it had made a commitment to deal with environmental issues, DJJ needed to determine how these were going to affect its future, and the way it was doing business. Regardless of which direction it decided to go, DJJ was aware of the far-reaching implications each of the strategic options could have for both the direction the company took and the way it was managed.

14.2 THE DAVID J. JOSEPH COMPANY

In 1863, Joseph Joseph, a native of Germany, settled in Cincinnati, OH, and began a hide and wood trading operation. Subsequently, his brother Samuel joined the company and in 1885, with the advent of growing industrialization and soaring demand for scrap metal, the Josephs turned to scrap iron trading, abandoning the hide and wool business.

The scrap iron company grew quickly and other businesses were added. After Joseph Joseph's death, his oldest son David, then manager of the scrap trading operation, ran seven plants and branch offices, covering almost every major steel center in the United States. With the beginning of World War I, demand for scrap iron increased significantly and by the end of the war, the Joseph Joseph and Brothers Company was one of the largest scrap metal firms in the nation, a position it again held in 1988.

In 1921, David Joseph, Sr. renamed the company as The David J. Joseph Company. His son became president in 1945. Consolidating its strength through the 1960s and 1970s, DJJ, in 1975, began the sale of its operations to SHV Holdings NV, a privately-held company located in Utrecht, the Netherlands. SHV had more than $2 billion in assets and operated world-wide businesses in energy, the trading of raw materials, and retail distribution of a wide variety of consumer goods. The sale was completed in 1980. By 1988, DJJ, still headquartered in Cincinnati, OH, employed approximately 750 people at 26 locations, including 12 ferrous trading offices across the United States.

By the mid- to late-1990s, nearly 40 percent of all the steel produced in the United States was made from purchased scrap. As a ferrous scrap broker, DJJ provided an essential link between the various sources of scrap and the mills and foundries which used it in the steel making and casting processes at this time. There were approximately 3,000 scrap dealers in the United States, operating mostly on a regional basis. Overall, the industry was fragmented and highly competitive, although this varied by region. Companies such as Cozzi, Luntz, and National Material Trading operated in Chicago and Cleveland. On a national level, competitors such as Luria Bros. and Tube City ranged from 50 percent to 90 percent of DJJ's size. Most of DJJ's competitors were brokers and processors, and many, especially those operating on a regional basis, were, much like DJJ in its early days, owned and managed by families.

DJJ employed about 40 traders with a variety of commercial and academic backgrounds. After an initial training period they were assigned to a district office and immediately became responsible for their own clients. They often spent 30 to 40 percent of their time in the field, realizing that much of their business depended on information obtained from customers. Most trades done by DJJ ranged from 50 to 5,000 tons, although some transactions exceeded 50,000 tons and dollar values of $5 million or more. Shippers of scrap often selected DJJ because of its financial strength. The practice in the industry was for buyers to advance a percentage of the purchase price at the time of shipment, with the balance to be paid later.

Besides the Ferrous Department, which represented 87 percent of DJJ's business, DJJ had two other operating departments. The Non-Ferrous Department was involved in the sale of non-ferrous scrap by-products generated at DJJ's processing plants, as well as in the domestic and international trading of non-ferrous scrap metals. The department represented approximately 6 percent of DJJ's $500 million in revenues.

The Railroad Equipment Leasing and Marketing Department represented 7 percent of DJJ's revenues. It was involved in all facets of used railroad equipment, including the purchase of equipment for scrapping at DJJ's facilities, the marketing of reusable equipment, and the owning and leasing of railroad equipment. This department was growing and adding significantly to the company's business, as well as enhancing some of the company's other activities through developing new markets and new transportation approaches.

An important aspect of the brokerage business has been transportation. Transportation often is a significant part of the total cost of any scrap transaction. DJJ's Transportation department offered the broker information on applicable rates on rail, truck, and barge shipments and also negotiated and monitored administration of contract freight rates. The Department also coordinated intermodal movements with a range of transportation carriers.

To complement its brokerage business, DJJ operated 12 specialized scrap processing and mill service facilities. Included in six of these plants were automobile shredding machines, each capable of processing approximately 600 automobile bodies per day. Some of these plants, which were often located near DJJ's major customers, were involved as well in scrapping retired railroad freight cars and track.

In 1985, DJJ established an International Department. This was DJJ's return to international trading of ferrous scrap after an absence of over 40 years. DJJ was now involved both in exporting U.S. scrap to overseas markets, and in importing to the United States high-quality scrap items and other ferrous units, such as pig iron.

DJJ ventured into a number of areas of innovation. For instance, it developed the first medium-sized shredders for the purpose of upgrading the quality of automobile scrap. And, in response to the increasingly stringent environmental regulations, incineration equipment was built (and patented) to facilitate the reduction of wood-lined boxcars to metal components.

Although financial information is incorporated into SHV's annual data, DJJ's 1988 net worth was estimated at $100 million. With operating margins between 3 percent and 4 percent, DJJ was typical of other national scrap metal brokers. DJJ dealt with several thousands of suppliers, the largest of whom represented less than 2 to 3 percent of DJJ's total volume. However, because of regional differences, a single supplier could provide as much as 10 percent of the volume for an individual yard. By contrast, the top 25 percent of its 207 customers represented approximately 90 percent of its total sales volume.

14.3 THE SCRAP BUSINESS

"Everything made from metal eventually becomes scrap—frequently many times over," explained Joe Hirschhorn, Senior Vice President, and member of DJJ's Executive Management Committee. "Old cars and trucks, railroad cars and tracks, steel beams from dismantled buildings and bridges are all raw material for the scrap processor," he continued. The process of ferrous scrap recycling started with the discarding of an item made from or containing iron or steel. The next activity involves collecting and processing. This includes sorting and reloading the material for shipment or cutting, shearing, baling, or shredding. After that, the scrap is melted and refined in the manufacture of new iron or steel for new products for consumer use. Using electromagnets, shears, shredders, and balers, scrap processors are in a highly-mechanized, capital-intensive manufacturing business. Scrap also is generated by many industrial companies as a by-product of manufacturing.

Ferrous scrap could be divided into three major categories: home, prompt industrial, and obsolete. Home scrap and prompt industrial scrap are involuntary by-products of manufacturing processes. Home scrap originates in a steel mill and consists of such items as ingot and slab ends and trimmings from the production process. Prompt industrial scrap includes clippings from an auto body manufacturer and turnings from lathes at a machine tool shop. Prompt industrial scrap also results from normal machining, stamping, and fabricating operations in the production of products made from steel. Obsolete scrap, the third major category, arises when a product made of iron or steel has served its useful life and is discarded, such as railroad cars or automobiles (see Figure 14.1).

As indicated earlier, and in spite of the general decline of the U.S. steel industry, there has been a strong trend in steel making to use more purchased scrap per ton of steel made. Purchased scrap receipts averaged slightly more than 27 tons per 100 tons of raw steel produced during the years 1954 to 1970. By 1984, this ratio had risen to 45 tons per 100.

This trend has closely tracked the growth of electric furnace steel making. In 1954, only about 5 percent of U.S. steel production was in electric furnaces. In 1983, that proportion had grown to 31.5 percent and had been running just over 37 percent during 1986. Electric furnace steel making became increasingly attractive for many reasons. Compared with an integrated, iron-ore based steel plant, the modern electric furnace shop could be constructed and started up faster, required only one-sixth the capital investment per ton of capacity, used only about one-fourth the energy per ton produced, offered greater operating flexibility, and required fewer man-hours per ton of output. In 1986, only 4 percent of total raw steel production was produced by plants using open hearths, down from 91 percent in 1954. Continuous casting has risen from 7 percent of steel made in 1973 to over 60 percent in 1987. Other technologies that had been developed included the use of outside-the-furnace ladle metallurgy stations to bring the melt to precise temperature, oxy-fuel burners, and consteel scrap preheating techniques. In part, these technological changes were a reaction to increased legislation and public policy with respect to the environment.

In 1986, steel mills used 72 percent of the 38 million tons of ferrous scrap that were purchased by U.S. customers that year. Iron and steel foundries accounted for the rest. Export of ferrous scrap

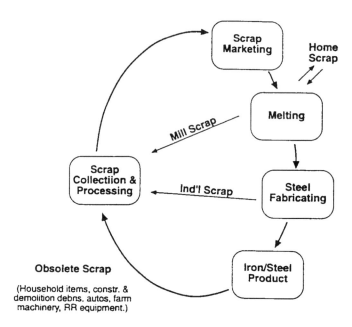

FIGURE 14.1 Pictorial showing relationships between iron and steel products and the scrap collection, marketing, and fabrication processes. (*Source:* David J. Joseph.)

from the United States reached 11.9 million net tons in 1986, representing nearly 24 percent of the total market for U.S. scrap.

One direct result of the conversion to continuous casting was a sharp reduction in steel mill home scrap generation and, therefore, in the proportion of scrap charges made up of revert home scrap. Tied to the reduced home scrap availability as well as the expanding role of electric furnaces, the amount of scrap purchased per 100 tons of steel made in the United States followed a trendline showing over 3 percent annual increases since 1970. Given this rate of growth, it was predicted that during an 80 million ton raw steel production year, a 600,000 to 700,000 ton increase in domestic purchased scrap would result from each additional one point share for electric furnaces.

"The scrap business is a demand-driven business. Scrap is bought, not sold, and is only worth what people are willing to pay for it," explained Hirschhorn. "Melting in a furnace is the only use for ferrous scrap and because of its heterogeneity it is not fungible."

Mills and foundries bought scrap when they needed it. If they did not need scrap to feed their furnaces, they would not buy scrap regardless of offers made by the broker. Prices of scrap were volatile and responsive to changes in the steel industry's operating rate. Industrial plants selling scrap had to dispose of the material each month and, therefore, would accept the highest bid offered from interested buyers. Scrap dealers did not need to sell each month and could resist lower prices, sometimes choosing to hold scrap when prices were below their costs to purchase and process. Industrial scrap supply tended not to be price elastic in that higher prices did not generally expand supply. The typical price for a ton of ferrous scrap metal was $100, although this varied depending upon which of the 60 to 80 different grades of scrap metal were involved. Also, depending upon general market conditions, the price of a ton of ferrous scrap metal could fluctuate between $75 and $150.

The scrap broker's role was to connect the unorganized and complex industry, which contained several thousand dealers, tens of thousands of manufacturing plants, hundreds of mills and foundries and scores of different scrap grades, into a market. The scrap broker traded scrap iron and steel between producers and consumers. About 40 percent of the scrap sold by brokers and large processors was prompt industrial scrap. These suppliers queried steel mills and foundries before the end of each month to determine their scrap requirements for the next 30 days. Based on this information they offered bids to industrial sellers. On the other hand, about 20 percent of the purchased scrap tonnage originated from the 7 to 9 million cars and trucks that were scrapped in the United States each year. Of the scrap purchased by U.S. consumers, 5 percent originated from railroads. The remaining 35 percent of U.S. customers' scrap purchases were other obsolete iron or steel products, such as demolition material, old water pipes, manhole covers, and kitchen sinks.

Discards of iron and steel products have exceeded ferrous scrap recovery each year since 1956 and the resulting additions to the inventory of ferrous scrap available for recovery had reached staggering volumes. In the United States, supply was not a constraint to increasing the recycling of ferrous scrap. Rather, level of demand, economic conditions, costs of processing, and transportation defined the limits of the industry.

The marketplace for ferrous scrap was not national and was comprised of numerous marketing regions. There were approximately 15 distinct marketing regions for scrap in the United States and many show supply-demand imbalances. Inter-territory shipments, grade substitutions, and increase or reduction in scrap collection could bring about equilibrium. For instance, the available supply of ferrous scrap in the northeastern United States exceeded regular demand from domestic consumers in the area by such a wide margin that approximately 40 percent of all scrap exported from the United States moved from points between Baltimore and Maine.

Over the years, variability of the scrap and steel industries have been driven, in part, by the significant changes the automobile industry was experiencing. Given the importance of the automobile industry to its business, DJJ kept a close watch on the changes that occurred in that industry.

14.4 AUTOMOTIVE INDUSTRY AND VEHICLE SCRAPPAGE

It was clear that the twentieth century was the century of the automobile. The first motorized vehicle was built in the mid-1880s, but by the turn of the century, there were only 8000 vehicles registered in the United States. The production of motor vehicles became a distinct and significant industry within the first decade of the twentieth century. The industry grew even more after Ford's introduction of the "Model T" in 1908. By 1920 registration had increased to over 9 million. Nearing the end of the twentieth century, there are over 190 million vehicles registered in the United States alone.

The automotive industry has been dominated by several major purchases. The Ford Motor Company, founded in 1903, emerged as a leader early in the industry's history. In 1921, it produced 58 percent of the world's automobiles. Around the middle of the century, half of the world's output was manufactured by three American corporations: Ford, General Motors (established in 1908), and Chrysler (established in 1925). Many of the smaller manufacturers such as Packard and Studebaker could not compete with the industry giants and stopped production during the 1950s.

In the late 1980s and early 1990s, leadership in automobile production had shifted. In 1993, Japan out-produced the United States, contributing 23.4 percent (11.2 million cars, buses, and trucks) of the world total. The United States manufactured 22.7 percent (10.9 million), while the combined countries of Western Europe produced 29.0 percent (13.9 million) of the world vehicles. The remaining 24.9 percent were produced by all other countries.

By 1993, the three top passenger car manufacturers in the United States were still Ford, Chrysler, and General Motors, although the top producers had shifted. Between 1987 and 1991, General Motors was the top producer with 2.5 million cars in 1991, followed by Ford with 1.2 million cars, and Chrysler producing 0.5 million cars. Production of passenger cars decreased between 1987 and 1991 by all three manufacturers.

The automobile industry relied on the steel industry for the supply of structural materials. The average passenger car manufactured in the United States in 1990 contained an estimated 1560 pounds of steel (conventional, high strength, stainless, and other steels). In 1978, however, the estimated steel total was about 2090 pounds. The twelve year trend showed a decrease in the use of steel in U.S.-built passenger cars.

Even though the use of steel in cars had decreased, the automotive industry consumed 14.0 percent of the U.S. production of all steels in 1989. The Motor Vehicle Manufacturers Association of the United States, Inc. (MVMA) 1991 report indicated that this figure was low because the shipments for steel service centers, distributors, and imports were not included.

Thus, in 1995, there was a greater reliance than ever before on the steel scrappage industry to supply needed steel. As the supply of ore in the United States had decreased, and the production of vehicles had grown, steel scrappage had been considered a major source of steel. The United States scrapped about 10 million passenger cars and trucks in 1990, which was equivalent to 94 percent of the new passenger cars registered that year (see Table 14.1). The number of new cars registered is about equal to the number of cars scrapped each year.

14.5 DESIGNING FOR THE ENVIRONMENT

Probably one of the most significant changes that has affected the automobile industry is the way cars are being designed. In many industries "green products" are being developed in response to both consumer desire for a cleaner environment and regulatory and advocacy groups' pressure. The automobile industry is no exception and since the mid-1980s a concerted effort had been underway to redesign the way cars are designed, the materials that are being used, and thus, the way cars can be discarded. Clearly, these trends directly impact DJJ's core scrap business.

Traditionally, as described earlier, the burden of recyclability was at the back end of the waste stream with the shredding company. However, because of the new materials being used, and their dependence on steel from scrap, car manufacturers needed to design for back-end recyclability, which meant that materials needed to be used that could be recycled, parts needed to be labeled, and an infrastructure for maximizing recovery had to be put in place. In short, the dynamics of automobile recycling were changing rapidly.*

TABLE 14.1 Yearly Passenger Car Scrappage in the United States

Year	No. Cars Scrapped (in mil.)	Scrapped as % of New Cars Registered	Scrapped as % in Use
1960	4.3	71.0	7.5
1965	6.2	67.1	9.0
1970	6.0	69.1	7.5
1975	6.8	72.6	7.2
1980	7.5	85.5	7.2
1985	8.4	76.4	7.4
1990	8.6	99.4	6.9
1992	7.4	91.2	6.1

Source: Ward's Automotive Yearbook, 1994, 234.

---

*The title of an article in the *Wall Street Journal*, July 10, 1995, underscored the significance of these changes to DJJ: "As auto companies put more plastics in their car, recyclers can recycle less."

A number of published reports offered further insights into the changing automobile recycling industry. For example, *Automotive Engineering,* a national trade journal, published yearly updates on automobile recyclability. In addition, a study performed by Arthur D. Little, commissioned by the Automotive Applications Committee of the American Iron and Steel Institute, reported on the state-of-the-art in automobile recycling in the United States. Some of the findings reported in this study were:

1. More than 90 percent of all passenger cars taken out of service were shredded through an established infrastructure for automobile recycling.
2. Between 75 percent and 80 percent of the weight of each automobile was recycled.
3. During the 1980s, the number of automobiles in operation grew at an annual rate of over 2 percent.
4. The average life span of a car was eleven years and climbing.
5. More than 9 million cars were scrapped each year.
6. The level of automobile recycling almost universally met or exceeded state and regional recycling goals.
7. Until the mid-1970s, environmental policy makers focused primarily on manufacturing operations. Between the late 1970s and early 1980s regulatory interests shifted to hazardous waste disposal and clean up, and in the late 1980s and early 1990s the focus broadened to include post-consumer solid waste disposal.

The ADL report concluded that with the use of new lightweight materials, and the pressure from RCRA to use recyclable materials, the *non-ferrous* industry was going to play a much larger role. It projected that by the end of the century 40 percent of a car's fluff would be plastic, up from 34 percent.

Thus, the automobile design and materials industries were finding themselves in a quandary. Design engineers not only had to meet customer and regulatory needs, but also environmental, cost, and styling criteria, and material suppliers had to produce lightweight, cost-effective, and recyclable materials. Plastic showed the most promise for meeting the need for lightweight materials (which also aided an automobile's fuel efficiency), and efforts were underway to deal with the recyclability issue. Of course, as the amount of plastic used in the manufacture of an automobile increased, the amount of metal recovery decreased.

Closely linked to the recyclability issue was the disassembly issue. It was clear that the disassembly of cars had to be done in a cost-effective and timely manner. A new computerized dismantling facility, such as one operated by Maryland Automotive Reclamation Corp. (MARC), determined electronically which parts of the car were reclaimable for hundreds of domestic and foreign cars. Once these parts were selected, coolant and fluids were extracted and the remaining hulk was then compressed and bundled for scrap resale. Plastics were separated both manually and mechanically and the facility could recycle 60,000 automobiles per year.

The plastics industry was actively involved not only in materials recycling and energy recovery from post-consumer automotive plastics, but also in the development of the necessary infrastructure to manage the process of reclaiming these materials. In the early 1990s there were over 12,000 dismantlers and wrecking yards in the United States, and about 200 shredders which produced over 50,000 metric tons of fluff each year.

Studies by the American Plastics Council suggested that less plastic would be sent to landfills if more resin types could be repaired. Barriers to increased repair and use included lack of economic incentives, the fragmented nature of the repair industry, and a lack of technical repair expertise. It was clear that vehicles designed for easier disassembly were going to help the situation.

Recovery economics for plastic parts or materials were directly related to ease of disassembly, volume of material recovered per part, time to decontaminate, and value of the material. The big challenge to the plastics industry was to design ways to recover resins in sufficient volumes for the development of commercially viable infrastructures. Clearly, an opportunity existed to recycle significant quantities of automotive polymers from scrap vehicles, a growing niche in the automobile recycling industry.

Steel and plastics, however, were not the only materials being used in the manufacturing of automobiles. Aluminum, which was used primarily for sheet body structure, offered rigidity and energy absorption to a vehicle's space frame structure. In the early 1990s, 50 to 60 percent of the aluminum in a car was recycled from old scrap. Recyclers were looking at ways to more effectively separate aluminum from other materials; more aluminum could be recycled if different aluminum alloys were selected.

Car manufacturers also were developing products and processes in response to the need to market "green cars." For example, recycled in-plant plastic scrap was being used in the fascia of the Dodge Caravan and Plymouth Voyager mini-vans. Ford Motor Company also used recycled plastic by converting bumpers into taillight housings, and was actively creating other new markets for post-consumer plastics. Its goal was to have 83 percent of the 1998 vehicles recyclable, with an average of 25 percent post-consumer plastic content or better. The 1995 Mercury Mystique was as environmentally friendly a car as Ford has ever built, with 80 percent recyclability by weight.

A number of companies had developed and patented new products and processes that were used in the manufacture of automobiles. Dow, for example, developed a process to chemically reclaim polyurethane foam (used in car seats). The Woodbridge Group developed and patented a process to recycle foam, tires, and fascia, and Dupont produced Dupont XTC, a thermoplastic composite sheet from recycled PET bottle resin.

Collaborative efforts between auto manufacturers, suppliers of parts and materials, and recyclers were also underway, both nationally and internationally. For example, VW, Audi, and Preussag Recycling GmbH, in Bremem, Germany, developed a nationwide recycling network that could dismantle 11,000 wrecks per year by 1997. In Britain and Germany, several scrap processors were working on the development of systems to convert residue to power through incineration, although the creation of emissions remained an obstacle. Also in Europe, an agreement between BMW, Fiat, and Renault called for the development of a European Union Auto Recycling network (open to all automobile manufacturers) to recycle nine million automobiles that were scrapped each year in the EU member nations (except Ireland, Greece, and Portugal). And, Peugeot-Citroën and Renault also developed a recycling system for up to 200 junked cars a day. Finally, in 1993, Daimler-Benz and Mitsubishi formed a joint venture to develop and market a metallurgical recycling process to replace traditional methods of shredding used cars.

In the United States, similar efforts were underway. In cooperation with the University of Detroit's Mercy's Center for Excellence in Environmental Engineering and Sciences, General Motors sponsored research for studying the recyclability of mixed thermoplastics. In 1992, BMW of North America unveiled a pilot plan for recycling junked cars in conjunction with members of the Automotive Recyclers Association. And, in the summer of 1994, the big three automakers started a joint venture called the Vehicle Recycling Development Center. This disassembly operation is aimed at teaching the three U.S. automobile manufacturers to design cars for easier dismantling and, thereby, to close the production loop (see Figure 14.2).

The push for automobile recycling that had intensified both in the United States and abroad was supported in many cases by new legislation. For example, in the Netherlands, a recycling fee was imposed on new car buyers to offset the cost of recycling. Evidence showed that by levying this environmental tax, the Dutch had successfully reduced the amount of fluff that ended up in landfills. However, in spite of the apparent success of the Dutch tax program, an MIT report argued that government regulators should not interfere with recycling, because such a large percentage of each automobile was being recycled successfully without the need for penalties or incentives. Other countries also were looking at levying an environmental tax on new automobiles.

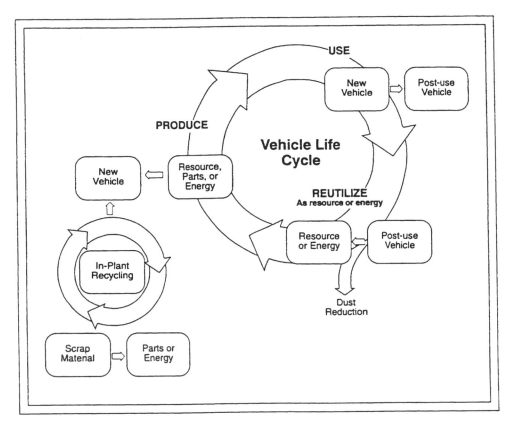

FIGURE 14.2 Motor vehicles can be seen in terms of recycling during vehicle production as well as reutilization during the life cycle of the vehicle. *Source: Automotive Engineering,* October 1992.

Regardless of the source of pressure, however, general agreement emerged among most members of the automobile industry that the need to establish an infrastructure and a demand for recycled material had to be put in place before recycling could be even more successful.

14.6 ENVIRONMENTAL ISSUES

It became clear to DJJ that during the 1980s and early 1990s, environmental issues were becoming more and more important to its business. Given that a portion (usually, about 20 percent) of the shredded automobile (the fluff) was being shipped off to landfills, DJJ was affected by an increasing number of city, state, and federal regulations that addressed waste disposal. For example, DJJ learned that the U.S. Environmental Protection Agency intended to investigate a monofill site used by one of its shredders located in Newport, KY. High concentrations of hazardous materials were being found at this landfill. In part, the EPA's interest was driven by a number of auto shredding operations in Massachusetts that had closed voluntarily because those companies did not send their nonmetal residues to landfills, as DJJ did. The EPA had found unsafe levels of PCBs (polychlorinated biphenyl) and waste oils in the samples it took in Massachusetts.

Traditionally, auto shredders had not been regulated, but because of the possibility of generating hazardous wastes, the EPA had started to look into problems. Resulting new regulations (see Box 14.1) required landfill owners to install double linings to prevent leakage into ground water. These developments were particularly relevant to DJJ because it preferred to have its own (monolithic) landfill in order to have more control over the day-to-day operations and adherence to regulations.

And, given the "cradle-to-grave" responsibility for discarded materials, regulatory agencies would eventually hold DJJ responsible regardless of whether it owned the landfill or not. Because many of its shredders could generate 1000 tons of fluff per month, DJJ's management was acutely aware of the importance of regulations and the long-term impact of its operations on the environment. Partially in response to the emerging significance of environmental regulations, DJJ's management had appointed Skip Rouster, who was a regional operations manager, to head the environment division. He reported to the Vice President of Operations. His task was to deal with environmental regulatory authorities whenever DJJ was confronted with environmental issues.

Box 1   Major Landfill Regulations

Prior to the 1970s, legislation affecting the environment was primarily of municipal or local concern. The trend began to change, however, with the Clean Air Act Amendments and the Clean Water Act of 1970. The Federal Water Pollution Control Act of 1972 began to affect landfill operations and planning when "consideration of the disposal on land of pollutants that affected water quality" was required for water quality control programs. This government action, however, had minimal observable effect.

In 1976, the passage of the Resource Conservation and Recovery Act (RCRA), brought waste management out from primarily local jurisdiction. This act was an attempt by Congress to "avoid potential problems posed by shrinking landfill capacity and the improper disposal of waste" instead of relying on remedial action. The act began to focus the issue not so much on landfill as on who will pay for the cost of improper disposal methods.

The main objectives covered hazardous waste management, solid waste management, and the procurement of materials made from recovered wastes. Intentions were to transfer costs (including health and safety) of disposal to those benefiting from the production of such wastes. That is to say, those individuals and organizations that either produced waste and made profit, or obtained benefit from use without cost. The increasing costs should have encouraged industry to move toward processes utilizing more recovered materials.

RCRA provided for the identification of hazardous wastes and the establishment of standards for the generation, transportation, treatment, storage, and disposal of such wastes. The only standard was the protection of human health and the environment. Non-hazardous waste regulations were intended to promote environmentally sound methods, maximize resource recovery utilization, and encourage conservation. Federal financial and technical assistance would be available to state and local units for the development of waste management plans. RCRA also established cradle-to-grave regulations and both civil and criminal penalties for violators. Cradle-to-grave implies that a producer of waste is liable for damage throughout the existence of said waste including post-disposal. States were allowed, even encouraged, to develop their own waste management plans. If the plan met minimum EPA standards, the states would retain primary jurisdiction, so enforcement of RCRA rested with the states, with EPA approval. Also any citizen could bring a lawsuit against someone failing to comply with regulation both past and present, including generators and handlers. RCRA may have fallen short in that it did not provide for enough enforcement. The decentralized system created inconsistencies and slack.

The Comprehensive Environmental Response, Compensation, and Liability Act (CERCLA) of 1980 (better know as Superfund) created a fund from taxes on crude oil and petroleum products to be used to remedy environmental problems caused by hazardous waste, especially chemical waste. The CERCLA continued the work of RCRA in that it is provided with the funding and authority to support the regulations of RCRA. The two are complementary and cover hazardous waste from point of generation to treatment or disposal. In addition, CERCLA broadened that applicability of regulation and liability by including the release of any hazardous substance into the environment.

continued

> continued
>
> CERCLA had five provisions:
>
> 1) The National Contingency Plan (NCP).
> 2) The National Priority List (NPL).
> 3) Reporting of regulated amounts of hazardous substance discharges.
> 4) Reporting of releases greater than the regulated reportable quantity.
> 5) Establishment of emergency response arrangements through the National Response Center, with immediate responsibility resting with the U.S. Coast Guard or USEPA.
>
> The act established a $1.6 billion, five-year fund allocated to clean up certain sites the NPL judged are more dangerous. Estimated costs of cleanups reached into the tens of billions of dollars. The liability provisions were extensive in that the only defenses were an act of God, an act of war, or an act or omission of a third party involved. However, no federal cause of action to individuals injured was established. Instead, CERCLA confirmed that primary responsibility for this aspect of hazardous waste management belonged in the private sector, and liability settlements were to be decided by evolving litigation. The NCP called for the President to establish some plan for the removal of oil and other hazardous substances from the environment. It required inclusion of provisions for the discovery and investigation of sites, methods of analyzing costs of cleanup, criteria for the extent of cleanup, and criteria for determining the priority for cleanup. The NPL was to be published each year under this provision.
>
> The Superfund Amendments and Re-authorization Act (SARA) of 1986 provided $8.5 billion over five years for the cleanup of abandoned and inoperative waste sites. Title III, Emergency Response and the Community Right-to-Know, provided a framework for emergency planning and required inventory information to be provided to the public. The emergency response and preparedness capabilities of states and local communities was to be enhanced. For example, facilities that released reportable quantities of hazardous substances were now required to notify local emergency committees and the state emergency response commission immediately. Also, facilities were required to submit a hazardous chemical inventory form to the local emergency planning committee, the state emergency planning commission, and the local fire department, of specified materials and minimum quantities present during the previous year. In addition, OSHA was required to establish standards to protect the health and safety of workers engaged in hazardous waste operations and emergency response activities.
>
> In 1988, parties involved in all aspects of waste were anticipating new EPA regulations, then long overdue. These were expected to include new restrictions on landfill site locations, new facility operating requirements, and increasing financial responsibility. The EPA intended these new regulations to support and integrate solid waste management systems that included reducing the amount of solid waste generated, recycling as much as possible, incinerating some refuse with appropriate controls, and continuing safe landfilling practices.

14.7 CONCLUSION

Having analyzed some of the market dynamics in the automobile, steel, and scrap industries, it had become clear to DJJ that the issues regarding alternative strategic directions were complex. For example, a decision to establish new alliances with automobile manufacturers would require significant human as well as monetary resources and would have long-lasting consequences for DJJ, even impacting its core business. Joint ventures also could be developed with other members of the automobile industry, such as steel, aluminum, or plastic producers. And, DJJ could even consider adding an increased ability to recover other recyclable materials from one or more of its scrap metal shredders. The costs associated with each of these alternatives were unknown, although SHV, DJJ's parent company, had encouraged DJJ to "think broad and deep." Money always was an issue, but the future of DJJ's operations was at stake and it was critical that DJJ develop a set of clear strategic goals that would position it to enter the twenty-first century successfully.

15 Environmental Hazards in Transportation: The Response of Grand Trunk Railroad to a Derailment Case Study Number 7

O. Homer Erekson and Timothy C. Krehbiel

CONTENTS

15.1 Introduction 291
15.2 Industry Setting and Environmental Challenge 293
    15.2.1 Company Description 293
    15.2.2 Accidents and Environmental Challenges in Transportation 294
        15.2.2.1 Environmental Hazards and Chlorine Release 296
15.3 A Systemic Approach to Emergency Preparedness 297
    15.3.1 The Downriver Mutual Aid Task Force 297
    15.3.2 The Environmental Management Action Committee 298
    15.3.3 The Transportation Community Awareness and Response Program 298
    15.3.4 Total Quality Management 299
15.4 Emergency Response Plan 300
    15.4.1 Emergency Response Plan in Action 300
        15.4.1.1 Emergency Response 300
        15.4.1.2 Cause of Derailment 301
15.5 References 302

15.1 INTRODUCTION

On Monday morning, May 22, 1995, Detroit was experiencing a pleasant spring morning with clear skies. Grand Trunk Railroad's train #385, which was traveling from Toronto to Flat Rock, MI approached an overpass to interstate I-696 in the Mt. Clemens Subdivision, approximately 10 miles northwest of downtown Detroit. It was 8:15 a.m. and rush hour traffic on I-696 was extremely busy as usual. Many motorists watched in horror as the train derailed while passing over them. Two train cars came crashing down onto the busy interstate and a third appeared to be dangling over the side of the overpass with one end coming to rest on the interstate and the other end resting against the top of the overpass (see Figure 15.1). Miraculously, all the motorists were able to avoid the fallen train cars. While the train was deraling, one motorist peered out her window trying to gauge where the falling car would land and was able to swerve at the last moment and avoid being crushed. Almost

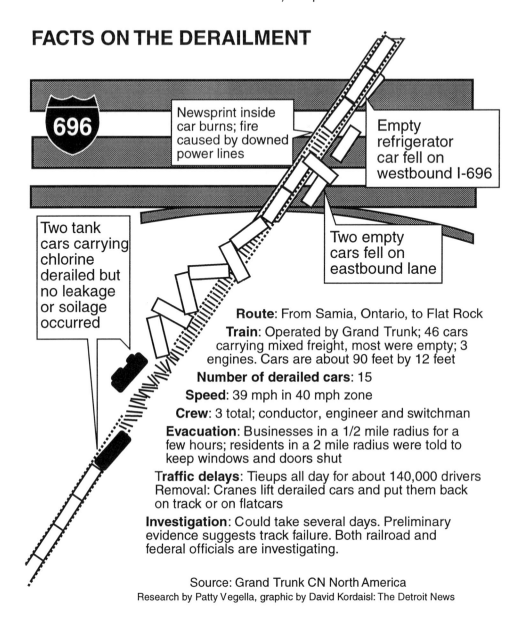

FIGURE 15.1 Facts on the derailment.

immediately a fire erupted in a derailed car which was sitting on the overpass. Its cause appeared to be a downed power line. The downed power line also draped itself across a car on the freeway below which had stopped a few feet short of a fallen train car. The train's crew, which consisted of a conductor, engineer, and switchman, radioed the dispatcher's office to report the incident.

The train consisted of 3 engines and 45 cars, 28 with loads and 17 empties. It had been traveling in a westerly direction at 40 mi/h, the speed limit for that section of track as set by the Federal Railroad Administration (FRA). In total 17 cars derailed. The train car in the eastbound lane and the dangling car were identified as empty hopper cars. The car in the westbound lane was an empty tanker car. The car on fire contained paper for newsprint. Of greatest concern, was the derailment of two Hazardous Materials (HazMat) cars containing liquid chlorine. One of the derailed HazMat cars was upright near the tracks right before the tracks entered the overpass. The other car, which was directly behind the other HazMat car on the train, had fallen onto its side into a ditch. Both of the HazMat cars appeared to be intact and not leaking.

Emergency crews responded quickly and the local fire department was able to contain the fire within the burning paper car. The two loaded cars containing liquid chlorine became the center of attention for both the emergency response personnel and the media owing to danger which might occur if the liquid chlorine were to escape. These two cars were shipped by ICI of Canada. To remove the cars, new rails would have to be laid and then a crane would have to lift the cars up and rerail them. Safety personnel believed that this operation would not cause any release of the highly toxic material, but they acted with extreme caution.

The sensational nature of the accident generated a huge amount of media attention. The local affiliates of all three major networks cut into normal programming to give the public dramatic, live shots from the derailment site. Grand Trunk's public relations office responded quickly. Public Relations Manager Gloria Combe knew that her role was to ensure that the media's coverage was accurate and balanced. In other words, she knew that the coverage needed to truthfully portray any potential dangers to the public but, at the same time, it need not excite fear or panic in the public by overstating or sensationalizing the potentiality for disaster.

Ms. Combe spent most of the day talking to TV, radio, and newspaper reporters informing them of the facts concerning the derailment and the huge efforts being made to correct the situation. This included several live, on-site interviews with local television. She stressed that no one was in any immediate danger from the derailed HazMat cars. In response to repeated media concerns, Ms. Combe interjected that over the past 20 years, the entire North American railroad industry had experienced only one death owing to transport of hazardous materials. David Wilson, Vice President and Senior Operating Officer in the United States, also gave interviews to the media while on-site.

At approximately 2:00 p.m. the Grand Trunk rerailment team prepared to lift the tanker cars using a large crane. The Warren Fire Department ordered an evacuation for a one half mile radius of the derailed HazMat cars. Schools and business were ordered to shut down and to evacuate. Residents were advised to leave, but were not forced from their own homes. The Red Cross set up shelters where local residents could find beds and food. After giving the local officials time to execute the evacuation, the two cars were successfully rerailed and transported from the area. The HazMat cars survived the entire derailment and rerailment episodes as designed—no punctures or leaks of any kind. At midnight, the evacuation was canceled.

During the entire rerailment, Interstate I-696 was closed in both directions, causing tie-ups for an estimated 140,000 drivers. Rush hour traffic on Monday afternoon was extremely slow as motorists tried to find alternate routes to I-696, the major east-west commuter route in Detroit. The highway reopened at 5:30 a.m., May 23, 1995, just in time for Tuesday morning rush hour. Rail service resumed Tuesday afternoon at 1:15 p.m. During the entire incident there were no injuries to the public, train crew, and response teams, nor was there a release of any hazardous materials.

15.2 INDUSTRY SETTING AND ENVIRONMENTAL CHALLENGE

15.2.1 COMPANY DESCRIPTION

Grand Trunk Western is the U.S. District of Canadian National (CN), operating 925 miles of track in Michigan, Ohio, Indiana, and Illinois, with approximately 2,300 employees. Canadian National itself operates Canada's largest railway system, supplying customers with freight rail transportation and related services.* CN's rail operations account for 90 percent ($4.2 billion) of its revenues and employed approximately 30,500 persons in 1994. CN describes its competitive advantages as "coast-

---

*As of 1994, Canadian National was composed of CN North America (its rail operations) and CN Enterprises (which includes CN Real Estate, CN Tower, CANAC International, AMF Technotransport, and CN Exploration).

to-coast service in Canada, highly efficient access to U.S. markets, the most balanced and diversified traffic mix of any railway in North America, and leading-edge information systems."[1]

Grand Trunk provides rail service to automotive, chemical, utility, steel, forest products, food, and other industries in its territory and connects with other railroads to provide services to the entire continent. Grand Trunk is the connection to the Chicago gateway for CN's international traffic. Figure 15.2 shows the operating revenues for both the Canadian and U.S. rail operations of CN by industry group.

15.2.2 ACCIDENTS AND ENVIRONMENTAL CHALLENGES IN TRANSPORTATION

A significant portion of the freight handled by railroad companies involves products from the chemical industry. Many of these products are hazardous materials, whose movement is governed by regulations issued under the Hazardous Materials Transportation Uniform Safety Act. Although the expected number of rail accidents involving HazMats in a given year is expected to be very small, it is important to have an appreciation for the actual and perceived risk and degree of preparedness within the rail industry and the communities it encounters.*

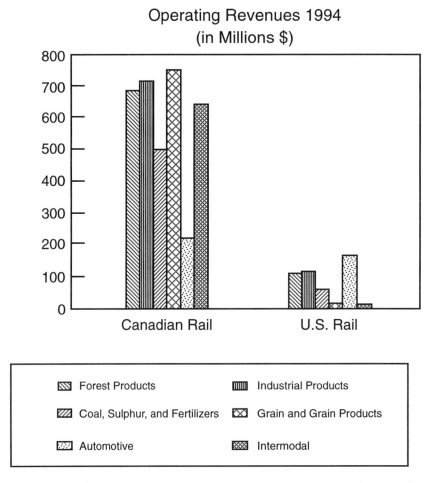

FIGURE 15.2 Canadian National operating revenues for 1994. See Reference 1.

---

*Meyer has offered an interesting hypothesis that because accidents are quite rare, and because most accidents have a negligible or minor impact on human health, industries and communities can become complacent about achieving sufficient safety standards and/or emergency preparedness. However, in the particular case he examined, provision of accurate and trusted information on risks before an event raised the sense of danger and minimized complacency. See Reference 2.

Environmental Hazards in Transportation

At a National Press Club press conference on rail safety, Edwin L. Harper, President and CEO of the Association of American Railroads, stated, "Never in the history of railroads has their safety record been as good as it has over the past three years, and 1994 was the safest ever." In fact, the rail safety record has been improving dramatically since 1981.[3] As shown in Figure 15.3, the train accident rate per million train miles decreased from 8.25 in 1981 to 3.82 in 1994.

These improvements in the accident rates of rail transport have been attributed to several sources, including technological advancements involving intensive computer applications for modeling and testing new freight car wheel designs and for improving train-handling techniques and monitoring potential causes of derailments before they occur. Resulting new freight car wheel designs reduced the number of accidents caused by wheel failure by approximately 80 percent from 1981 to 1994. In addition, technology has played a significant role in improving the communication of important information during an emergency situation.

Beginning in the 1980s, the industry began to tag all rail cars for electronic reading. The system finally adopted automatic equipment identification (AEI), and requires that all rail cars be equipped with transponders coded for each car, which can be read electronically as the cars pass certain

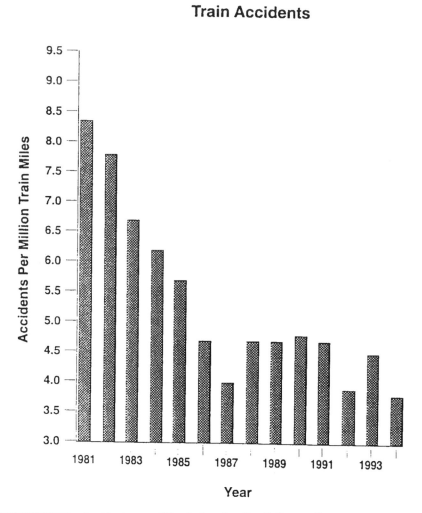

FIGURE 15.3 Accidents per million train miles. See Reference 9.

scanner points. This information feeds to computers which assist in keeping track of the cars as they pass these points. Information regarding the contents of all cars also is maintained by computer.

Federal law requires that all trains carrying hazardous materials have an accurate listing of all cars in the train (by their order in the train) and identifying those containing hazardous materials, together with specific information dealing with the handling of such materials if they should be involved in a derailment. Train crews are required to carry this listing, called a *train consist*, with them. It also must be available through the railroad's offices.

Harper argues that much of the improved railroads' safety performance was due to replacement of command-and-control safety regulations with performance standards, "Since enactment of the Staggers Act and the resulting increase in railroad cash flow, freight railroads have made capital expenditures in excess of $50 billion to improve track, signals, equipment, and information systems."[3]

15.2.2.1 Environmental Hazards and Chlorine Release

Similar patterns in reduction of accidents are found when looking at HazMats. From 1980 to 1988, accidents involving trains with hazardous materials were down 42 percent, while accidents resulting in the release of HazMats decreased 63 percent. Since 1988, as shown in Figure 15.4a, the percentage of accidents for the United States involving damage to or derailment of a HazMat car where there was a release of hazardous materials has declined from 21.9 percent in 1989 to 11.1 percent in 1993. Figure 15.4b shows comparable data for Michigan alone. From 1980 to 1988, Grand Trunk moved more than 80,000 carloads of HazMats and had just three measurable spills, each involving less than 10 gallons of product.[4] Moreover, according to data from the Association of American Railroads, only one rail worker from 1975 to 1995 has been killed while transporting hazardous materials in the United States and Canada.[5]

Regardless of the excellent overall safety record of the railroad industry and Grand Trunk in particular, citizens in the Detroit metropolitan area were faced with a potential release of a serious HazMat, liquid chlorine, in this instance. Chlorine is a greenish yellowish gas with an irritating odor, in this case present in a liquid form. It is used to make solvents, many chemicals, disinfectants, chlorine bleach cleaners, and plastics. Exposure to chlorine can cause irritation of the eyes, nose, and throat, including tearing, coughing, sputum, bloody nose, and chest pain. Higher levels can cause pulmonary edema (a buildup of fluid in the lungs) and death. Chronic (long-term) health effects can include cancer and reproductive hazards. In addition to human health effects, chlorine has high acute toxicity effects to aquatic life, and chronic ecological effects including shortened lifespan, reproductive problems, and changes in appearance or behavior for animals, birds, fish, and plants.[6]

Year	Total Number of Accidents	Accidents in Which a HazMat Car was Damaged or Derailed	Accidents in Which There Was a Release of Hazardous Materials	Percent Released Due to Accident
1989	517	251	55	21.9
1990	466	236	35	14.8
1991	525	293	47	16.0
1992	482	230	27	11.7
1993	559	262	29	11.1

Source: Association of American Railroads, June 5, 1995.

FIGURE 15.4a Train accidents involving transportation of hazardous materials in the United States.

Year	Total Number of Accidents	Accidents in Which a HazMat Car was Damaged or Derailed	Accidents in Which There Was a Release of Hazardous Materials	Percent Released Due to Accident
1984	10	7	1	14.3
1985	10	20	0	0.0
1986	13	10	1	10.0
1987	15	26	3	11.5
1988	9	14	0	0.0
1989	10	23	3	13.0
1990	3	2	0	0.0
1991	5	6	0	0.0
1992	5	9	0	0.0
1993	5	2	1	50.0

Source: Robert J. Chaprnka, President, Michigan Railroads Association, testimony, June, 1995.

FIGURE 15.4b Train accidents involving transportation of hazardous materials in Michigan.

Moreover, the costs of a derailment are significant for a rail company. A minor derailment involving only a few cars, none of them carrying hazardous materials, may easily approach $100,000. In 1989, Grand Trunk showed the following estimated costs for a typical minor derailment:

- Equipment damage $20,000
- Track damage   6,000
- Freight and signal damage 35,000
- Cleanup costs   10,000

In most cases, railroads must pay claims for loss, damage, and delay of freight because they are largely self-insured. To the extent that there is insurance coverage, these cost estimates do not include effects on insurance premia or loss of coverage arising from serious accidents. In derailments involving HazMats, these costs are significantly higher because of the greater threat to the population and the need to retain more specialized emergency response teams.

15.3 A SYSTEMIC APPROACH TO EMERGENCY PREPAREDNESS

The following quote from *Prevention... Preparation... Response: a report on rail safety by Grand Trunk Corporation*[4] summarizes Grand Trunk's philosophy on emergency preparedness: "The perfectionist side strives for an accident-free environment by focusing on preventative steps ranging from education to track inspections to equipment improvements. The realistic side recognizes that accidents can occur and stresses thorough preparation and prompt response." The company's commitment to this philosophy is evidenced by Grand Trunk's involvement with associations like the Downriver Mutual Aid Task Force and the Transportation Community Awareness and Response Program.

15.3.1 THE DOWNRIVER MUTUAL AID TASK FORCE

The heavily industrialized area just south of Detroit is known as the Downriver area. In 1968, 15 Downriver communities established The Downriver Mutual Aid Task Force as a co-operative for sharing police, fire, medical, and other emergency response personnel in the event of a civil disturbance. Over time, the focus of the task force has broadened to include virtually any potential disaster including a hazardous material spill. Preparation includes continual training, education, sharing of equipment, and maintaining accurate up-to-date information regarding materials and shipments.

15.3.2 THE ENVIRONMENTAL MANAGEMENT ACTION COMMITTEE

In 1987, Grand Trunk believed that to be able to deliver on its commitment to environmental excellence, Grand Trunk would have to become more pro-active on environmental issues. In addition, improvements were needed in communications between departments. It was agreed that a cross-functional team needed to be formed. Under the leadership of John Baukus, an engineering and planning officer at Grand Trunk, the Emergency Management Action Committee (EMAC) was formed. Mary P. Sclawy was named Recorder and Maker of the Agenda for EMAC. Each of the eight major divisions of Grand Trunk agreed to contribute a decision-maker to EMAC. EMAC would meet once a month and the representatives would report back to their respective departments. The departments involved were Safety, Transportation, Law, Real Estate, Engineering, Mechanical, Purchasing, and Marketing.

The committee kept an eye on any and all environmental issues. The agenda typically consisted of discussions concerning new regulations, pro-active measures for possible problems, and remedial measures for existing problems. EMAC was an excellent forum for discussion of how policy in one department would affect the other departments. This increased communication resulted in more of a systems approach to managerial decision-making. Like most cross functional teams using a systems approach, attention became more focused on the processes and less on the final outcomes. This focus on the processes reinforced the interconnectedness of the departments.

A good example of pro-active management involving EMAC deals with purchasing requirements. The purchasing department requires a cradle-to-grave analysis for all new purchases. Products are purchased only when waste streams are identified and acceptable disposal procedures are in place. The purchasing department has an approved list of items from which departments can order. All items not on the list can be purchased only after sufficiently documenting that the product waste stream can be disposed in a safe and environmentally sound manner and the disposal method is in place.

15.3.3 THE TRANSPORTATION COMMUNITY AWARENESS AND RESPONSE PROGRAM

The Transportation Community Awareness and Response (TRANSCAER) program is a cooperative effort involving the Michigan Railroads Association, the Chemical Manufacturers Association, the Michigan Department of State Police, and the fire and police departments in communities through which hazardous materials travel. The unofficial slogan for TRANSCAER is "Replacing fear and indecision with knowledge and action." In 1991, Lansing, MI became the site of TRANSCAER's Hazardous Materials training center. The center, which was funded primarily by the Michigan Railroads Association and the Chemical Manufacturers Association, is administered by the Michigan Department of State Police. Grand Trunk joined in the effort by donating several HazMat tank cars. The center trains emergency response personnel through classroom instruction and hands on training involving tank cars, trucks, and storage tanks. TRANSCAER also conducts workshops and informational meetings across the state of Michigan. In 1993 and 1994, over 1500 public officials attended TRANSCAER meetings. TRANSCAER also helps individual communities develop their own emergency response plans.

The TRANSCAER Implementation Flow diagram given in Figure 15.5 stresses the systemic approach needed to manage the transportation of hazardous materials. The first step to develop a

Environmental Hazards in Transportation

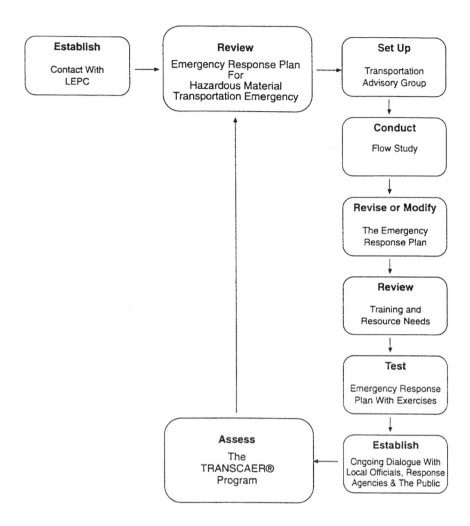

FIGURE 15.5 TRANSCAER implementation flow. See Reference 10.

TRANSCAER Project is to contact the Local Emergency Planning Committee (LEPC). After this step, the model suggests a never ending loop of planning and continuous improvement. The model incorporates dialogue from many different parties including local officials and the general public. This dialogue includes factual discussion of the dangers of transporting hazardous materials, open responses to public concerns, and participation in community activities, civic events, and educational programs. Companies and communities adopting this TRANSCAER model will confirm those areas of emergency preparedness that are working effectively as well as identify and improve those areas which need more attention.

15.3.4 TOTAL QUALITY MANAGEMENT

In the early 1990s, Canadian National undertook a major effort to become a more quality conscious organization. Quality Action Teams were formed and allowed to apply creative thinking to process improvement. In 1994, the Performance Management Process (PMP) was introduced in order to give employees a better understanding of how they can contribute to the overall success of the company. PMP linked employee performance directly to corporate goals and replaced the traditional annual employee assessment with an ongoing planning and evaluation process.

Supervisors for Grand Trunk Western attended 16 hour seminars on Total Quality Management (TQM). The classes were taught by an internal TQM specialist. The workshops were typical of many executive short courses on TQM. The workshop focused on identifying customers, providing quality service to those customers, and encouraging departments to work together to accomplish common goals. Topics included the quality improvement cycle, teamwork, benchmarking, and systems thinking. After completion of the workshops, quality teams were formed to address many different issues. The teams consisted of supervisors and scheduled employees.

Jack Buysse, Supervisor of Hazardous Materials, was one of the supervisors who attended one of the workshops. A long time employee of Grand Trunk, he managed with a philosophy of doing it correctly initially and then continually improving. He believed that the objective of the railroad was to fulfill customers' needs. To do this he believed that the company needed to continually garner new support with customers and internal staff. Buysse believed that the workshops and quality teams were a success. He felt that the approach could be very helpful in improving communication between departments.

15.4 EMERGENCY RESPONSE PLAN

In the event of an actual emergency, a member of the train crew calls the dispatcher. The dispatcher then calls the people on a calling list. The list contains both the work and home phone numbers of corporate executives, as well as predetermined emergency response personnel. Each of these initial contacts then follows a set of predetermined procedures.

The emergency response plan contains the following steps:

1. Clean up derailment.
2. Get track back together.
3. Find cause (one of the following: act of God, track, mechanical or human failure, or vandalism).
4. Reports to state agencies (Michigan Department of Transportation, Michigan Department of Environmental Quality).
5. Report to Federal Railroad Administration (FRA).

At times public hearings also are held, though not a standard part of the Emergency Response Plan.

At the conclusion of this process, the data and analysis of the derailment are evaluated to determine if the Emergency Response Plan (see the model in Figure 15.5) adequately provides for contingencies encountered and amendments made if it does not.

15.4.1 Emergency Response Plan in Action

15.4.1.1 Emergency Response

Within minutes of the I-696 derailment, a member of the train's crew called the dispatcher. The dispatcher then called the people on the calling list. In addition to the Grand Trunk personnel, nine emergency response teams responded to the accident: Warren Fire Department, Roseville Fire Department, Warren Police Department, Roseville Police Department, Michigan State Police, Dow Chemical Emergency Response Team, Michigan Department of Transportation, National Transportation Safety Board, and the Federal Railroad Administration. The rerailment was conducted by a Grand Trunk rerailment team specifically trained to handle such incidents. The Dow Chemical response team also was present to represent the shipper's interest and to act as an advisory. The response of the local fire, police, and rescue personnel was quick, professional, and their efforts were applauded by Grand Trunk Railroad. The effective response by the local safety officials was due in part to the extensive training provided to local officials by Grand Trunk, including the TRANSCAER program. Grand Trunk agreed to pay for all costs incurred by local officials.

The orchestration of the various emergency response teams throughout the entire episode was outstanding. The success of the operation was largely owing to Grand Trunk's commitment to its stakeholders. Grand Trunk had long realized that if it were transporting hazardous materials, it had to view the local residents as stakeholders. To be responsive to the stakeholders needs, it had to have extensive training programs in place for their response teams. Grand Trunk also realized that it needed the support of the local emergency response police and fire departments. During the I-696 incident, the derailment response systems had worked exactly as planned. Mary P. Sclawy, Corporate Attorney for Grand Trunk remarked, "At a derailment, watching the response workers is amazing. Everyone knows exactly what to do."

15.4.1.2 Cause of Derailment

Preliminary cause of the derailment, as determined by on-site Grand Trunk workers, was a pair of broken joint bars. As shown in Figure 15.6, steel bars are used on either sides of railroad rail sections to hold the rail sections together. Although it is not unusual to have an occasional broken joint bar, it is highly unusual to have both pairs of joint bars break and for this break to lead to a derailment.

The company took full responsibility for the derailment, but believed that the cracks in the joint bars could not have been detected prior to the accident. David Wilson, Vice President of Operations for Grand Trunk stated, "I don't think any inspector would have seen that. We don't feel we had an employee failure."[7] The company felt very lucky that no one was injured in the incident, and was proud of the fact that the emergency response system worked so well.

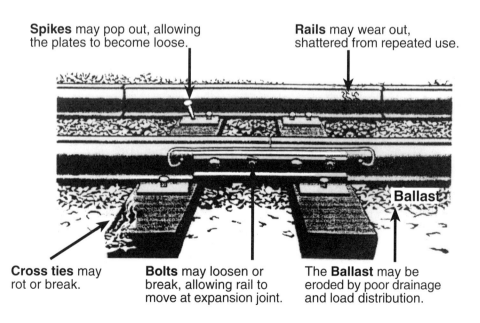

FIGURE 15.6 Potential track defects.

At the same time, the transportation unions were making headlines with their claims that cutbacks in the number of rail inspectors by Grand Trunk and other Michigan railroads were threatening rail safety. Grand Trunk has one person assigned to inspect 50-plus miles of track twice a week, while the Federal Railroad Administration (FRA) has one inspector, based in Detroit, to cover 4000 miles of track in Michigan. According to Rich McLean, Michigan Vice Chairman of the Brotherhood of Maintenance of Way Employees claimed that Grand Trunk officials were putting profits in front of safety. "I know for a fact they've cut in half the number of inspectors and doubled the amount of track they have to inspect." In response, Wilson claims, "There's been no reduction in safety. As technology comes in, we can reduce the amount of people."[7]

The broken joint bars were sent to a metallurgical lab at the Federal Railroad Administration (FRA) to determine the cause of the breaks. In August, the FRA released its initial report of the incident. The report indicated that the breaks in the joint bars were consistent with what one would expect from repeated stress under normal use. Although the report was only preliminary and did not try to pinpoint the cause for the derailment, nothing in the report contradicted the company's belief that the derailment was caused by the broken joint bars. Bob Cerri of the National Transportation Safety Board indicated that no track safety standards were violated: "It was one of those things you couldn't detect. Bars broke inside under the rail—not visible at the time of inspection. The break occurred after the derailment—the inspector could not see the break as a result of normal track inspection."[8]

15.5 REFERENCES

1. Canadian National Annual Report, Montreal, Quebec, 1994.
2. Meyer, P. B., Responding to accidents posing environmental risks: walking the fine line between panic and complacency, *Sustain*, Spring, 24, 1996.
3. Harper, E. L., Railroads achieve safest year ever, Remarks at a joint Association of American Railroads and Department of Transportation Press Conference on Rail Safety, National Press Club, May 25, 1995.
4. Grand Trunk Corporation, *Prevention . . . Preparation . . . Response: a report on rail safety*, 1989.
5. Chaprnka, R. J., President, Michigan Railroads Association, and Such, A. J., Executive Director, Michigan Chemical Council, personal correspondence to Representative Paul Hillegonds, May 25, 1995.
6. U.S. Department of Transportation (DOT), Chlorine Report No. UN 1017, 1989.
7. St. John, P. and Schabath, G., Feds, railroad dispute accident cause, *The Detroit News*, May 26, 1995.
8. Personal communication with Gloria R. Combe, Manager, CN-U.S. public affairs, August 12, 1996.
9. Federal Railroad Administration, Accident/Incident Bulletins, Nos. 150–163, Tables 511, 1, 7, 13, 16, 1994.
10. TRANSCAER®, Transportation Community Awareness and Emergency Response, Chemical Manufacturers Association, Washington, D.C., 1995.

16 The Procter & Gamble Company: Reducing Packaging Waste in the United States and Germany, 1987–1994 Case Study Number 8

Jan Willem Bol, Orie L. Loucks, David W. Rosenthal, and Steven Skeels

CONTENTS

16.1	Introduction	303
16.2	Part I: Lenor Pouch and Downy Refill Packaging	304
	16.2.1 The Procter & Gamble Company—A Brief History	304
	16.2.2 The Procter & Gamble Company, 1989–1990	305
	16.2.3 Development of the Lenor (Germany) and Downy (United States) Refill Packages	308
	16.2.4 Environmental Concern and Eco-labeling in Germany	309
	16.2.5 Level of Environmental Concern in the United States	310
	16.2.6 The Fabric Softener Market—the United States and Germany	311
	16.2.7 Test Results for the Lenor and Downy Refills	313
	16.2.8 Conclusion for Fabric Softeners	316
16.3	Part II: Recovering Plastic Waste for P&G's Plastic Bottle Products	317
	16.3.1 P&G's Market Incentives for Plastic Recovery and Use in the United States	318
	16.3.2 Closing the Recycling Loop in the United States and in Germany: Comparisons	319
16.4	Conclusion	321
16.5	References	321

16.1 INTRODUCTION

During the late 1980s, Procter & Gamble, a large multinational organization headquartered in Cincinnati, OH, was implementing a number of consumer-driven changes to its product line. Extensive marketing research had shown that more and more consumers were expressing interest in "environmentally friendly" products, and Procter & Gamble, eager to respond to consumers' needs, as well as be ahead of its competitors, was investigating profitable ways to respond.

This case describes some of Procter & Gamble's marketing and production decisions during the period 1987 to 1993 in both Germany and the United States. Part I details a successful change in Germany at about the time a project team in the United States was considering changes in their approaches to a well-established U. S. brand. The test marketing data offers the reader an opportunity to make a recommendation as to what to do next. Part II provides insight into post-consumer plastic solid waste, and options for its reduction. This part of the case also considers similarities (in the issues) and differences (in the options) between the U. S. and German markets.

16.2 PART I: LENOR POUCH AND DOWNY REFILL PACKAGING

In June 1990, Brad Irwin, Advertising Manager of Fabric Conditioner Products, examined the data from the first six months of test market results for Downy Fabric Softener's refill package. The results appeared to be very positive, but Irwin faced the decision of extending the test market, modifying the test, or recommending that the new package be made available nationally.

The introduction of the new package into the test market, the prior November, had attracted considerable attention in the press because of the package's reduction of solid waste. The package reduced the amount of solid waste by approximately 75 percent vs. the existing packaging. The positive tone of the press coverage had apparently matched public sentiment about improving the environment because Downy Refill test market sales exceeded expectations.

Procter & Gamble management had focused time and attention on environmental improvement as an ongoing part of their corporate culture. John Smale, Chairman and Chief Executive Officer until January 1990, had made the commitment to "get in front of the curve" on environmentalism. The employees of the company were urged to make environmental stewardship a daily part of the business.

The Downy Refill package followed an earlier decision in Germany, where, in 1987, a plastic refill pouch had been successfully introduced for Lenor Fabric Softener, Downy's sister product. The German initiative, based on market research that indicated consumers would be receptive to such an initiative, reduced plastic by 85 percent. Additionally, consumers felt good that they could do something to help the environment. In Canada, similar plastic pouches were in the process of being introduced. Still, Irwin worried over the acceptance of the new package in the United States. Was the U.S. market ready for the inherent inconvenience the new package entailed?

Beyond the environmental issues, there were differences in commercial issues between the two countries as well. Acceptance by the retail trade in the United States was critical to the new package's success. The amount of shelf space available for the new package was limited, and Irwin did not want to lose existing space. The appropriate positioning for the product was still at issue, the balance of economy and environmentalism was yet to be determined in the United States, despite the success of Lenor in Germany. A related issue was that of pricing. Some cost savings would be generated by the new package, but the degree to which that should be reflected in the pricing was still in question.

16.2.1 THE PROCTER & GAMBLE COMPANY—A BRIEF HISTORY

Unknown to each other, William Procter, an Englishman, and James Gamble, an Irishman, had left their home countries to pursue their fortunes in the new western frontier of America. They individually found their way to Cincinnati where health problems and the rigors of travel forced them to remain. Cincinnati also carried the nickname Porkopolis from its role as a center for the slaughtering and processing of hogs. The oils and tallow from the meat processing generated other business opportunities, and Procter quickly began producing candles. Gamble went to work for a maker of soap.

Their ties were established by marrying sisters, and in 1837 they combined their efforts and created the new Procter & Gamble Company. Shipment of their goods was largely by water, up and down the Ohio River. By the 1850s the wharf hands who loaded the river boats had begun to use the moon and stars symbol on the side of the P&G boxes to distinguish boxes of Star Candles.

During the Civil War the company supplied soap and candles to the Union Army, thus expanding the company's business and at the same time creating a high level of brand awareness. In 1879 Ivory Soap was developed and by 1882 was advertised nationally for the first time. The product typified what was to become the P&G practice of business: ". . . to provide products of superior quality and value that best meet the needs of the world's consumers." Table 16.1 lists some of P&G's popular brands and their dates of introduction.

16.2.2 THE PROCTER & GAMBLE COMPANY, 1989–1990

In 1989, the Procter & Gamble Company (P&G) ranked number 14 on the Fortune 500. The company had $21.4 billion in sales worldwide, assets of $16.4 billion, and net earnings after taxes of $1.2 billion. The company employed 79,000 people worldwide. Table 16.2 provides a financial summary for the years 1988 to 1990.

P&G manufactured and marketed consumer products in the categories of laundry and cleaning, personal care, foods and beverages, and pulp and chemicals. P&G brands included such well-known names as Crest, Pampers, Tide, and Downy. By 1990, they marketed more than 250 brands in over 140 countries worldwide. Additionally, in 1990, P&G's international operations accounted for almost 40 percent of its world-wide sales and from 1986 to 1989 grew almost five times as much as its U.S. domestic sales (See Table 16.2).

The company had major operations in 46 countries outside the United States. Tables 16.3 and 16.4 show business results by geographic area and business segment. One of its operations was P&G GmbH in Germany, which was established in 1963 and, by 1990, had become one of that country's premier marketers selling more than thirty brands, employing almost 7000 people, and sales revenues of more than DM 1 billion.

The company's world-wide business was summed up by their statement of purpose:

"We will provide products of superior quality and value that best fill the needs of the world's consumers.

"We will achieve that purpose through an organization and a working environment which attracts the finest people; fully develops and challenges our individual talents; encourages our free and spirited collaboration to drive the business ahead; and maintains the Company's historic principles of integrity, and doing the right thing.

TABLE 16.1
Procter & Gamble, Familiar Brands

Brand:	Date:
Ivory Soap	1882
Crisco Shortening	1911
Camay	1926
Tide	1946
Prell	1946
Crest	1955
Charmin	1957
Downy	1960
Pampers	1961
Folgers	1963
Bounce	1972
Pert	1986
Pantene	1992

TABLE 16.2
Consolidated Statement of Earnings

Years Ended June 30 (Millions of Dollars except Per Share Amounts)	1990	1989	1988
Income			
Net sales	$24,081	$21,398	$19,336
Interest and other income	561	291	155
	24,642	21,689	19,491
Costs and Expenses			
Cost of products sold	14,658	13,371	11,880
Marketing, administrative, and other expenses	7,121	5,988	5,660
Interest expense	442	391	321
	22,221	19,750	17,861
Earnings before income taxes	2,421	1,939	1,630
Income taxes	819	733	610
Net earnings	$1,602	$1,206	$1,020
	—	—	—
Net earnings per common share*	$4.49	$3.56	$2.98
Net earnings per common share assuming full dilution*	$4.27	$3.47	$2.96
Dividends per common share*	$1.75	$1.50	$1.38
Average shares outstanding* (in millions)	346.1	334.4	338.6

* Adjusted for two-for-one stock split effective October 20, 1989.

TABLE 16.3
Comparison of Sales and Earnings by Geographic Component

Millions of Dollars		United States	International	Corporate	Total
Net sales	1988	$12,423	$7,294	$ (381)	$19,336
	1989	13,312	8,529	(443)	21,398
	1990	14,962	9,618	(499)	24,081
Net earnings	1988	864	305	(149)	1,020
	1989	927	411	(138)	1,206
	1990	1,304	467	(169)	1,602
Assets	1988	8,346	4,751	1,723	14,820
	1989	8,669	5,260	2,422	16,351
	1990	9,742	6,516	2,229	18,487

TABLE 16.4
Comparison of Sales, Earnings, and Expenditures by Product Group

Millions of Dollars		Product Groups					
		Laundry and Cleaning	Personal Care	Food and Beverage	Pulp and Chemicals	Corporate	Total
Net sales	1988	$6,668	$8,676	$2,963	$1,532	$(503)	$19,336
	1989	7,138	10,032	3,029	1,778	(579)	21,398
	1990	7,942	11,767	3,318	1,666	(612)	24,081
Earnings before	1988	699	888	32	248	(237)	1,630
income taxes	1989*	754	1,031	(14)	362	(194)	1,939
	1990*	781	1,314	304	307	(285)	2,421
Assets	1988	2,852	7,114	1,721	1,410	1,723	14,820
	1989	2,964	7,511	2,023	1,431	2,422	16,351
	1990	3,296	8,786	2,726	1,450	2,229	18,487
Capital expenditures	1988	285	483	120	117	13	1,018
	1989	273	510	101	138	7	1,029
	1990	383	586	131	197	3	1,300
Depreciation,	1988	149	375	88	79	6	697
depletion	1989	151	428	90	90	8	767
and amortization	1990	170	464	117	101	7	859

* Adjusted for two-for-one stock split effective October 20, 1989.

> "Through the successful pursuit of our commitment, we expect our brands to achieve leadership share and profit positions and that, as a result, our business, our people, our shareholders, and the communities in which we live and work, will prosper."

In January 1990, Ed Artzt became Chairman of the Board and Chief Executive as John Smale stepped down after nearly 16 years at the head of the organization. The new top management team believed that globalization was key to the company's growth in the 1990s and beyond. In the 1990 annual report Artzt described the principles of globalization at P&G:

> "Globalization is making products and services competitive with anyone's, anywhere, whether that competition takes place at home or abroad. Globalization means doing a better job than competitors at satisfying consumers' needs, and their demand for quality, no matter where they live. It means creating the network and infrastructure to efficiently compete in the increasingly homogeneous worldwide marketplace.

> "Globalization has special meaning within Procter & Gamble. It means that we will continue to change from a U.S.-based business that sells some of its products in international markets into a truly world company, a company that thinks of everything it does—including development of products—in terms of the entire world. We will increasingly plan the growth of our business, and our technology investments, on a worldwide basis."

P&G also was strongly committed to maintaining and improving the quality of the environment. Following Smale's lead, in 1990 the company published *P&G's Environmental Quality Policy*, which outlined their environmental position:

> P&G continually strives to improve the environmental quality of its products, packaging, and operations around the world.

The company had taken a number of avenues to meet this commitment (See Box 16.1). Steps being investigated include: use of recycled plastics to make household cleaner and detergent bottles; composting or recycling pulp and plastic from disposable diapers; and reformulation of products in more concentrated versions to save space and packaging. The new packaging for Lenor fabric softener in Germany and Downy in the United States was part of this ongoing transition. Although these initiatives were developed within a framework of different market conditions and independent decision-making of the U.S. and German divisions, both were based on hearing the needs of consumers, worldwide.

16.2.3 Development of the Lenor (Germany) and Downy (United States) Refill Packages

The 1987 introduction of Lenor doypack pouches into the West German fabric softener market was keenly observed by P&G executives throughout the organization.[1] This line extension achieved an 85 percent reduction in packaging materials as compared to Lenor's standard plastic container. At the time, P&G personnel in Germany had debated whether to introduce the concentrated version of Lenor in a refill pouch, a bag in a box, or in a cardboard carton.

Lenor had been launched in Germany in 1963 and by the late 1980s had grown to DM 180 million in sales. The concentrated form accounted for 30 percent of Lenor's sales volume even though consumers perceived this as less effective than the diluted version. In 1990, Lenor had become the leading fabric softener in Germany with a 37 percent market share. Priced approximately 10 percent above its competitors, it had battled fierce price competition with aggressive promotion, which in part had led to significant profit and share erosion. For example, during the late 1980s Lenor's sales had declined by 7.5 percent per year and Lenor's marketing team had looked for opportunities to turn the situation around. Moreover, unlike in the United States, German marketing laws prevented P&G

Box 16.1   Procter & Gamble's Approach to Environmental Responsibility

Our Policy
P&G's Environmental Quality Policy publicly affirms our commitment to operating in an environmentally responsible manner, with emphasis on continuously improving the environmental performance of our products, packages, and processes. The policy also serves to unite P&G people around the world in a common commitment to environmental responsibility.

Environmental Quality Policy
In accordance with the above commitments, it is Procter & Gamble's policy to:

- Ensure our products, packaging, and operations are safe for our employees, consumers, and the environment.
- Reduce or prevent the environmental impact of our products and packaging in their design, manufacture, distribution, use, and disposal, whenever possible.
- Meet or exceed the requirements of all environmental laws and regulations.
- Continually assess our environmental technology and programs, and monitor progress toward environmental goals.
- Provide our consumers, customers, employees, communities, public interest groups, and others with relevant and appropriate factual information about the environmental quality of P&G products, packaging, and operations.
- Ensure every employee understands and is responsible and accountable for incorporating environmental considerations in daily business activities.
- Have operating policies, programs, and resources in place to implement our environmental quality policy.

from using coupons and refunds, and contests were tightly regulated by governmental agencies. Advertising on the two state-owned national channels was also restricted in terms of the number of available time slots.

After careful consideration of a number of options the Lenor marketing team had decided to launch the refill package, and when Brad Irwin traveled from the United States to Brussels, Belgium, in the spring of 1988 to attend meetings with his European counterparts, one of the topics of discussion was the success of the German introduction and the ability to transfer that technology to other markets. Irwin recalls:

> "From an international standpoint, one of the marketing issues we talked about was the cross-cultural transfer. We have a tendency to think about a product's life cycle, taking the product from here in the United States to someplace else. Here we were talking about taking it the other direction. The Germans were further along in their thinking than we were in the United States."

Irwin commented on the internal forces driving his division of the U.S. company toward a new packaging decision for the Downy product in the United States:

> "There were two sources, really. One was the technical community, the packaging and product development group which are somewhat integrated, and they were watching the German situation very closely.

> "The marketing community, including finance, sales, and advertising, was also looking at the German situation more from a commercial aspect. We all work for the same company and share information across borders. P&G personnel in Germany were watching from their standpoint to see if this was a consumer acceptable idea. Was this an opportunity to make our packaging less burdensome to the environment?

> "There was a third influence which was growing consumer interest developing around [the environment]. There was also a pragmatic business issue; we had a product called Triple Concentrate Downy which was in the market, albeit not successful, and which was actually the product that we were going to put in the refill and dilute. Simply put, we had the product on the market. We were marketing it in a different way and so we had an infrastructure in place. It was more of a packaging issue. We had the concentrate developed, there was no challenge in that. There was a question of how we could make this concentrated product, which we liked a lot, more successful because it was a good product. We had spent years developing it and frankly it hadn't been terribly successful in the market. Before the Europeans were actually doing the pouch deal there was an eagerness to find a way to commercialize this great technology because the path that we had chosen wasn't terribly successful."

Thus, two years after the introduction of Lenor's plastic doypack in Germany, the U.S. Downy project team wondered if an environmentally improved package could help the introduction of the Downy concentrate and refill in the United States.

16.2.4 ENVIRONMENTAL CONCERN AND ECO-LABELING IN GERMANY

Before answering the U.S. packaging question, management felt it needed to understand more about environmental issues in Germany and the political circumstances that had favored the Lenor pouch refill package. Germany was the European equivalent of California as a consumer market: the continent's biggest, richest country with the largest amount of waste and the strongest environmental lobby.[2] German legislation, passed first in 1975, was intended to reduce waste, increase recycling, and allocate costs based on the polluter pays principle. Under the German application of this principle, the industry generating the waste (or pollutants) is assessed the full costs of assimilating or burning the waste, as opposed to the cost being borne by society in charges for landfill disposal or as environmental degradation. One example evident in Germany was communities that provided inexpensive landfills which industries took advantage of, instead of finding production methods to reduce waste generation. However, the early German legislation lacked enforcement provisions, mandatory targets for emissions or waste reduction, or recycling quotas. Consequently, waste generation increased, landfills reached capacity, and incineration prices continued to climb as there were no government or market penalties for failing to reduce waste.[3]

Eco-labeling was one of the proposed solutions. Such labels were meant to be a prominent identification of a low waste or low pollution product, and the government hoped these would stimulate industry to produce less environmentally burdensome products. The first national eco-labeling program was established in Germany by 1977. The government sponsored a program to designate environmentally friendly products with the Blue Angel trademark. This trademark indicated to consumers that the product met certain national standards. Manufacturers who earned the label, also paid a usage fee, intended to cover program review and operating costs. Generally, about 15 percent of the products in a given category could meet the requirements for the label.[4]

The Blue Angel program issued 500 labels in 1984, 2,000 in 1987, and 3,100 in 1989. Proposals from manufacturers for additional product categories were received each year by the program administrators who approved a limited number of additions annually. The Blue Angel was recognized by 80 percent of German consumers and was credited by some government officials and environmentalists for "stimulating consumer awareness" of environmental issues, inspiring industrial innovation, and reducing the environmental burdens created by consumer products.

With respect to fabric softeners, a small but vocal number of German consumers believed softeners to be superfluous products. This view was held more strongly in Germany than in many other countries in Europe or the United States. Half of the Germans believed that protection of the environment was critical, and this sentiment had grown significantly during the 1980s, in part owing to the success of die Grünen, the Green party. Almost half of the German consumers reported using fewer environmentally harmful products. P&G even weathered a boycott in 1984 of one of its products, Top Job, targeted by some activists as environmentally harmful. Yet, price and convenience prevailed, contradicting the fact that German consumers claimed to be willing to pay more for a product they perceived to have a lesser environmental impact.

16.2.5 Level of Environmental Concern in the United States

Although Irwin and his counterparts agreed that U.S. consumers were less concerned than the Germans about the environment, there was mounting evidence that in 1989 the environment was becoming an important issue in the United States. P&G had purchased, subscribed to, or directly commissioned a number of public opinion studies which addressed the importance of the environment to consumers. In 1982, when consumers were asked to name the two most important problems in the United States, the environment tied for 13th out of 16. Yet, when the test was repeated in 1989, the environment ranked 6th. The clear trend showed increasing concern. Table 16.5 shows the results of a national survey conducted in the fall of 1989.

The findings of the research compiled by P&G indicated that the shift in attitudes among Americans was dramatic and widespread. At the same time, most consumers were poorly informed of the facts, especially scientific facts about environmental issues. Regardless of their level of understanding, however, consumers found environmental issues to be highly-charged emotionally. A significant segment of the population believed that more money should be spent on the improvement and protection of the environment and claimed that they were willing to make trade-offs to benefit the environment. Box 16.2 shows the results of a P&G study.

Early in 1990, the Roper organization, a well-known U.S.–based market research firm, conducted an extensive survey on environmental attitudes and behaviors in the United States. Based upon 1413 surveys, they categorized American consumers into five groups, each group exhibiting a different level of commitment to the environment (Box 16.3). Overall, the study concluded that the majority of Americans were non-environmentalists and that there was a fundamental unwillingness to pay more for environmentally friendly products.

TABLE 16.5
What Do You Think Are the Two Most Important Problems Facing the United States Today?

Third Quarter	1989	1988	1987	1982
Drugs	39 percent	32 percent	14 percent	2 percent
Other social problems	21	23	28	10
Foreign affairs	21	29	47	25
Government spending	17	28	16	5
Poverty	16	11	8	1
Environment	16	8	5	2
Crime	14	10	9	10
Unemployment	9	17	15	47
Economy	9	10	10	37
Moral decay	7	6	8	7
Inflation	6	8	7	33
Government	6	8	15	6
AIDS	5	8	—	—
Energy	1	1	3	3
Other	6	5	7	8
Don't know	3	5	6	2

Table combines first and second choices.

The Roper study identified the degree of commitment consumers felt toward the environment, particularly when there would be additional costs for friendly products. Some consumers indicated they were willing to pay slightly more while others were unwilling to pay any amount extra. The Roper results added perspective to what seemed to be a significant movement among consumers toward environmentalism.

16.2.6 THE FABRIC SOFTENER MARKET—THE UNITED STATES AND GERMANY

In Germany as in the United States, fabric softeners came in two forms: liquid or woven sheets. Liquid fabric softeners were used in the washing machine, usually added during the rinse cycle.

Box 16.2 Results of Questions on the Environment

Respondents agree that:	1986 percent	1989 percent
"I would be willing to give up convenience products and services I now enjoy if it meant helping to preserve our natural resources."	72	81
"I would be willing to spend a few hours a week of my own time helping to reduce the pollution problem."	58	66
"I would be willing to pay as much as 10 percent more for grocery items if I could be sure that they would not harm the environment."	47	64

> **Box 16.3   Attitudes vs. Behaviors**
>
> - Using cluster analysis, respondents were examined individually to see how the behavior of each was similar to and different from other respondents. Three demographic variables—income, education, and gender—correlated most strongly with environmental activism. The more affluent and better-educated, and women, more than men, are likely to be involved.
> - Based on the analysis of similarities and differences, five different groups of consumers were identified from this research, as follows:
>
> True-blue Greens   11 percent
>   Committed environmentalists
>   Strong support for regulations, regardless of cost or
>   inconvenience
>   Engage regularly in pro-environmental practices
> Greenback Greens   11 percent
>   Match true-blue greens in many feelings
>   Willing to pay substantially higher prices for green products
>   Prefer solutions not requiring high involvement on their part
> Sprouts   26 percent
>   Concerned but uncertain about environmental-economic
>   tradeoffs
>   Less willing to make personal sacrifices
>   Important "swing group" in policy debate
> Grousers   24 percent
>   Critical of all players in game
>   Unwilling to pay higher prices for green products
> Basic Browns   28 percent
>   Virtual absence of pro-environmental activities
>   Lack support for regulation
>
> The Roper study says, "There are five distinct groups of Americans when it comes to environmental attitudes and behaviors. Two of them—which represent just under a quarter of the national population—are environmentalists. Two others—representing slightly more that half of all Americans—are not. A fifth group—accounting for the remaining quarter—is a key 'swing' group on environmental matters."

Sheets were used in the dryer where the heat of the drying cycle released the softening agents. Liquid fabric softeners were either diluted or concentrated and were packaged in high density polyethylene (HDPE) plastic containers. In the late 1980s the total German fabric softener market was DM 346 million, but had been declining by 1 to 3 percent per year. Lenor had experienced a 7.5 percent annual decrease in sales, in part owing to the fierce competition it received from Colgate Palmolive (Softlan), Unilever (Snuggle), Henkel (Vernel), and Generic/Private Label.

The decline in Lenor's sales had also been driven by some consumers who had stopped using fabric softeners because of environmental concerns. For example, more than 20 percent of German fabric softener users were aware of the product's alleged harmful effects on the environment and most of those users had stopped using the product category.

The U.S. fabric softener market, including both liquid and sheets, was about $1.1 billion in 1989, continuing a modest growth trend. Despite the size and continued growth of the U.S. market, Irwin was concerned for the future of the business as softeners were not strictly necessary in doing laundry:

> "An important context is the discretionary nature of the product. Your laundry will live without fabric softener, although I hate to say that. The fact that people will readily stop using your entire category cre-

ates a whole new attitude. Laundry detergents . . . People will wash their clothes. They'll try to find a less environmentally burdensome way to do it, but with fabric softeners they can simply stop. In Germany [some] did. That adds some urgency to being out in the lead [in environmental friendliness].

"We know that only 68 percent of households use softener as opposed to 94 percent on detergents. The key reason for not using softener is 'don't need it.' So the environment could feed as a part of 'don't need it,' and after all, 15 percent of households somehow get by without this important additive."

Roughly 68 percent of U.S. households used some form of fabric softener. Some 35 percent used sheets only, 23 percent used liquid only, and 10 percent used both. Fabric softener users tended to use the products for about 85 percent of laundry loads. Heavy users, which tended to be larger households and households with several young children, purchased softeners about once every 2 to 3 weeks. Light users, on the other hand, purchased softeners about once every 5 to 6 weeks. Box 16.4 provides a summary of category statistics.

Downy commanded nearly 30 percent volume share of market (equivalent units vs. competitors). The nearest competitor was Snuggle liquid with roughly an 11 percent share. Table 16.6 shows market share data. Irwin described the positioning of the brands in the market:

"Downy liquid and Snuggle liquid are the two premium-priced brands. Final Touch and StaPuf are a mid-priced segment, about twenty percent lower. Private labels and other liquids are about fifty percent below (Downy). There is a slight pricing difference between Snuggle and Downy, in about the five percent range depending on the geographic location. So, within the premium tier there is a slight tiering between Downy liquid and Snuggle. [Snuggle's] strategy is softness that is a little less expensive. That's what they have on the packaging, that's what the advertising says. They want to set the stage for an incentive to be competitive at a five to ten percent lower price than Downy liquid."

16.2.7 TEST RESULTS FOR THE LENOR AND DOWNY REFILLS

In order to address Lenor's declining sales, the German marketing team had considered two options, both aimed at capitalizing on consumers' environmental concerns. Some on the team believed that relaunching the Lenor concentrate would be best, whereas others felt that a refill pack would be better. The refill pack could be a laminated cardboard carton, a bag in a box, or a soft plastic "doypack" pouch.

German consumers seemed to like the laminated cardboard carton better, but the results of the consumer tests were close. Both designs offered an 85 percent reduction in package material volume, but each had its advantages and disadvantages. For example, Germans had experience with the cardboard carton and perceived it as more environmentally benign, even though the plastic-coated box was harder to recycle.*

Given the German experience, Irwin and the Downy team were convinced that the concentrate and refill concept was worth testing in the United States, and in 1988 began a series of concept and use tests. By early 1989 the tests were complete and the carton was chosen as the new Downy Refill (US) package. Table 16.7 presents the test results. Irwin stated:

"We took the carton and pouch to a concept and use test. The consumers see the concept, they rate the product at that point in time, and they tell us whether they would buy it or not. We give them the product to take home and use and then we interview them about four weeks later. They give us a rating of 'not very good' to 'excellent.'

---

*Yet, vis-à-vis the United States, the laminated carton was easier to recycle in Germany because a fairly extensive paper recycling infrastructure existed. Virtually no plastic recycling could be done, however, owing to the failure of an earlier attempt to set up plastic recycling facilities. In the United States, at that time, neither of these packages could be recycled to any significant extent nor could the original HDPE plastic bottles.

> **Box 16.4   Fabric Softener Study — Category Background**
>
> 1. Market size ($ Millions)
>
> | | 1987–88 | 1988–89 | 1989–90 |
> |---|---|---|---|
> | | 975 | 997 | 1,054 |
>
> 2. Consumer habfits (percent of U.S. households)
>
> | | 1989–90 |
> |---|---|
> | Use washer | 91 |
> | Use dryer | 82 |
> | Use any fabric softener | 68 |
> | Use liquid only | 23 |
> | Use sheet only | 35 |
> | Use both | 10 |
> | Non-users | 23 |
>
> 3. Loads per week
>
> | | 1989–90 |
> |---|---|
> | Average loads per week | 7.7 |
> | Average loads per week softened — liquid users | 6.9 |
> | Average loads per week softened — sheet users | 6.6 |
>
> 4. Purchase frequency
>
> | Average: | 3–4 Weeks |
> |---|---|
> | Heavy users: | 2–3 Weeks |
> | Light users: | 5–6 Weeks |
>
> 5. Benefit ranking
>   - The most important benefits for liquid users are:
>   1. Softness
>   2. Freshness
>   3. Static
>   - The most important benefits for sheet users are:
>   1. Static
>   2. Freshness
>   3. Softness

"What we found was that, frankly, both the carton and the pouch were acceptable. It wasn't a clear choice between the two, but the carton had a slight advantage in purchase intent. We also had some other data that suggested that the inconvenience of the doypack was an important component and that [U.S.] consumers appreciated the handleability of the carton. At this point we decided internally to go ahead with the carton.

"Another component of a decision like this is how easy it would be to get into the business. The carton had advantages in the sense that there is machinery that you can buy that fills milk cartons. It is relatively easy to buy on the market."

There was some concern, primarily in the United States, that the cardboard carton looked like a milk carton and could lead to accidents.* In order to address this concern, P&G undertook a number of actions. They consulted with poison control centers about accidental ingestion. Package artwork and dimensions were used which differed significantly from milk and juice cartons. The relative non-toxic nature of the product was also taken into consideration in the U.S. packaging decision. There

---

* This was, as was discovered later, less of a concern in Germany than in the United States.

TABLE 16.6
U.S. Fabric Softener Market Shares—Food*

	Volume Equivalent Units 1989–1990
Downy liquid	29.7
• Bottles	29.7
• Refill	NA
Snuggle liquid	11.1
Final Touch liquid	4.6
StaPuf liquid	0.9
Private label liquids	5.4
All other liquids	2.1
Total liquids 53.8	
Sheets 46.2	

* Does not reflect other distribution outlets such as discount stores, etc.

TABLE 16.7
Refill Concept And Use Results—1989

	Carton	Pouch
Overall rating percent	72	73
Definitely would buy—percent	34	30
Favorable comments—Percent		
Odor	52	52
Softness	46	39
Convenience	33	31
Economy	12	16
Environment	12	15

also were some negatives associated with the pouch package design. The pouch, when tested, seemed messy because it had a tendency to spill. It also had technical difficulties, although it was felt that those technical problems could easily be overcome.

Following the concept and use tests, the company undertook a series of focus groups and supporting quantitative research to determine the best approach to present the new package to the public. The approaches being considered were economy, environment, or a combination of the two. The environment only approach quickly proved to be the weakest of the three, and was dropped from serious consideration. Table 16.8 presents the test results. Irwin stated:

> "One of the big debates was that there were two viable alternatives that came up in positioning. One was economy, and one was environment and economy. When you offer Downy in a carton there was a clear expectation that it ought to cost less. So, the consumers understood the cost concept in a heartbeat. Plus, we had this situation where we had a competitor at a 5 to 10 percent discount [vs. Downy], so it became very appealing for us to give the consumer a less expensive way to buy Downy.

> "So, there was quite a bit of debate about whether we only focus on economy or add the environmental issue. The environment was a double-edged sword. We were concerned about doing it badly, and one thought was let's just talk the economic aspects of it and let the consumer get the environment for free."

TABLE 16.8
Refill Concept Only Results—1989

	Alternative	
	Economy Only Concept	Environment/Economy Concept
Definitely would buy (Percent)	46	48
Favorable Comments—Percent		
Convenience	21	22
Economy	72	59
Environment	31	48

The economy message derived from cost savings associated with both the concentrate and the refill package. The expected manufacturer's selling price per case for the new package was 24 percent below that of the existing plastic bottles. Manufacturing costs were 19 percent below that of the existing Downy bottles. Reduction in marketing and promotional expenditures brought the refill profits to parity with the current bottles. Irwin expected the reduction in marketing and promotional costs to come largely as a result of a "low everyday pricing" policy which limited the variation in pricing brought about by promotional dealing at the retail level. Refill economics are presented in Table 16.9.

In September 1989, the doypack concept was introduced in Canada, and in November the Downy Refill began test marketing in the Washington, D.C./Baltimore area. As usual, the test market was supported by print and television advertising as well as other promotional activity at nationally equivalent levels.

The test market results appeared to be very positive. After six months, distribution had slightly surpassed its target and sales had nearly doubled forecasts. Concerns over cannibalization appeared to be unfounded as total sales for Downy Liquid, both bottles and refills combined, were up for the period. Test market results are presented in Tables 16.10, 16.11 and 16.12.

16.2.8 CONCLUSION FOR FABRIC SOFTENERS

Brad Irwin couldn't help but wonder if things weren't looking "too good" for the Downy Refill project. Throughout the testing process the results had been consistently positive. The consumers were receiving a product that cost less while also providing an environmental benefit. Trade acceptance and consumer purchasing patterns had exceeded P&G's forecasts. The next step should be national distribution.

Still, there were questions that concerned Irwin. The test market had been conducted only in the Washington, D.C./Baltimore area. If the market were somehow unrepresentative of the national market, going national could be a mistake. The test market had only run for six months. Traditionally, P&G had run test markets for a year or more in order to allow buying patterns to stabilize. The inconvenience of refilling the Downy bottle with concentrate and mixing with water could become bur-

TABLE 16.9
Refill Economics—1991 (Index vs. Bottle)

	Average bottle	21.5 oz. refill
Price/case	(100)	(76)
Cost/case	(100)	(81)
Marketing support/case	(100)	(41)
Profit/case	(100)	(100)

TABLE 16.10
Refill Test Market

	Refill Test Market Share Results		
	Base (11/88–10/89)	Test (1/90–5/90)	Difference Points (Percentage)
Baltimore/Washington			
Refill	—	7.4	+7.4 (—)
Liquid Downy bottles	25.6	21.5	−4.1 (16%)
Total Downy liquid	25.6	28.9	+3.3 (13%)

TABLE 16.11
Consumer Ratings of Downy

	Pre-Refill	6-Months Post-Refill
Being good for the environment—percent	62	81
Being economical to use—percent	59	69
Softness—percent	78	76
Freshness—percent	79	76

TABLE 16.12
Demographics of Downy Refill (Index to Total Households)

	Snuggle Liquid	Refill	Downy Bottle
Female head of household 18–24	23% (116)	27% (132)	20% (97)
4 + household size	35% (121)	35% (121)	30% (104)
Children 6–11	22% (117)	23% (119)	20% (104)

densome over time, which would be demonstrated by consumers resuming old buying and use habits. However, repeat purchase behavior did not yet show such a problem.

Irwin wondered if the publicity surrounding the environmental friendliness of the Refill package might backfire. Since fabric softeners weren't strictly necessary, might not consumers take the attitude that P&G was making too much of the new packaging?

The debate over the price positioning still lingered in Irwin's thoughts. If the price were raised just a little, the economy message could still be used, and the profitability of the product could be improved dramatically. The test market would be the place to determine the outcome of such a decision.

Irwin felt some urgency in pushing the new product to go national. The technology which enabled Downy to be produced in a concentrated form was unavailable to the competition, thus creating a window of opportunity for the company. The longer he waited in test, the more time the competition would have to develop and market their own concentrated formulation.

16.3 PART II: RECOVERING PLASTIC WASTE FOR P&G'S PLASTIC BOTTLE PRODUCTS

Coincident with the development of refill concentrates in both Germany and the United States, Procter & Gamble personnel in the United States determined that the public's concerns about solid waste included plastic and plastics packaging. On the one hand, consumers liked plastic packaging's light weight, shatter resistance, and low cost. But on the other hand, they believed it could not be safely incinerated, and would remain forever in a landfill. In addition, for many activists, plastic packaging was symbolic of what they described as a throwaway society. Therefore, P&G personnel in the United States identified plastic packaging as the main material providing the opportunity for use as a recycled material, and that participation in utilizing post-consumer plastic recycling was an appropriate goal. While some agreed that packages made from recovered plastics packaging would never surmount the technical and infrastructure challenges without prohibitive capital and research expenditures, others among P&G's suppliers were willing to work with the company to determine feasibility. P&G concluded that its primary contribution to the company's plastics recycling would be to take a leadership role in demonstrating that past-consumer plastics could be recycled into plastic packaging.

Years earlier, the bottle deposit legislation had been introduced in some states to reduce roadside littering, while increasing reuse of glass bottles. However, plastic bottles cannot be cleaned thoroughly, and other issues such as very high costs and inefficiency precluded the bottle deposit approach from stimulating novel, market-place solutions to solid waste. The problems are compounded when consumers discard the reusable containers on which a deposit is paid, even as manufacturers rely on the returns for their production processes.

16.3.1 P&G's Market Incentives for Plastic Recovery and Use in the United States

P&G managers in the United States, such as Tom Rattray and Jack Sneller, who were considering packaging options for the new Downy product, recognized that, in pursuit of their company's environmental policy, their goal would be to design an innovative and effective alternative to conventional plastic bottles. P&G had begun using recycled material in metal, glass, and fiber packaging many years earlier, and those decisions had been made based on consumer needs, cost and technical performance issues, as well as on conservation. The same would have to apply now to the new Downy packaging. For all practical purposes, the standard practice for new packaging with recycled materials had to be the meeting of previously established performance standards. Rattray of P&G's Corporate Packaging Division explained:

> "We really did go out there and set the standards and say, 'We're not going to have recycled bottles unless they're just as good quality as our virgin bottles.'"

In 1988, P&G had been approached by the firm that made the Spic & Span Pine bottle and P&G agreed to explore the use of 100 percent recycled polyethylene terephthalate (PET) to produce some of its plastic bottle requirements. This was a technical decision. After the test bottles were shown to meet specifications, P&G management made the decision to let the public know how well plastic could be recycled, done at a formal press conference in November 1988. The success in using post-consumer plastic waste in this one product led P&G to consider incorporating some recycled plastic in the manufacture of bottles for other cleaning products, shampoo, paper product wraps, and bags.

The primary source for recycled plastic for most P&G products initially would have to be post-consumer milk bottle resin (high density polyethylene HDPE). P&G also considered incorporating detergent bottle resin (copolymer HDPE) into new detergent bottles, but the technology was uncertain. Another challenge existed in achieving sufficient color sorting of the recovered plastic waste to ensure package color consistency. PET from a wide variety of packages (soda bottles, peanut butter jars, salad oil bottles, etc.) could be used for clear bottles, but the latter could not be made using any of the recovered colored plastics. Rattray explained:

"Our problem is that you don't make blue bottles, you don't make yellow bottles, or you don't make fluorescent Tide bottles if you've got [waste plastic] that's black. So, what we did very quickly, which is typical of what we do in this kind of industry, [was] we established parameters that become a standard for the country and sometimes the world. Today there is very good, clean post-consumer recycled material that you can use as you use virgin material, with a standard level of colorence to get to where you want to be. I think it's only because we turned a lot of trucks around in the early days. We said, 'Nope, that's not good enough. We're not going to use it.' "

Fortunately, these new processes did not constitute a major financial decision for P&G, as these developments were considered to be "just another material qualification," part of the responsibility of the various groups handling product packaging. The development team determined that modest capital investments would be required on the part of their supplier companies, as different types of plastic bottles, some with 25 percent recycled content and others 100 percent recycled, could be made in the same plant as those from virgin material, with only minor input modifications. But those decisions would be made by bottle suppliers, not P&G. The suppliers would build their development costs into the price of bottles sold over an extended period of time, if P&G could assure a long-term purchase contract. That would be a significant strategic decision internally for P&G, but not one involving great risks for the company. The biggest issue would be to ensure adequate quality and quantity control in the supply of recovered post-consumer plastic waste.

16.3.2 Closing the Recycling Loop in the United States and in Germany: Comparisons

To meet consumer demand for environmentally friendly products worldwide, and comply with government regulations on waste disposal in many countries, P&G constantly sought technologically advanced production methods and environmentally sound recycling techniques. Dr. Deborah Anderson, Vice President, Environmental Quality—Worldwide saw the problem as steps in a loop they would have to close:

$$\text{Households} \rightarrow \text{Recovery processes} \rightarrow \text{End-use markets} \rightarrow \text{Households.}$$

To be a major player in solving the solid waste problem, P&G would have to enter this loop, ideally by becoming an end-use market, she said. In the Unites States, as in other countries, the company had begun to use Life-cycle inventory (LCI) extensively to identify priority areas for improvement in production processes. Life cycle analysis (LCA),[5] an alternative approach, establishes not only the use of resources and resulting discharges to air, water, and solid waste, but also the net impact on the environments they affect. On the other hand, LCI stops at the determination of resource consumption and discharge volumes, a technology that is better understood and agreed upon by the scientific and environmental community than LCA, which is more complex.

An early example of the significance of change in the cycle was seen in Germany where source reduction for the consumer was achieved through pouch packaging of the new Lenor container. This step alone reduced packaging volume for Lenor in Germany by 85 percent. Large plastic bottles purchased in Germany for P&G's Ariel, for example, were being repeatedly refilled in the home with diluted concentrate from the Ariel concentrate purchased there. However, P&G management in Germany believed that, in their market, it would be economically and environmentally feasible to reuse only 25 percent of post-consumer plastic in new bottles, because of material contamination and lack of uniformity. An option being considered in Germany was to use plastic as an industrial chemical feed-stock, rather than as relatively low-value input to the manufacture of plastic bottles, reflecting the culture and government of that market.

Al St. Clair of P&G's Worldwide Environmental Quality Coordination Division described some of the pressures which drove the German commitment to recycling:

> "The first factor is that if you look at Germany vs. the United States, they are at about 3.8 percent of the land area that we are. When you take a look at the German people, they have a much different ethic around conservation of land. The second factor is that they have 8 times the population density that we do; they have about 80 million people crammed into an area the size of Nevada. So, when someone says to these folks, 'land fill crisis,' they really believe land fill crisis. The third factor is that when you go out to the dump in Germany, you'll find that people are paying 4 times as much as we are [in tipping fees] and that those costs have been escalating at about 50 percent per year."

In the United States, P&G's decisions regarding waste reduction through recycling were complicated by still other issues. Among them were the differences in the level of environmental concern shown by consumers from region to region within the United States. The economics of small volume recycling were such that the cost of recycled material had been, at times, more expensive than virgin material. Further, collection channels for post-consumer waste would have to be developed to a much greater extent in order to provide the quality and volume needed for P&G to use recycled material consistently in their production. Some individuals among P&G's suppliers wondered whether a sufficient, high quality supply would be available to P&G in the United States to make bottles with a range of recycled content, up to 100 percent.

While public concern about the environment was growing in strength, research had shown little evidence that consumers were willing to purchase everyday products based upon the environmental friendliness of their packaging. Rattray commented:

> "For the general public, so far, I haven't found any tangible evidence I can confirm that says recycled content sells products.

> "That's a concept that I think we just haven't gotten the public educated on yet. They want to recycle, but looking for recycled content is not a big deal. "When you think about it, I don't think it'll ever sell soap.

> When you go to buy that can of corn, you really don't think about how much of the steel in the can has been recycled. You want to come out with the corn. Ditto Downy or Tide or whatever. I have been looking unsuccessfully for any example of a recycled content product that gained in sales because of that, and I know of [only] two categories where products could do that. One is paper products and the other is trash bags.

> "A real question in people's minds [is] whether this [interest in recycling] is for real. Is this a quickie thing, a fad that in four years is all gone away?"

There were reasons, also, to wonder whether the cost of recycled material for plastic bottles in the United States would be greater than the cost of purchasing virgin material. Supplies depended on uncertain technologies (especially cleaning), consistency, and volume. Costs of collecting, sorting, and cleaning the recycled materials, combined with the growth in production capacity for virgin material, contributed to the possibility of a pricing imbalance. Further, there were concerns that once a commitment had been made to generating a continuing stream of recycled material, production of recycled products would need to develop quickly or the demand for recycled material could exceed the supply, making the cost imbalance even worse. Rattray stated:

> "Virgin resin is cheaper or at best at parity with recycled, and it's pretty hard to talk people into paying more for a used car rather than a new car.

> "It's the whole evolution of the system. Of course, as soon as you get cheaper, you get a whole new set of dynamics. So far that hasn't happened, if you look at recycled fiber in cartons . . . Our detergents cartons have been made out of recycled fiber for years and years because it's cheap. You never even thought

about it. That's just the way you made them. But [with plastic] you can't do in 5 or 6 years what has taken 100 years for other industries."

16.4 CONCLUSION

According to P&G's life cycle inventory, new approaches, including source reduction and large-scale recovery of recycled plastic for new packaging, would benefit the environment. There was some agreement that long-term commitments by P&G could help to define an overall market *approach* to plastic waste reduction. At the same time, different markets and conditions in the United States and overseas made the issues and choices complex.

Anderson summed up the company's approach:

"We don't have a One Right Answer. We haven't said that the Downy refill is what we have to go with everywhere in all countries and in every part of the United States; that we have to go with bag and box everywhere, or we have to go with a recycled-content package everywhere. What we are trying to do is confirm that all of these are steps in the right direction, and then we will tailor to the local infrastructures, local consumer preferences, and local economies.

"An example: In Canada we have pouches for Downy. In Canada the consumers were really used to using pouches because there they buy milk in pouches. In the United States, the consumers said, 'Yuk, I won't buy this.' So, it made sense to go with source reduction through use of pouches in Canada. They didn't have that much plastic recycling set up then, so source reduction was a good move.'

"The pouches were also a good move in Japan, because there they (Japan) have a high rate of incineration and they preferred a pouch over a paper carton because it gave them higher energy recovery when incinerated."

16.5 REFERENCES

1. Quelch, J., Drumwright, M., and Yao, J., The Procter & Gamble Company, Lenor Refill Package, Harvard Business School Case No. 9-592-016.
2. Megan, R., Packaging a revolution. *World Watch,* 6(5), 28, 1993.
3. Cairncross, F., How Europe's companies reposition to recycle, *Harv. Bus. Rev.,* 70(2), 34, 1992.
4. Bureau of National Affairs (BNA), Cabinet approves amended waste law despite widespread industry objections, *BNA International Environment Daily,* April 2, 1991.
5. Kuta, C. C., Koch, D. G., Hildebrandt, C. C., and Janzen, D. C., Improvement of products and packaging through the use of life-cycle analysis, in *Resources Conservation and Recycling,* Elsevier, New York, 14(3/4), 185, 1995.

17 Electricity Load Management and Stakeholders: Cincinnati Gas & Electric Company Case Study Number 9

H. Gregory Hume, Alison Ohl, Raymond F. Gorman, and Orie L. Loucks

CONTENTS

17.1	Introduction	323
17.2	A Changing Industry	324
	17.2.1 Load Management and Demand-Side Management	324
	17.2.2 Industry Response to Conservation Programs	326
17.3	The Cincinnati Gas & Electric Company	327
	17.3.1 Regulatory Setting	327
	17.3.2 Integrated Resource Planning	327
	17.3.3 Cost-Effectiveness	329
17.4	Public-Private Consensus-Building for DSM	331
	17.4.1 The DSM Collaborative	332
17.5	Industry Deregulation and DSM	334
17.6	Conclusion	335
17.7	References	335

17.1 INTRODUCTION

On a Monday morning in early August 1996, the commissioners of the Public Utility Commission of Ohio, PUCO, knew they had a long day ahead of them. They had completed several days of hearings on continuation of the demand-side management program (DSM) of Cincinnati Gas & Electric Company, CG&E. The Company had been working since 1989 on establishment of DSM, but with the prospect of major changes in the electric industry, steps toward deregulation and open competition for markets, the company felt that the benefits of DSM would now be less than the costs, and not in their customers' best interest. The Commission had to decide whether to continue with the existing programs, the result of years of planning, or to cut back on DSM in light of the deregulation taking place in the electric utility industry.

To assist with its deliberation, the PUCO had contracted with several industry consultants. They recommended that the Commission should start the deliberations by reviewing the history of the industry and the demand-side management program.

17.2 A CHANGING INDUSTRY

Traditionally, successful and profitable operation of electric utilities was accomplished by investing in capacity growth of electricity production at the best price possible for their service area, thereby attracting more expansion. The nature of the business had been such that utilities increase sales and revenue by making investments in new capacity, while petitioning the regulatory commission to allow them to recover those costs by adjusting rates to provide a fair rate of return on their investment. At the same time, a growing demand for electricity in all aspects of our society created a further impetus to increase generating capacity. Costs remained low in areas where economies of scale were achieved, encouraging industrial and commercial consumption, thereby reinforcing the cycle. During the 1970s, these forces created strong incentives for growth on the part of utilities, and eventually led to overbuilding of electric generating capacity and concomitant pressure for substantial rate increases. As rates began to rise, owing largely to fuel price increases and other events of the 1970s, discontent on the part of energy consumers began to be vocalized.

In conjunction with increased concern for environmental protection, attention turned to problems of global dependence on fossil fuels and environmental damages from their use. Ideas for conservation and load management (C&LM) were proposed as a partial solution. Load management refers to programs that shift demand from one time to another, typically in an effort to reduce peak demand during the workday.[1] Litigious conflicts between utilities and intervening environmental groups, reinforced by the traditional rate setting process, tended to result in more dollars being spent on C&LM programs, but without the full support of the utilities who continued to think in terms of capacity growth. Utilities were apt to take the defensive position that only utility people knew what was going on, and that the environmental concerns were unreasonable. As the economy climbed out of a slump in the mid-1980s, attention fell away from conservation ideas and a focus returned to the costs of compliance with new regulations, adding more pressure for rate increases or load management. The improving economy also brought growing interest in environmental issues worldwide.

17.2.1 Load Management and Demand-Side Management

In the mid-1980s, Least-Cost Planning (LCP) emerged across the United States as a pro-active regulatory tool promoting the use of excess capacity available at that time. LCP is a model of utility planning that opens up options for capital investment in specific solutions that would provide the needed energy at the least-cost.[2] LCP could use conservation programs to slow down the growth rate of electricity demand, in support of the idea that, even though there was a current excess, it would be wise to milk the excess as much as possible to delay building new capacity as long as possible.[3] By encouraging decreased growth, or no growth for the present, new building and its capital costs could be postponed for years. This postponement was attractive in terms of reducing near-term overall costs. Thus, a drive among regulators sought to reduce the need for expansion of energy generators. It was likely that the prospect of increased costs for environmental compliance at those facilities had an impact as well. For example, when comparing investment patterns, an option such as a new plant could lead to electricity costs of $0.05 per kWh, whereas a conservation program such as promotion of efficient refrigerators, could, effectively, increase available power at a cost of as little as $0.02 per kWh.[2]

From these beginnings, a trend toward demand-side management (DSM) developed. Until the energy crisis of the 1970s, this demand-side management was almost exclusively of the load-building variety, such as convincing homeowners to replace gas stoves with electric ones.[4] Since then, DSM has become widely known as a method that allows the electric companies to reduce or even eliminate the need for building new capacity.[5] It relies on the proposition that, while electricity to meet both current and future demand is typically increased by the conversion of supply resources (fossil fuels), it also can be increased by decreasing the daily use of energy by current consumers.[6] Changes can occur in the pattern of use as well as in overall level of use.

On any given day, the demand for electricity within a utility system follows a predictable path that peaks at some point (mid- to late-afternoon in the summer air conditioning season, for example) and reaches a minimum during the late night of that 24-hour period. The utility must install sufficient generating capacity to meet the maximum demand or peak load, but 50 percent or more of this capacity will not be required during off peak hours. During peak hours, industrial customers may be charged a demand charge in addition to their kilowatt consumption, based on how much they contribute to that peak demand, reflecting the cost of needing that extra generating capacity. Residential customers usually do not have this charge.

Utilities have developed a number of load management strategies to help alleviate the strains of peaks and the unused capacity valleys (i.e., trying to level the load over the 24-hour period.) These include peak clipping, load shifting, strategic conservation, valley filling, strategic growth, and flexible load. Figure 17.1 illustrates each of these load-shaping strategies. The horizontal axis represents 24 hours of the day beginning at midnight. The vertical axis is the level of energy demand placed on the utility at that time. Some of these options, like strategic load growth, do not suggest the use of demand-side or strategic conservation programs. In fact, such a strategy may encourage increased electricity generation. However, those strategies or incentives that aim to decrease the peak level are consistent with, and often obtained by, DSM programs. They are effective in reducing the need to install new generating capacity and are encouraged by regulatory commissions as an effective means of lowering customer utility bills.

Still, not all of these programs necessarily mean that the utility will reduce overall energy consumption, because consumption may increase during off-peak periods to equal or exceed the consumption avoided during peak periods (i.e., the load shape is leveled, but the total consumption can be greater than before).

Electricity conservation programs are another approach to improving the efficiency of basic electric power utilization. Some studies have suggested that 30 to 75 percent of current electrical

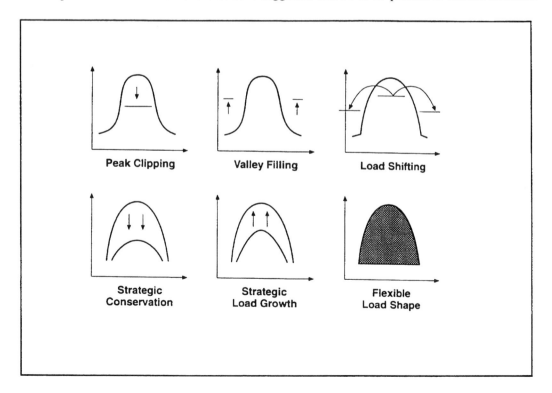

FIGURE 17.1 Pattern of daily electricity loading and strategies for shifting the peak.

consumption might be eliminated at costs between $0.006 and 0.026 per kWh.[7] Savings derive from proven technologies such as replacement of tungsten filament lamps with compact fluorescent bulbs, improved insulation, industrial motor drive controllers, and occupancy sensors, but the computation is complex owing to the multiple cost elements of electricity production described under the Regulatory Setting, below. The savings noted here may be comparable to current retail electricity prices, . . . averaging about $0.07 per kWh.[8]

In 1992, a study suggested that potential incremental savings from DSM programs, between 1991 and 2000, could range from 2 to 19 percent of all electricity sales in 2000, and 8 to 19 percent of peak demand in 2000.[1] A separate 1990 report suggested that forecasted DSM savings for CG&E could reach 0.8 percent in the year 2000 compared to other Ohio utilities whose forecasted savings ranged from 0 to 2.76 percent.[2] Another study in the state of New York estimated that power from new plants would cost $0.05 to 0.10 per kWh (including the energy supply component and the demand component, see later sections), while power from DSM programs, the demand component by itself, may cost $0.014 to 0.050 per kWh. Power from many existing plants begins at $0.03 per kWh and runs higher.[1] Managers in electricity production point out, however, that these prices should be structured to reflect *cost per kW of generating capacity*, which would be appreciably different from the market prices just cited. In any case, estimated costs or savings from these conservation programs yield a wide range of potential benefits, and one goal of DSM was to create ways by which there could also be a benefit to the utility itself.

17.2.2 Industry Response to Conservation Programs

Since the mid-1980s there also has been a substantial change in attitudes toward conservation or efficiency measures on the part of both regulatory commissions and utilities. As one observer has put it, "During the energy crisis of the 1970s, consumers were responsible for energy conservation; today a large part of the burden has shifted to the utility."[4] There are many explanations for this shift. One influence is the success of the conservation measures adopted in the 1970s, albeit forcibly, owing to higher prices and a tight fuel supply, a lesson that has not been forgotten.

In addition, an early industry trend toward low-pollution nuclear power had not proven prosperous. Nuclear power seemed like an excellent source of electricity, even considering the waste disposal problem. However, encounters with out-of-control construction costs, extended completion times, systems that simply did not work, and vocal, local citizens opposing projects, dampened the prospects. These difficulties meant not only that utilities would be short of projected capacity and might require more traditional capital investment, but also that some had incurred huge costs for failed nuclear start-ups that either customers or shareholders would have to cover.

Federal and state pollution regulations begun in the 1970s also continued to be a cost consideration for operation of fossil-fuel-based generating plants. The potentially more restrictive regulations implied in the Clean Air Act Amendments of 1990 also made the large investments in new plants and pollution prevention technology unattractive, especially in comparison with lower-cost DSM programs. A utility might have invested in a proven pollution control technology and find shortly thereafter that the regulations and standards could have changed. DSM programs were designed to be relatively flexible, not requiring a great deal of risk on part of the utility. They were characterized by a low initial investment cost with a fairly high return. Often, these programs could be used even as a marketing tool through customer relations.*

With rising costs for regulatory compliance (e.g., flue-gas scrubbers), DSM initiatives such as conservation or efficiency programs can appear to be a more cost-effective way for a utility to be increasing overall demand, as long as it is distributed satisfactorily. Thus, conservation and load

---

*By 1990, DSM programs yielded a 1 percent decrease in U.S. electricity consumption and a 4 percent reduction in peak demand. According to utility plans, DSM programs should yield a reduction of 4 percent (consumption) and 6 percent (peak demand) by the turn of the century. See Reference 6.

shaping were inextricably linked. Changes in technology and pressures from regulatory bodies also contribute to how each option is considered. Regulatory agencies have cooperated in the recent past by modifying cost-recovery mechanisms in the rate-determining process, to allow recovery by the utilities of some avoided costs. Furthermore, avoiding construction of new plants usually means avoiding rate cases before the Commission that ask for increases and open the utility up to public purview.

17.3 THE CINCINNATI GAS & ELECTRIC COMPANY

In 1992, the Cincinnati Gas & Electric Company found itself dealing with all these changes in the industry. The company then served 699,000 electric customers and 416,000 gas customers over a 3,000 mi^2 service area. It owned a peak generating capability of 5,044 megawatts, generating an electricity peak load of 4,002 megawatts. Its annual report showed a total operating income of $260 million, $236 million of which was from electricity produced. Its 1991 heat rate of 9,886 Btu per kWh ranked the company as the second most efficient electricity producer among 100 of the largest investor-owned utilities in the nation and first among the coal-burning utilities.* It was expected that for the period 1993 to 2013, "over 90 percent of the annual system energy requirements would be supplied by coal. . . . The majority of coal acquired by the [company would originate] from the states of Ohio, Kentucky, and West Virginia."[10] This region typically provides high-sulfur coal that creates high pollution considerations.

17.3.1 REGULATORY SETTING

The CG&E Company is regulated by the Public Utilities Commission of Ohio (PUCO). This organization, among other responsibilities, oversees and sets utility rates for CG&E customers. Rates are comprised of two parts, consisting of a fuel-based cost portion and a base rate. The first accounts for the daily costs of fuel used to provide a given amount of electricity. The base rate is a regulatory-determined base dollar amount that allows the utility to recover costs of its capital investment (e.g., a new plant) and a fair rate of return (see Box 17.1). CG&E can file, at its own discretion as needed, for a rate case to adjust its base rate. It is required to file annually as to the cost-of-fuel portion, but usually the parties agree there is no need for a hearing procedure. In a full rate case, the company provides an application with projected revenues and an appropriate rate of return. The Public Utilities Commission staff reviews the documentation, and then files their own judgment. Any interested party, or intervenor, may respond to that report. Then a commission hearing is held where the staff report becomes the focus of testimony.[3] The Ohio Consumers' Counsel (OCC), an office funded by a surcharge on utility bills, serves as a representative for residential customer concerns on rate issues with non-municipal utilities. The OCC frequently prepares a response to the report. Other typical intervenors include environmental groups such as the Sierra Club.

17.3.2 INTEGRATED RESOURCE PLANNING

In the late 1980s, a small group of people close to Ohio regulatory activities were proposing changes in PUCO standing rules consistent with the ideas of least-cost planning, although there were varying degrees of support among the five commissioners.[3] "[It was] mostly an internally driven idea,

---

*In a move consistent with industry trends, CG&E began taking steps to merge with PSI Resources, Inc., of Indiana in December 1992. The two companies are now operating divisions of the new company, CINergy Corporation. CG&E's subsidiary, The Union Light, Heat and Power Company of Kentucky, and the non-regulated businesses are subsidiaries of the new corporation. The merger created the 13th largest investor-owned electric utility in the nation (based on generating capacity), and is expected to have decreased revenue requirements and reserve margins. Cost savings are expected from lower fuel costs, reduced operating and maintenance expenses, and delayed capital investments. See Reference 9.

> **Box 17.1   Elements of the Rate-Making Process**
>
> The rate-making process looks at data for a test-year of expenses. This test-year is a twelve-month period, not necessarily a calendar year, three months of which have to be actual numbers over which revenues and expenses have been determined. The required revenue is equal to operation and maintenance expenses, depreciation of capital, taxes, and a rate of return. The rate of return is based simply on the product of the rate base and a weighted cost of capital. This capital includes common equity, preferred stock, and long-term debt. A disincentive for DSM programs can occur if a program does not include a capital expenditure that could be included in the above calculations. The equation can be seen as follows:
>
> $$\text{required revenue} = (\text{operation and maintenance expense}) + (\text{depreciation}) + (\text{taxes}) + (\text{rate of return}),$$
>
> where
>
> $$\text{depreciation} = \text{the portion of utility plant consumed in providing utility service,}$$
> $$\text{rate of return} = (\text{rate base})(\text{return on investment}),$$
> $$\text{rate base} = (\text{net plant}) + (\text{working capital}) - (\text{deferred taxes}),$$
> $$\text{return on investment} = \text{weighted cost of capital including common equity, preferred stock and long-term debt,}$$
> $$\text{operation and maintenance} = \text{documentation from plant records.}$$

[derived] from keeping track of what was going on elsewhere [in the nation]".[3] Steve Puican, an economist in the Utilities Department of the Commission, also noted the advantage of DSM:

> "to avoid the need to build additional capacity in the future or defer it to the extent possible. DSM means different things to different people. We look at it in terms of the least cost plan. If you can reduce the growth rate in demand for less than what it would cost you to have to build something to meet that demand, then rate payers will be better off, because in the long term they will have lower rates."[3]

The state of Ohio passed legislation in 1983 requiring utilities to file annual 20-year forecasts. These reports were intended to help determine whether utilities, in the absence of a truly competitive marketplace, were overbuilding and overforecasting demand for their electricity. In 1988, the Commission opened up its rulemaking to modify the forecasting rules and require integrated resource planning (IRP). It required companies to look at demand-side management programs as a way of developing resources, an alternative to their traditional supply-side option.

Initially, there was hesitancy on the part of utilities to use demand-side resources for electricity supply, and the initial rules contained several financial disincentives. Additional hearings produced guidelines outlining what costs could be recovered in the rate base, and how they were to be recovered. By statute, base rate adjustments, which would include the DSM program effects, were not allowed in the absence of a rate case. In an effort to avoid the necessity of legislation, the Commission tried to arrange suitable incentives within its own authority by allowing utilities to put relevant costs owing to DSM project outlays into deferred accounts for ultimate recovery in a subsequent rate case. Thus, utilities could invest in DSM programs without the lengthy negotiations with the Commission and apply later for recovery of the costs. Although communications between the utilities and those that would approve the costs were encouraged, the process was not risk free.

Electricity Load Management and Stakeholders

The purpose of the new integrated resource planning was to help prepare CG&E to meet its 20-year forecasted electricity demand. In the long run, forecasted demand regularly exceeds existing load capacity. The IRP process (See Box 17.2) included consideration of supply-side resources, demand-side resources, and legislative compliance. Taken together, these emphasize the most cost-effective plan that also meets the need.

17.3.3 Cost-Effectiveness

Mr. Puican, the Commission's economist, explained that "if the DSM is not cheaper than the supply-side, we don't then consider that conservation is a societal good. So conservation may have to be given the benefit of the doubt initially, as opposed to supply-side alone. If the supply-side option is cheaper, then we are going to go with the supply-side option; it is hard-core economics," he said.[3]

Box 17.2   Outline of Integrated Resource Planning (IRP)

The IRP process consists of six steps:

1. Electricity load forecasting.
2. Review of the existing systems.
3. Screening of supply-side options.
4. Screening of demand-side options.
5. Environmental compliance planning.
6. PUCO filing.

At CG&E, the IRP Committee had the key functional areas of the company represented, including marketing with responsibility for DSM program evaluation and development. The IRP plan selected should represent the lowest present value total cost and lowest present value rate subject to the Company's willingness to accept the risks and uncertainties inherent in the plan. In addition, the plan selected also should contain the lowest present value of utility costs. The process employed to develop the IRP involves integration of the forecast, generation options, and demand-side options. In addition, it is important to select the best way to conduct the integration, while incorporating interrelationships with other planning areas, for example, fuel procurement and transmission/distribution planning.

The utility's planning objectives included understanding and responding to customers' needs, having a high quality investment for shareholders, and conducting field operations with responsible concern for the environment. In addition, the Company, through its DSM program options, expected to alter customers' energy use patterns and efficiencies, while improving customer service. By focusing on reducing peak demand, using cost-effective DSM, the Company could lower the costs customers pay for electricity, reducing expensive (in both capacity and energy production) peak costs.

In this filing, three load-shaping strategies are considered: peak clipping, strategic conservation that includes an associated peak reduction, and load shifting. Each of these objectives has the potential to reduce or shift peak demand. The preliminary load forecast, preliminary supply-side options, and demand-side options were loaded into PROVIEW, a proprietary product of Energy Management Associates, for development of the IRP. This model conducts a dynamic optimization that is capable of evaluating thousands of alternate plans to determine the plan with the lowest present value total resource cost, lowest present value rate, or lowest present value utility cost. PROVIEW utilizes a dynamic programming technique that is useful for making a sequence of interrelated decisions, and a systematic procedure for determining the combination of decisions that maximizes the desired outcome (least cost) within the given constraints. However, for this filing, once it was determined that the DSM programs would enter the plan, DSM was not allowed to compete against supply-side options, but was constrained to be in the plan.

> **Box 17.3   The Total Resource Cost Test (TRC)**
>
> Program assumptions and data:
>
> | | |
> |---|---|
> | Incremental Equipment Cost | $200.00 |
> | Rebate | 50.00 |
>
> | | |
> |---|---|
> | Utility administration costs: | 50.00 |
> | Equipment life | 15 years |
>
> Program impacts:
>
> | | |
> |---|---|
> | Energy savings | 300 kWh/year |
>
> Demand savings: $0.2 per kW
>
> Utility avoided supply costs:
>
> | | |
> |---|---|
> | Energy | $0.03 per kWh |
> | Capacity | 60.00 kWh per year |
>
> Program costs (illustrative computation):
>
> | | |
> |---|---|
> | Utility rebate | $ 50.00 |
> | Incremental participant cost | 150.00 |
> | Utility administrative cost | 50.00 |
>
> TOTAL Program Costs $250.00
>
> Program benefits (avoided costs):
>
> Energy: (300 kWh/year)($0.03/kWh) = $9 per year
>
> ($9 per year)(15 years) = $135.00
>
> Capacity: ($0.2 per kW)(60.00 kW per year) = ($12 per year)
>
> ($12 per year)(15 years) = $180.00
>
> | | |
> |---|---|
> | Total program benefits: | $315.00 |
> | Net benefits equals (total costs minus avoided costs): | $ 65.00 |

Thus, to be considered in the plan, demand-side options had to be considered in a total resource cost (TRC) and pass two tests: a Commission test for cost-effectiveness and a financial attractiveness test. As an illustration, consider a program with the TRC assumptions and sample calculation shown in Box 17.3.

The total resource cost test illustrated in Box 17.3 was designed by the Commission as a simple cost-benefit analysis intended to provide a societal perspective. The benefits side consists of the two avoided costs:

1. Energy development and operations costs, a savings associated with the use of existing facilities.
2. Avoided capacity development, a savings associated with not having to expand facilities.

The cost side contains the cost of program administration and incremental equipment costs. Administrative costs include overhead and advertising. Incremental equipment cost is the difference between the baseline equipment or state cost, and the more efficient technology. If a rebate is offered

Electricity Load Management and Stakeholders   331

to participants as an incentive to participate in the program, it is included explicitly as a cost, but lsubtracted from the incremental cost to the utility. Hence, it does not matter to the result who (utility or participant) pays what portion of the investment in efficiency. A TRC ratio (avoided costs/costs) greater than one suggests a net benefit, while values less than one indicate a net loss to society. The TRC is not a test of the effect on rates; even programs that pass the TRC may increase rates.

The program would be attractive to CG&E if the costs that are incurred by the company can be recovered in rates. Recall that the rate base is dependent on capital expenditures. However, investments in DSM measures are not really capital investments. The rules established by the Commission for recovering DSM program costs created a neutral-plus incentive. For these purposes, the cost components of a DSM program are defined as:

1. Program expenses (rebates, administration, promotion, etc.).
2. Lost revenues (revenues lost owing to the successful decrease in demand from the program).
3. Shared savings (10 percent for the typical participant).

"A shared-savings incentive . . . reward[s] the utility based on the net benefit of the program."[4] Thus, all parties reap benefits of cost-effective energy efficiency: The participating customers have lower bills, all ratepayers observe a relatively lower cost of providing electric service, and the utility is allowed to retain a fraction of the net-savings as earnings.

There is some risk to the utility, however. Although the operating plan containing the proposed DSM program receives approval from the Commission, and the special account for DSM expenses is established, actual program implementation occurs before the extent of cost recovery is known. The Commission retains the authority to rule on the prudency and cost-effectiveness of the DSM program expenditures at the base-rate proceedings after the program has been implemented and during the life of the program.

17.4 PUBLIC-PRIVATE CONSENSUS-BUILDING FOR DSM

In a May 1992 rate case (Case No. 91-410-EL-AIR) the Commission called upon CG&E to involve community-based organizations in a voluntary, non-adversarial decision-making process capable of advancing DSM.[11] The result was establishment of the DSM Collaborative, a step toward a new relationship among stakeholders in the industry.

Just as events of the 1970s and 1980s complicated planning and decision processes for the industry, the increased number of rate cases, facility siting cases, and rulemaking hearings also increased the contentiousness of each proceeding during the 1980s and 1990s. The previous laissez-faire path of the electric industry development and regulation was now somewhat in disarray. The pattern of public interventions forced regulators to hear expert views from all sides, and to seek acceptable solutions to difficult problems, solutions that often pleased no one. Unintended incentives and disincentives were hard to avoid. Dissatisfaction and distrust also developed during this expensive and lengthy adversarial process. As it happened, many parties began trying to settle controversies outside of the established process. The DSM Collaborative was to be the beginning of a new paradigm of problem-solving that produced solutions carefully and deliberately constructed as a focused outcome from the stakeholders involved.[12]

Experiments with consensus-based processes, or alternative dispute resolution (ADR), also became acceptable within the utility industry during the previous ten years. Applications included both rule-making and rate-making. Various mechanisms for consensus building were being practiced, demonstrating a range of levels of participation and dialogue.[13] Information on the eight principles used in the process can be found in the report by Raab.*

17.4.1 THE DSM COLLABORATIVE

CG&E'S DSM Program Manager, Van Needham, points out that, "Historically, [the] company was leery of outside participation in company planning because, previously, [they] had had hostile intervenors in rate cases as their only contact."[13] Now there were to be other non-traditional intervenors involved. Jim Dugan, another company manager said,

> "Initially the company recognized the need to do the Collaborative in order to get the rate increase. There was also a feeling that the Commission felt this was an important thing, so the company wanted to accommodate. The company was going through some self-examination at about the same time. The company began to feel it was important to get close to customers and create a relationship. It was time to recognize the customers' perspective of the utility."[14]

At this time CG&E's DSM Collaborative could be described as a Prospective Collaborative, a forum for DSM initiatives that proposed a high level of participation and dialogue.[15] This type "involve[s] designing future plans or programs for utilities, or policies for [regulatory commissions]."[12] Collaboratives typically exist outside of a particular controversy or proceeding, but help to inform analyses of those proceedings. They produce changes in utility plans that usually are subjected later to regulatory approval. Processes like the Collaborative, that are intended to reduce time, cost, and uncertainty can be used in both litigation and policy setting circumstances. CG&E Collaborative's Strategic Plan describes the situation as follows:

> "The collaborative is both assertive and cooperative. It is the opposite of avoidance. Collaborative means full involvement with another party to attempt to work with that other party to find some solutions which fully satisfy the concerns of both parties. This involves digging into an issue to identify the underlying concerns of the parties and to find alternatives which satisfy both sets of concerns."[16]

The CG&E Collaborative was unique in its size and scope of participants. It was open to all interested parties, not only the company, regulatory agencies, and consumer advocate representatives, as is generally the case of other DSM collaboratives in Ohio (See Figure 17.2). In fact, PUCO and the Ohio Consumer Counsel membership are excluded explicitly, although representatives attend as observers and advice-givers.† The commissioners had much to do with setting up the CG&E Collaborative in a form rather different from others in the state:

---

* Eight principles for consensus:

  1. Initiate consensus building as early as possible.
  2. Include all stakeholders.
  3. Secure direct involvement of the PUC whenever possible.
  4. Provide adequate resources.
  5. Do not exclude contentious or sensitive issues from consensus-building efforts.
  6. Consider assisted negotiation.
  7. Structure consensus-building processes to supplement traditional adjudicatory and rule-making procedures.
  8. Modify traditional procedures to better accommodate consensus-building opportunities.

See Reference 12, page 223.

† Other collaboratives in Ohio exist, but few have the character of CG&E. Others may be less broad or more political. There is some evidence that the relationship between company and customers in the CG&E case transcends rate case issues and political levers.

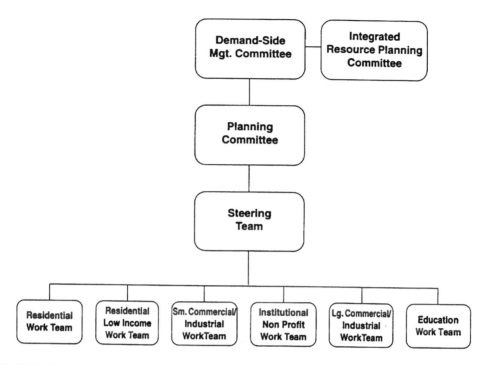

FIGURE 17.2 Organization structure for the demand-side management collaborative at the Cincinnati Gas & Electric Company.

"It was the thought that you get community organizations involved and those community organizations, maybe knowing their clients better than maybe the utility does, might be better able to advise the utility on how to best implement some of these programs. CG&E is different because the scale of the collaborative is far larger than that of any other collaborative."[3]

The first meeting of non-utility participants in the Cooperative was held in the early fall of 1992. There were 12 to 14 people present, including low-income and environmental group representatives. Many of these individuals were selected by the company's Planning Committee to serve on the initial Steering Committee.[17] The first group of participants were selected through a company survey of potential participants. The first meeting of the entire Collaborative was shortly thereafter, and monthly meetings began in January 1993. Each full Collaborative meeting dedicated time to DSM education and information sharing, as well as time for a general community billboard, part of the informal team-building and communication goal. CG&E's CEO Jackson Randolph met with the Steering Team in May 1993 to reaffirm the company's commitment to the Collaborative. He reminded them of the company's expectation of cost-effectiveness, broadened communications with customers, advocacy for cost recovery and mutual cooperation.

The Collaborative was seen to serve many purposes. It "provides a forum for The Cincinnati Gas & Electric Company and its customers to work together in order to develop and implement DSM programs."[16] It was to result in increased communication between the utility and its customers, education in DSM and energy issues, and "advocacy for developed programs."[16] It provided a process for ongoing exchanges of information, expectations, objectives, and goals among customers, the Commission, and CG&E during the planning process. It was intended to be a long-term, dynamic, continuous process that would be utilized routinely in DSM program planning and other activities involving energy-related issues. The four stated objectives, consistent with the mission, include:

1. Program Development—the Collaborative will facilitate DSM programs.
2. Communication—the Collaborative will provide an interactive communication process.
3. Education—the Collaborative will educate customers.
4. Advocacy—the Collaborative will promote the collaborative process as the preferred method of problem-solving in an effort to minimize litigation.[16]

17.5 INDUSTRY DEREGULATION AND DSM

By August 1996, the role of the DSM Collaborative seemed much less significant for CG&E and the Commission, however. The federal Energy Policy Act of 1992 had opened up interstate wholesale electricity markets, taking a major step toward a free market in the supplying of electricity to a service area such as Cincinnati. Some electric utility companies had surplus capacity, and might provide electricity without new capital investment, and possibly at a lower cost than conservation or demand-side management. This federal law, and subsequent changes in Ohio's Revised Code guiding electric utility operations, and the Commission's adoption of the Ohio Energy Strategy, all raised doubts as to the validity of long-term DSM assumptions and measures (such as the Total Resource Cost measure, see Box 17.3).[17] The issue now before the Commission, therefore, concerned whether to retain the processes set up through CG&E's DSM Collaborative, or accept a whole new approach to reducing the cost of electricity through deregulation and competition in the marketplace.

At least two of the intervenors before the Commission had argued for the inherent worth of DSM, the Collaborative, and the measures (such as TRC) used to allocate benefits and costs. The dialog and engagement with all stakeholders was important, they said, and the improved understanding would save money in the long run. TRC, of course, made assumptions well into the future as to where the electricity supply and related expense would be allocated, and although it would probably be from CG&E, that was not necessarily the case.

On the other hand, the Company pointed to the prospect of serious flaws in the TRC out-year assumptions (see Box 17.3 and its discussion), and argued that, in a deregulated environment, their customers could be hurt by erroneous indicators and flawed DSM decisions under those circumstances. In an open electricity-supply market for the Cincinnati area, all customers would, potentially, be able to find their own lowest cost supplier by shifting to lower-demand periods, with suppliers possibly other than CG&E (assuming power-line leasing also is available). In this view, demand-side management would be changed substantially. CG&E suggested to the Commission that they deal with these changes and uncertainties through a generic rate case involving all the utilities in Ohio.

Given the prospects for further changes in DSM, CG&E could not yet provide any specific outline as to how deregulation in their service area would actually operate or what it would mean economically for their customers. The functioning of an open, deregulated market for electricity would get underway in 1997 and 1998 only in California, and it would be several years before existing contracts would expire in other market areas, such as the Ohio Valley, allowing for the beginning of open competition. Stephen Puican, the Commission economist, in a published article the previous year, had noted that regulators probably should not change requirements "while waiting for the fog of uncertainty regarding competition to clear."[18] At the same time, the Commission did not want a major utility, and the Cincinnati service area, to be burdened with DSM cost analysis guidelines that would soon be out-of-date and possibly flawed.

17.6 CONCLUSION

The participants in the DSM Collaborative have embarked on what was expected to be a long and fruitful new road. The company was becoming much less of a faceless monolith, and had many stakeholders to listen to now. "These are customers that were not necessarily represented by traditional intervenors, even though intervenors claimed they were," said Van Needham.[13] Electricity conservation continues to be a major need, but there appear to be several competing ways to reach that goal. Participants in the rate-setting proceedings all agreed that it was still unclear, under deregulation, as to whether benefits would accrue to all participants as planned, but the Commission would have to make a determination.

17.7 REFERENCES

1. Nadel, S., Utility demand-side management experience and potential—a critical review. *Annu. Rev. Energ. Environ.*, 17, 507, 1992; Nadel, S. M., Reid M. W., and Wolcott D. R., *Regulatory Incentives for Demand-Side Management,* American Council for an Energy-Efficient Economy, Washington, D.C., 1992.
2. Morgan, R., Promoting energy efficiency through utility regulation, presented at a conference, Rebuilding Ohio: Energy, The Economy and The Environment, Columbus, OH, June 13, 1993.
3. Puican, S., Personal interview about Public Utilities Commission of Ohio, February 18, 1994.
4. Stoft, S. and Gilbert R. J., A review and analysis of electric utility conservation incentives, *Yale J. Regul.,* 11(1), 1, 1994.
5. Gellings, C. W. and Chamberlin, J. H., *Demand-Side Management Planning*, Fairmont Press, Lilburn, GA, 1993.
6. Roodman, D. M., Power brokers: managing demand for electricity, *World Watch* 6(5), 22, 1993.
7. EPRI study on 1992 DSM activity shows lighting as top technology. *Energy User News,* October 1993.
8. Joskow, P. L. and Marron, D. B., What does utility-subsidized energy efficiency really cost?, *Science,* 260, 281 and 370, 1993.
9. Cincinnati Gas & Electric Company, Annual Report, 1992.
10. Cincinnati Gas & Electric Company, *Electric Long-Term Forecast Report,* Vol. 1, submitted to the Public Utilities Commission of Ohio, Division of Forecasting, 1993.
11. Public Utilities Commission of Ohio, Annual Report, 1992.
12. Raab, J., *Using Consensus Building to Improve Utility Regulation,* American Council for an Energy-Efficient Economy, Washington, D.C., 1994.
13. Needham, V., The Cincinnati Gas & Electric Company, telephone interview, August 16, 1994.
14. Dugan, J., The Cincinnati Gas & Electric Company, telephone interview, August 12, 1994.
15. The Cincinnati Gas & Electric Co., The collaborative effort, organization structure, brochure, 1993.
16. Cincinnati Gas & Electric Co., Strategic plan: the collaborative effort, brochure, 1993–1994.
17. Public Utilities Commission of Ohio, Summary, opinion and order, Cincinnati Gas & Electric Company Case Nos. 95-203-EL-FOR, 1996.
18. Puican, S. E., DSM and the transition to a competitive industry, *Public Utilities Fortnightly,* June 15, 1995.

18 Community-Industry Dialogue in Risk Management: Responsible Care® and Worst-Case Scenarios in the Valley of the Shadow * Case Study Number 10

O. Homer Erekson and Pamela C. Johnson

CONTENTS

18.1 Introduction 337
18.2 Challenges to the Chemical Industry 338
    18.2.1 Regulatory Environment 338
    18.2.2 The Chemical Industry Responds 339
18.3 Context for Kanawha Valley 341
18.4 Worst-Case Scenarios in the Kanawha Valley 342
    18.4.1 Denial 343
    18.4.2 Terror 344
    18.4.3 Community Building and Technical Challenges 344
18.5 The Road to Safety Street: Communication 348
18.6 Epilogue 351
18.7 References 352

18.1 INTRODUCTION

Richard (Dick) Knowles came, as plant manager of the DuPont Belle plant, to the Kanawha River Valley near Charleston, WV from Niagara Falls. In his prior position, he served as DuPont plant manager soon after the infamous Love Canal episode involving a housing development built on a toxic waste site. In Charleston he faced a community still in deep grief because of mistakes from within the chemical industry. Dick's years in Niagara Falls were personally transforming. He says that it was there he learned in meetings with people in the community that "*listening* to people was

---

*This case is an expanded version of "Responsible Care® and worst-case scenarios in the Kanawha Valley, USA." A CD-ROM version, "In the Valley of the Shadow," is available from the Council for Ethics in Economics, 125 E. Broad St., Columbus, OH 43215-3605. See Reference 1.

really half of the process of healing . . . and that trust is hard to build—it is very fragile." He went on to say that his benchmark question became, "Would I put my most loved one at the plant fence and let them live there for 70 years?"*

The chemical companies in the Kanawha Valley were working with the community to develop appropriate accident prevention plans, trying to decide how much information to share, and how to disseminate it. The chemical companies were being asked to identify up to three compounds each and to convey to the community the worst conceivable accident that could occur with no safeguards in place. There were very few guidelines in place as to how to choose the chemical compounds to be included, how to define a worst-case scenario, and a more probable scenario, which technical model to use, and what assumptions to make in that model, how to elicit cooperation among the plants within the Valley, and how to communicate to the public in an informative, but non-alarming manner. Knowles noted, "We were faced with a challenge that could threaten the existence of our facilities, if we didn't do the communications openly, thoroughly, and well."

At a major plant managers meeting in August 1993, there was a proposal to reorganize the process to return control to the companies. Knowles maintained his belief in full disclosure and full participation by the public in the entire process, including the technical aspects. Some of the other plant managers replied, "You'll make us look bad. You'll ruin it for the rest of us." The meeting ended on a very tense note as Knowles was considering whether to break ranks and move ahead with full communication of the worst-case scenarios for the DuPont plant.

18.2 CHALLENGES TO THE CHEMICAL INDUSTRY

In December 1984, a Union Carbide plant in Bhopal, India, suffered a major leak of methyl isocyanate (MIC), "the single most astonishing and terrible event" in the history of the chemical industry, according to Union Carbide CEO Robert Kennedy.[2] This accident attracted international attention with more than 3000 people dying and an undetermined, but in the tens of thousands, number of individuals injured. Despite the international uproar and the resulting criticism of the chemical industry, for many persons living far distant from Bhopal, the reality of this industrial accident may have been understated.

Such was not the case for the chemical industry. At the 1984 annual meeting of the Chemical Manufacturers Association (CMA), board chairman William G. Simeral noted that "while the industry saw itself as a producer of high-tech jobs and exports, the public saw only leaking drums and hazardous waste sites—and that image meant trouble for the industry."[3]

Clearly the citizens of the Kanawha Valley, many of whom were employees of the chemical companies or relatives of employees, became increasingly concerned in the early 1990s as they discovered more about the risks to the community. In January 1992, Pam Nixon, chairwoman of People Concerned About MIC and a person hospitalized from an earlier chemical accident, made a written request that chemical companies release their worst-case scenarios.

The issue for the companies was how should a company and an industry respond to the need for public dialogue on the existence of toxic risk to a community?

18.2.1 REGULATORY ENVIRONMENT

After the December 1984 Bhopal incident, the U.S. Environmental Protection Agency (EPA) formed a Chemical Emergency Preparedness program, which was expanded in 1986 when Congress approved the Federal Emergency Planning and Community Right-to-Know Act. Although these actions increased company responsibility in risk management and public information, the requirements escalated in October 1993, when the EPA proposed a rule under the Clean Air Act calling for

---

* Throughout this case, quotations are used from personal interviews with Richard K. Knowles, on February 10, 1995, and March 7, 1995.

a Risk Management Plan. This plan would require companies to develop a Risk Management Program that includes a *hazard assessment* program with a five-year release history (including the size, concentration, and duration of releases) and consideration of a range of release (including *worst case scenarios*), a prevention program, and an emergency response program, along with a risk management plan documenting the results of the facility's risk management program. These were to be provided to the EPA, the Chemical Safety and Hazard Investigation Board, the state emergency response commission, the local emergency planning committee, and the *public*.

A major focus of the risk management programs would be the worst-case scenarios. They were defined as *the loss of all of the regulated substance from the process in an accidental release that leads to the worst off-site consequences*. But Craig Matthiessen, Director of Chemical Accident Prevention at the EPA, said "the worst case is not the objective. The goal is prevention and preparedness and response, and the worst case is the road to that goal. The purpose is to ensure that facilities take steps to reduce the likelihood and severity of accidental chemical release that could harm the public and the environment . . . and to ensure that the public and state and local governments receive facility-specific information on potential hazards and the steps being taken to prevent accidents."[4]

18.2.2 THE CHEMICAL INDUSTRY RESPONDS

The Chemical Manufacturers Association (CMA) is an industry-wide trade association that provides technical support in areas such as toxicology, biochemistry, engineering, and environmental affairs, lobbies the federal government on behalf of the association's chemical producers, and provides a forum for company executives to exchange views on issues of common concern.[5] As of June 1996, CMA consisted of 192 members who represented about 90 percent of the U.S. basic industrial chemical manufacturing capacity.

In April 1985, in response to the Bhopal accident, CMA developed the Community Awareness and Emergency Response (CAER) program. Initially, this was a voluntary program designed to help assure emergency preparedness and to foster community right-to-know. Jon Holtzman, CMA's Vice President for Communications noted that, although plant managers initially balked at sharing potential risks at their plant sites with the public, "the plant managers learned that no one wanted them to leave town. They just wanted to know they and their families were safe."[3] As the CAER program was being implemented throughout the United States, four significant results began to emerge:

- The public voluntarily was given access to potentially damaging information.
- The EPA was persuaded that the industry was in earnest about making progress.
- The performance of local plant managers began to change.
- In drafting Title III of the Superfund Amendment and Reauthorization Amendment (SARA), legislators adopted whole sections of CAER.[5]

However, even with the early successes of CAER in the United States, CMA began to observe similar activities in Canada. The Canadian Chemical Producers Association (CCPA) was taking a more broad-based approach in formulating a program it termed "Responsible Care," that involved codes of practice for transportation, distribution, manufacturing, research and development, and hazardous waste operations. The program was a formal condition of membership in CCPA. Interestingly, in the Preamble to the CCPA Codes of Practice, it states "Member company performance must reflect the concerns, needs, and values [of the stakeholders] . . . this is ethical thinking, decision-making, and performance."[6]

In 1987, CMA convened a meeting of CEO's and other leaders of the U.S. chemical industry to consider development of a similar broad-based program. In September 1988, Responsible Care® was unanimously approved by CMA's Board and subsequently adopted by all member companies as an obligation of membership in CMA.

Responsible Care® is organized around a set of ten guiding principles (see Box 18.1) and six Codes of Management Practice.* It commits the chemical industry to responsible management of chemicals from development through production, distribution, and ultimate use. The six Codes of Management Practice are the Community Awareness and Emergency Response (CAER) Code, the Pollution Prevention Code, the Process Safety Code, the Distribution Code, the Employee Health and Safety Code, and the Product Stewardship Code.

Two codes are especially relevant for this case. The goal of the CAER Code is to "assure emergency preparedness and to foster community right-to-know. It demands a commitment to openness and community dialogue. The code has two major components: first, to assure that member facilities that manufacture, process, use, distribute, or store hazardous materials initiate and maintain a community outreach program to openly communicate relevant, useful information responsive to the public's questions and concerns about safety, health, and the environment; and second, to help protect employees and communities by assuring that each facility has an emergency response program to respond rapidly and effectively to emergencies."[7]

The Process Safety Code "is designed to prevent fires, explosions, and accidental chemical releases. The Code is comprised of a series of management practices ... based on the principle that facilities will be safe if they are designed according to sound engineering practices, built, operated, and maintained properly, and periodically reviewed for conformance.... The Code must be implemented with full recognition of the community's interest, expectations, and participation in achieving safe operations."[8]

In both of these codes, it is clear that efforts are made to develop close relationships between participating companies and their communities. Charles Aldag, Chairman of the Responsible Care® Coordinating Group, has noted that to be successful, Responsible Care® has to address performance,

---

Box 18.1   Guiding Principles of Responsible Care®

CMA member companies are committed to support a continuing effort to improve the industry's responsible management of chemicals. They pledge to manage their business according to these principles:

- To recognize and respond to community concerns about chemicals and our operations.
- To develop and produce chemicals that can be manufactured, transported, used, and disposed of safely.
- To make health, safety, and environment considerations a priority in our planning for all existing and new products and processes.
- To report promptly to officials, employees, customers, and the public, information on chemical-related health or environmental hazards and to recommend protective measures.
- To counsel customers on the safe use, transportation, and disposal of chemical products.
- To operate our plants and facilities in a manner that protects the environment and the health and safety of our employees and the public.
- To extend knowledge by conducting or supporting research on the health, safety, and environmental effects of our products, processes, and waste materials.
- To work with others to resolve problems created by past handling and disposal of hazardous substances.
- To participate with government and others in creating responsible laws, regulations, and standards to safeguard the community, workplace, and environment.
- To promote the principles and practices of Responsible Care ® by sharing experiences and offering assistance to others who produce, handle, use, transport, or dispose of chemicals.

---

*Responsible Care® also has a "Phase II" Implementation Plan that deals with various issues such as management systems verification and mutual assistance among CMA companies.

process, and dialogue issues. "The experience of the chemical industry in the 1980s showed itself as a performance problem. But the heart of the codes is process failure. This could only be remedied by involving the community in open and frank dialogue about the potential risks facing neighbors of chemical plants. Fundamentally, environmental responsibility is a matter of ethics."[9]

With the momentum of a set of well-articulated Codes of Management Practice endorsed throughout the chemical industry, CMA then set its course to work with local plants to begin to demonstrate the application of Responsible Care®. One significant window of opportunity was to encourage companies to take the lead by voluntary participation in a worst-case scenario community forum, well before the Clean Air Act rules would be promulgated.

18.3 CONTEXT FOR KANAWHA VALLEY

The Kanawha Valley is a narrow river valley bounded by 1200 ft mountains that run for about 30 linear miles. It has a population of approximately 250,000 living in a number of different communities, including Charleston, South Charleston, Institute, and Nitro. Spread along the valley are eight different chemical companies operating 15 different plants or technical centers (see Figure 18.1), providing economic livelihood for approximately 15,000 people. Almost everyone living in the area has family members or friends employed in the chemical industry.

That the Kanawha Valley is such a center of chemical production is not surprising because it is generally given the distinction of being the birthplace of the chemical industry in the United States. Soon after the Civil War, the advantages of this location for the infant industry were becoming apparent. It was convenient to ready sources of coal as a source of energy and to salt mines that provided a primary ingredient for the production of chlorine. Moreover, its proximity to the river facilitated transportation of supplies and finished products.

With the growth of the chemical industry, an important source of economic stability for the region was realized. Despite occasional accidents, residents generally supported the industry and accepted the inherent risks associated with chemical production. However, given the history of conflict between coal miners and management in the early part of the century, the relationship between management and labor in West Virginia was less than ideal. This also affects relationships between industry and the community in the Kanawha Valley to this day. Recent trends in downsizing in response to competitive threats have affected the work environment as well. Some of the companies

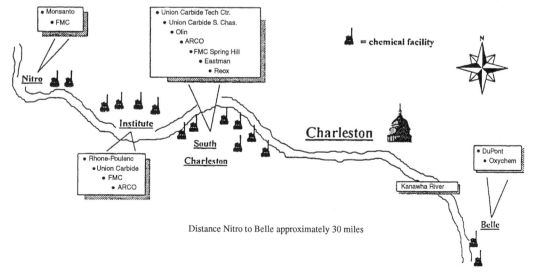

FIGURE 18.1 Chemical facilities in the Kanawha Valley of West Virginia. *Source:* National Institute for Chemical Studies.

have begun using contract labor, with one of the unions taking the position that use of contract workers, whose skills and experience vary, increases the risk of accidents.

Minor accidents (small leaks or releases) occur with some regularity. The chemical plants in the Valley listed over 400 small leaks and 32 reportable leaks from 1989 to 1993. Heightening the concern of the public was the revelation by the Rhone-Poulenc Institute plant (purchased from Union Carbide in 1986) that over that time period, there had been 43 small leaks of MIC.* Major releases were less frequent, but did occur occasionally. In August 1985, shortly after the Bhopal incident, 135 residents of the Valley were treated following a release of 500 gallons of aldicarb oxime from the Union Carbide plant in South Charleston.

18.4 WORST-CASE SCENARIOS IN THE KANAWHA VALLEY

"We know that you guys can kill us. Just tell me what you are doing to prevent that from happening."

The sentiment behind this haunting statement had lived in the Kanawha Valley for many years. In 1992, the chemical companies in the Valley were to begin facing its significance in a manner heretofore not seen in any industry, particularly their own.

Although there would be support in the form of workshops and input from the CMA in Washington, D.C., the burden of the work from the industry perspective would fall directly upon the plant managers and other employees. Their framework for reference would be the CAER Code. Dick Doyle, Vice-President for Responsible Care®, adds, "CMA can provide encouragement and support for our members' efforts in carrying out the Responsible Care® initiative, but in the end it's the men and women throughout the chemical industry that have the responsibility to see that it works. This is no small task, but is essential if we hope to improve the chemical industry's performance and ultimately earn the public's trust."[11]

Coming to the Kanawha Valley, Knowles found a community with vital chemical plants where "business was great, but you had to hold your noses when you went by our plant." Moreover, because of the long-standing mistrust alluded to earlier, he also found the situation between the plants and the community, especially the newspapers, as being very adversarial. It was in this environment that the chemical companies were to consider releasing information about worst-case scenarios to the public.

The worst-case scenario process was organized and directed by the Local Emergency Planning Committee (LEPC). The Kanawha/Putnam Local Emergency Planning Committee (LEPC) was established in 1987 as part of the Emergency Planning and Community Right-to-Know Act. Its membership included representatives from law enforcement, fire departments, emergency medical services, environmental groups, hospitals, industrial facilities, and local communities. While the formation of this group was mandated by regulation, the leadership of the LEPC were volunteers from the public. The resulting collaboration with the industry people provided a basis for the development of trust in the process. It was to this group that Nixon made her request for three worst-case scenarios from the local chemical plants.

In the fall of 1992, two committees were formed by the LEPC. Figure 18.2 provides a schematic of the Hazard Assessment/Risk Communication network. The Hazard Assessment Committee was charged with development of the technical description of the hazards that existed and the risks posed to the community. It was headed by Dr. Paul Hill, President of the National Institute for Chemical Studies. The Hazard Assessment Committee was to establish the process to allow the chemical plants to uniformly evaluate worst-case scenarios and associated risk information using common criteria. It was to involve citizens, industry, and government at all levels of planning and communication.

---

*Leaks above a certain level must be reported to the government. For instance, a release of 100 lb or more of anhydrous ammonia must be reported, whereas a release of 10 lb or more of methyl isocyanate must be reported. For a summary of the accidents reported at Kanawha Valley chemical plants from 1989 to 1993, see Reference 10.

Community-Industry Dialogue in Risk Management

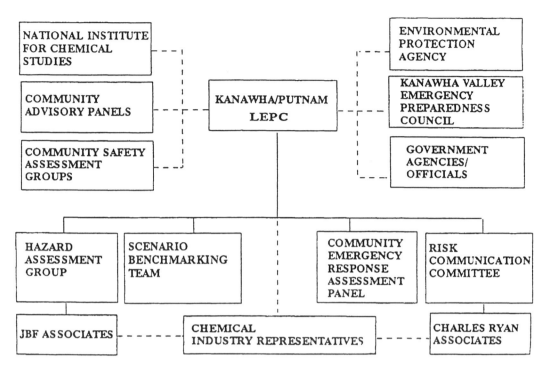

FIGURE 18.2 Kanawha Valley Hazard Assessment/Risk Communication Network.

The Risk Communications Committee was charged with developing a process to allow the chemical facilities to communicate worst-case scenarios and associated risk information to the community. This included producing written documentation compiling worst-case assessments in a clear, concise, and readable form, and development of a public forum by which worst-case information could be communicated to the public. It was co-chaired by two members of the public and one industry representative. One of the co-chairs of the committee, Mary Frances Bleidt recalled, "We were hoping to create a sense of community, not scare the hell out of people."[12]

Another important organizational feature was the hiring of JBF Associates, Inc. by the LEPC to serve as an independent third-party reviewer of the Hazard Assessment Project. This was a nationally recognized consulting firm who had performed hundreds of hazard analyses and risk assessments on chemical processes with a reputation for high-quality work and a "call-it-like-we-see-it" ethic. They were asked to provide technical advice, perform quality assurance reviews, ensure analyses were performed using generally accepted practices, and assist in interpretation of the results.

The organizational pieces were in place. The committees began their work. Still, the next few months saw the process struggle. Knowles described how the plant managers went from a point of *"denial* to *terror* to *community building."*

18.4.1 Denial

The local plant managers were under considerable pressure by local environmental groups and the newspapers. Nixon reflected the initial skepticism of many of the local citizens when she said, "There were a lot of engineers, and they had confidence in the safety measures—many were redundant systems—and so they said that accidents that assumed a total lack of safety controls couldn't happen. The work of the committee was slowed down because of plant reps and corporate people who would fly in to hear what assumptions we were making and why—then at the next meeting, there would be someone new from corporate, and we had to keep explaining over and over again."[13] Some citizens complained that meetings were set at inconvenient times to fully involve the public.

In reflecting upon the frustration of the plant managers at this early stage, Knowles offered, "Plant managers don't like getting beat up. But if you come in with an intense level of love, care, and authenticity, you can work through this with the community. If people feel trusted and part of the process, it can be beguiling. You must be careful not to manipulate the process. You have to be able to look at yourself in the mirror and ask if what you are doing is right. And while you're doing it, you'd better have your plant living up to the highest ethical standards, because if you have a double standard, your employees will talk."

18.4.2 Terror

During the latter part of 1992 and into 1993, two crucial organizational themes were at work. First, the plant managers decided to respond jointly. Knowles commented, "Perhaps, it was an issue of vulnerability. We probably felt there was safety in numbers. Also, it was a very complicated technology, and it was important to have consistency and coherence in how we presented the information. We probably felt that we could have more influence as a block within LEPC and in the community."

Second, during the late summer of 1993, many plant managers became increasingly concerned about the progress of the technical work in defining worst-case scenarios and developing the necessary models. There was a move to slow the process down, to move control of the process from the LEPC back to the companies. It was at this time, in August 1993, when Knowles was faced with whether to break ranks with the other plant managers and to continue the open process that was underway. He said, "I am concerned with what is the right thing to do, and I am also responsible to look out for DuPont stockholders. [Continuing the process was] right for the community and for DuPont. The only choice in my mind was between the *certainty* of failure if we took back control of the committees, and the *possibility* of failure if we did it in a more open and participative way."

The very next day, the Rhone-Poulenc Institute plant had a major explosion, killing one person and injuring several others. The community had to shelter-in-place, remaining inside a safe location until the episode passed. This event had a profound effect on moving the Hazard Assessment Project ahead. There was no longer a choice as to whether companies would participate fully. The community demanded the worst-case scenarios be developed and publicly distributed.

18.4.3 Community Building and Technical Challenges

With the new momentum in place, the Hazard Assessment Committee began to formally specify the method for determining worst-case scenarios. This process involved stakeholders from the entire community, including industry, government leaders, the emergency response community, and the general public. Even though defining risk involves inherently technical models and data, involving the public helped to humanize the process.

Hill, chair of the committee, remarked, "In the decision on how to determine how far down the standard should be for say flammability, most industry people said, 'okay, here's what the literature says. The 2.3 level is where eardrums burst and presents definite hazard; that is where the standard should be.' The wonderful thing about what community residents like Nixon added was to say, 'That may well be. As engineers looking at technical chemical data, it would be natural to stop at the technical standard. But as lay people, we will be more interested in an absolute standard—not just my eardrums, but will my home be left standing? Will all of our records be destroyed?' The perspective of the community people added a reality check to the technical data from industry. It was like a pyramid where you added piece by piece, establishing a baseline of information on issues, concerns, and literature searches until you got to an apex where consensus existed. And there was not a lot of fighting. My job in chairing was to help gather all of the pieces and pull all of the parties into consensus."[14]

Out of this process, the following technical issues were determined:

- *Determination of scope of analysis:* Each plant would prepare a worst-case scenario (with no safeguards) and a more likely accident scenario (with safeguards).
- *Choice of benchmarking model:* The CAMEO/ALOHA computer model was chosen as the benchmarking model for all plants. However, companies also could provide supplemental information using any additional software available.
- *Definition of models:* In addition to adopting the benchmark model, and supplementing with information as appropriate, each plant faced various other modeling issues and limitations. Each plant agreed to provide a worst-case scenario and a more probable scenario for up to three chemicals (see Table 18.1 for a list of the companies and chemical compounds). The assessment protocol included a description of the failure, mitigation systems in place, release rate, duration and form, meteorological conditions, level of concern, and the potentially exposed population.
- *Worst-case scenario:* Defined as the catastrophic failure of the largest single chemical vessel or pipeline which results in worst off-site consequences, releasing the contents over a ten-minute period under the most stable weather conditions (almost no wind). No passive controls (e.g., a dike), administrative procedures (e.g., maintaining storage tanks below maximum capacity), or emergency response capabilities were allowed in the scenarios.
- *More probable release scenarios:* Defined with smaller quantities of materials released under most stable weather conditions, with existing control procedures and systems in place; it was necessary that the plume resulting from the scenario would have to cross the plant fence line.
- *Measurement of health effects:* Following the EPA's Technical Guidance for Hazards Analysis "Green Book," health effects from the scenarios were calibrated using the Emergency Response Planning Guidelines (ERPG) levels two and three (See Figures 18.3 and 18.4). ERPG-2 is the

TABLE 18.1
Chemical Plants and Chemical Compounds in Kanawha Valley Worst-Case Scenarios

Plant	Chemical Compounds		
FMC, Nitro	chlorine	phosphorous trichloride	phosphorous oxychloride
Monsanto, Nitro	chlorine	anhydrous ammonia	hydrogen sulfide
FMC, Institute	methylene chloride		
Rhone-Poulenc	methyl isocyanate	chlorine	phosgene
Union Carbide, Institute	butadiene	ethylene oxide	
Arco Chemical, Institute	propylene oxide		
Olin, South Charleston	chlorine		
FMC, Spring Hill	liquid hydrogen		
Union Carbide, South Charleston	monomethylamine	sulfur trioxide	
Arco Chemical, South Charleston	propylene oxide	vinylidene chloride	acrylonitile
Union Carbide Tech Center, South Charleston	dimethylamine		
Arco Tech Center, South Charleston	methylene chloride		
DuPont, Belle	anhydrous ammonia	butyl isocyanate	dimethylamine
Occidental Chemical, Belle	chlorine	methyl chloride	hydrogen chloride

Source: Charleston Daily Mail, June 3, 1994.

maximum airborne concentration below which it is believed that nearly all individuals could be exposed for up to 1 hour without experiencing or developing irreversible or other serious health effects or symptoms which could impair an individual's ability to take protective action. ERPG-3 is the maximum airborne concentration below which it is believed that nearly all individuals could be exposed for up to 1 hour without experiencing or developing life-threatening health effects. Where ERPG data were not available for certain chemicals, alternative standards for determining health effects had to be identified, and in the case of flammability, the level at which there would be significant community impacts had to be determined.

EMERGENCY RESPONSE PLANNING GUIDELINES (ERPG)

Concentration at which severe, permanent, life-threatening health effects can occur if exposure exceeds 1 hour. — ERPG-3

Concentration at which serious health effects can occur if exposure exceeds 1 hour. — ERPG-2

Concentration at which increased irritation and odor effects can occur if exposure exceeds 1 hour. — ERPG-1

FIGURE 18.3 Emergency response planning guidelines. See Reference 15.

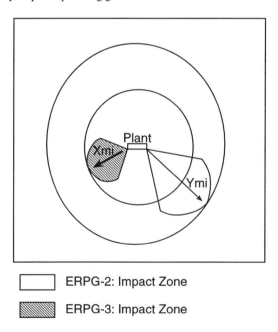

Figure 18.4 Plume model for worst-case and more probable scenarios showing radii for two levels of the EPA Emergency Response Planning Guidelines (ERPG) discussed in the text. See Reference 19.

In the fall of 1993, each plant then began the formal development of its scenarios using the LEPC parameters. The process was supported by Responsible Care® CAER Code workshops covering regulations related to the Clean Air Act and case studies on worst-case scenarios. DuPont Belle provided a model for anhydrous ammonia that was accepted as a format by all of the plants. The plume model, demonstrated in DuPont's model in Figure 18.5, was a relatively simple visual representation of a complex scenario that could be communicated effectively to the public, and provided a basis for comparison across plants. The inner circle closer to the plant showed the area subject to the EPRG-3 (life-threatening health effects) level. As shown, for DuPont Belle, the worst-case

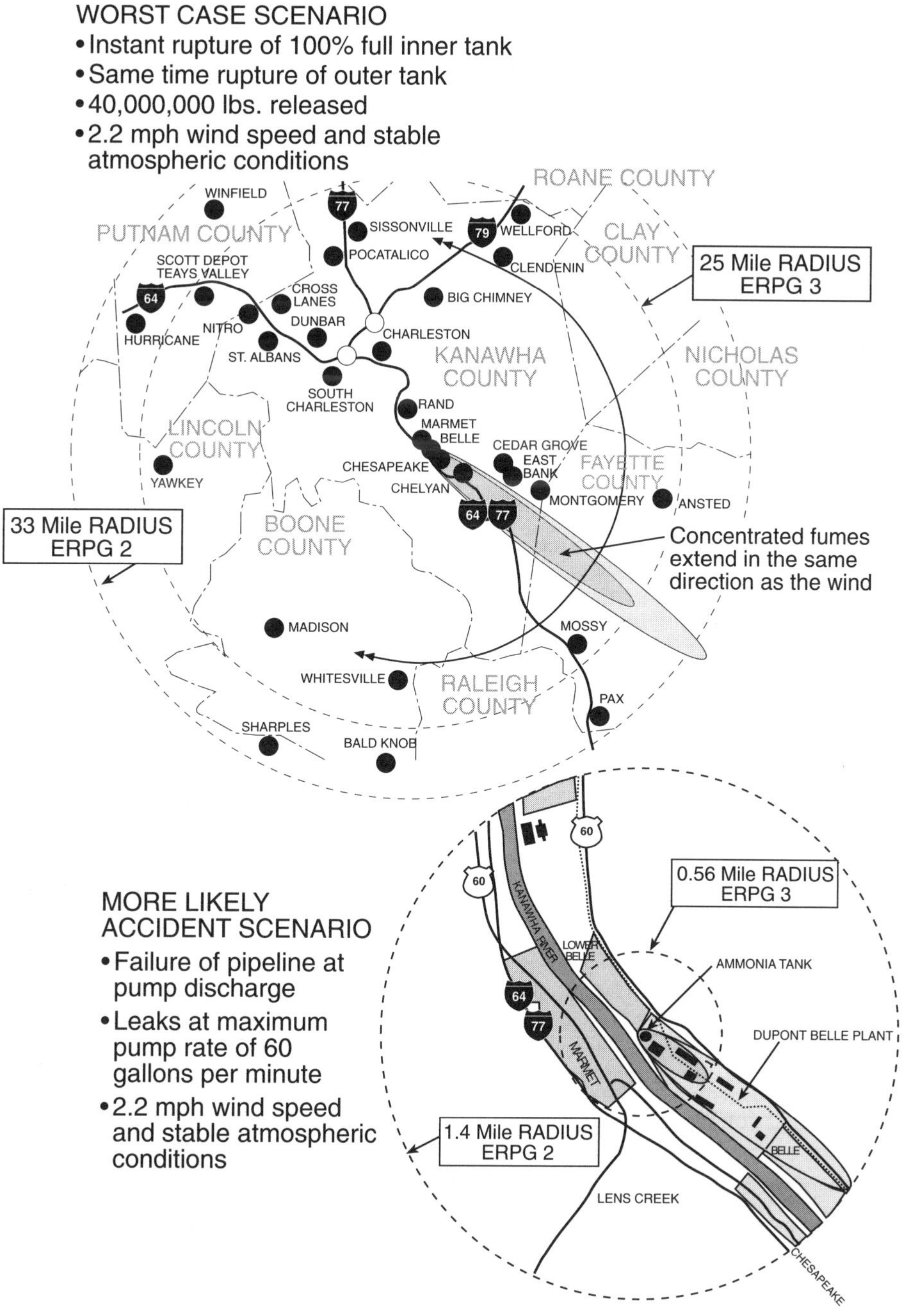

WORST CASE SCENARIO
- Instant rupture of 100% full inner tank
- Same time rupture of outer tank
- 40,000,000 lbs. released
- 2.2 mph wind speed and stable atmospheric conditions

MORE LIKELY ACCIDENT SCENARIO
- Failure of pipeline at pump discharge
- Leaks at maximum pump rate of 60 gallons per minute
- 2.2 mph wind speed and stable atmospheric conditions

FIGURE 18.5 DuPont Belle plant worst-case and more probable-case scenario. See Reference 20.

scenario for anhydrous ammonia reached a 25 mi radius extending in the same direction as the wind. The outer circle showed the area subject to the EPRG-2 (serious health effects) level, and extended 33 mi for DuPont Belle. The lower chart in the same figure showed the more likely accident scenario, with the EPRG-3 radius reducing to only 0.56 mi and the EPRG-2 radius extending only 1.4 mi. The dramatic differences regarding distance resulted from the various layers of protection and other mitigation factors in place (see Figure 18.6).

Ultimately, 72 scenarios were modeled. In reviewing these scenarios, the worst-case scenario hazard zones ranged from less than 0.5 mi to greater than 25 mi, with 52 percent of the ERPG-2 hazard zones being less than 6 mi (see Figure 18.7a). The more probable scenario hazard zones ranged from 0.2 to 6.2 mi, with 59 percent of the ERPG-2 zones being less than 2 mi, and 86 percent being less than 3 mi (see Figure 18.7b).[15] In addition to the plume models, the risk management plan summaries included a five-year history of the release of the chemical compound, uses for that compound, and prevention and emergency strategies of the plant.

Was the hard work finally over? Clearly, a lot of energy, patience, and technical consultation had resulted in a remarkable collaborative effort on the part of chemical companies in a competitive industry, along with extensive participation of multiple stakeholders, to agree upon the technical specifications of the worst-case and probable-case scenarios. But looking back on the process, those involved expressed the consensus view that what was ahead would be the most difficult steps to take.

18.5 THE ROAD TO SAFETY STREET: COMMUNICATION

In agreeing to reveal the detailed information on worst-case and probable-case scenarios, along with the five-year history of accidents, the chemical companies in the Valley had shown good faith in an open stakeholder-based process. Clearly, the involvement of the public at all levels was consistent with both the CAER Code and Process Safety Code's guidelines.

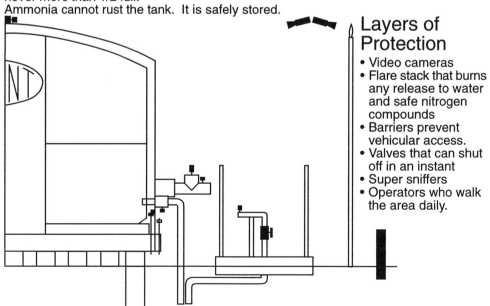

FIGURE 18.6 DuPont Belle safeguards with anhydrous ammonia. See Reference 21.

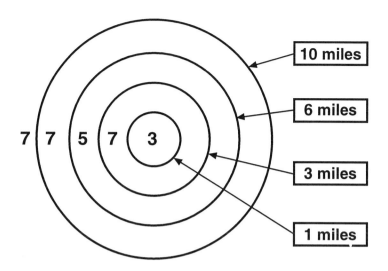

FIGURE 18.7a Frequency of worst-case scenarios (ERPG-2).

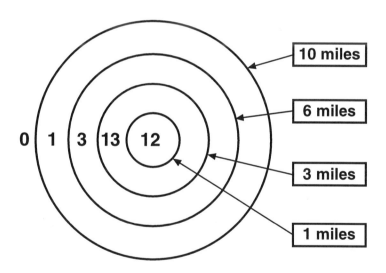

FIGURE 18.7b Frequency of more probable scenarios (ERPG-2). See Reference 15.

Part of the responsibility of stakeholders is then to use the information provided by industry in a fair manner that conveys full information to the public. The Risk Communication Committee was charged with developing the means of providing results of the risk management plans to the public. It was assisted with technical matters from the chemical industry by Knowles, Tom Neinhammer (Arco), Thad Epps (Union Carbide), and Tony LaLonde (FMC).

The greater challenge, however, was dealing with conflict and a lack of trust in the community. Knowles commented, "The situation with the community, especially the newspapers, was very adversarial." Reinforcing this position, Bleidt, committee co-chair noted, "Everyone is afraid of the Charleston Gazette because they do witch hunts, and we knew [one of their reporters] would crucify us. He had been going around the United States trying to dig up stories on danger from chemicals and relate them to here. He wasn't able to get the technical information both because we didn't have it, and because we wanted to communicate it all at the same time. He saw this as evidence of a cover-up. When I was talking with him at one point, I told him we were training people to talk to the media. The next day, the paper ran an editorial in which they quoted me and then questioned why training was necessary if the intent was not to lie and manipulate the press. And when I read the editorial the next morning, I was so upset that I just went back to bed. Several of the plant managers called and apologized for the paper criticizing me."[16]

But the basis for conflict was within industry itself. Because of their training and the nature of their work, plant managers approached communication from a technical orientation. Knowles said, "Most plant managers are not politicians. Most of them have worked in laboratories and grew through competence, not through relationships. When you open yourself to others, it can cause deep pain—and getting to this point was a deep personal journey for me. It is really a process of deep caring and love—and if it not based on this, it is nothing more than fraud."

But entering into dialogue with the public about what they perceived to be technical issues was not easy for plant managers. Bleidt noted, "I didn't realize how secretive the plant managers were — they were keeping information from the public. For example, one plant was keeping thousands of pounds of MIC—much more than they needed for their production processes. The plant managers had to give up autonomy, and this was traumatic for them. I was so mad. They were autonomous and pompous, and not trained to deal with the public. And yet, I also felt sorry for them. I bet you they lost sleep, lost weight, had ulcers all from working through this experience. No one ever had to give up to the extent the plant managers did. The only thing that we community reps gave was time. It was most perilous for the plant managers."

Knowles maintained his emphasis on inclusivity and full information. He even brought a set of junior high school students into the process, including the committee meetings. "Their presence truly changed the committee meetings—brought an openness and an authenticity." Working hard during the later fall of 1993, the Communication Committee proposed that there should be a very open public forum to communicate the results. The decision was made to hold a program entitled *Safety Street: Managing Our Risks Together*, not only to have a formal seminar at the Charleston Civic Center, but also to go where people naturally congregated in the Valley—in the mall!

This decision was not greeted warmly by industry officials. In the spring of 1994, the Responsible Care® Coordinator meeting of CMA was held, with about 180 people, some of whom were CEOs. Knowles reflected, "Some were not happy at all—especially when they heard that Safety Street was escalating to the mall. Everyone was worried—would '60 Minutes' show up? They called it a 'scare fair.' Back in the Valley, I could see the community coming together. But there was a lot of industry pressure and anxiety about the possible public reaction that would result. Everyone was getting tense. We had only one chance, and it had to be right. There were no dress rehearsals. So for planning purposes, many of us were on this thing like a blanket."

While the communication efforts were proceeding, JBF Associates, Inc. was performing quality assurance reviews and the Hazard Assessment Committee was working hard to secure remaining technical issues and to provide guidance for final data submission to the LEPC. In February 1994, a public relations firm, Charles Ryan and Associates was selected as communication project coordinator. By April 1994, it had initiated a communication plan.

Then on June 3 to 4, 1994, "Safety Street" was held starting with a seminar on the first day at the Charleston Civic Center. The purpose of the seminar was to present the worst-case and more-probable-case scenarios in the morning and then to provide unlimited access to plant personnel and their information in booths in the afternoon and in the mall the next day. Presenters included members of the LEPC, local citizens, the Director of Chemical Accident Prevention for the U.S. EPA, plant managers and other industry personnel, the President of JBF Associates, Inc., and other community leaders.

Approximately 700 people attended the first day session, with one third from out-of-town, one third from the community, and one third from local industry. Estimates of participation at Safety

Street events in the mall ranged from 3000 to 5000 persons. There were no reactive demonstrations, but rather genuine questions and dialogue. Leaders from industry and the public reported satisfaction with the event, and there was a climate of improved trust with increased expectations for further openness and dialogue. As part of its report, JBF Associates, Inc. concluded that the following lessons had been learned from the process of community-industry dialogue leading to Safety Street:

- *Consistency* is important.
- *Consensus* among analysts is elusive, but possible.
- *Coordination* among stakeholders is difficult, but essential.
- *Communication* is time-consuming, but necessary.
- *Credible* third-party involvement can be valuable.

18.6 EPILOGUE

Following Safety Street, the Community Emergency Response Evaluation Group (CEREG) was formed with the objective of evaluating the effectiveness of community emergency responses for each worst-case and more probable-case scenario presented by the chemical facilities. They have been in the process of assessing the internal emergency response plans of each chemical plant, and also will assess the emergency response plans of local emergency responders such as fire departments and ambulance companies. Ultimately, they will create a set of recommendations to bring cohesion to the various response plans.

As to the impact on the community and those involved in the process, it is useful to consider the survey results before and after Safety Street. A survey of 403 randomly selected adult head-of-households residing within the Valley was conducted by telephone in July 1994.[17] The results were compared to a community assessment survey conducted in March 1994 (see Table 18.2). The opinion towards chemical companies remained largely unchanged before and after Safety Street, with just over 60 percent of respondents continuing to have a favorable view of chemical companies, but with a decrease in unfavorable opinions from 35 to 31 percent. Moreover, there were significant differences among various demographic cohorts with respondents aged 55 and over, non-union households, upper-income households, and long-time residents having a more positive perception of chemical companies. Also it is interesting to note that, despite the Safety Street event and subsequent media coverage, there was little change in the perception of the respondents as to the potential for a major leak or explosion at a chemical plant. However, there was a significant increase in the percentages of respondents who believed post-Safety Street that other environmental issues (disposal of hazardous waste, air emissions from chemical plants, and automobile emissions) were no problem.

Does the relatively little change in the public's perspectives about chemical companies or environmental issues mean that Safety Street had little ultimate impact? Or does it mean that the process of community-industry dialogue about risk management and worst-case scenarios tempered what might have otherwise been very alarming information?

In reflecting about the overall experience, Knowles commented, "We identified three conditions necessary for self-organizing systems to emerge: there needed to be an open and free flow of information; we had to establish an environment of open relationships with mutual access; and we had to be self-referencing to what was going on—connecting our own values to it. You have to provide information and help people to understand and interpret the information. You have to understand that I was not special in all of this. There were a bunch of good people who did something extraordinary. The people involved were all instruments for the emergence of the worst-case scenarios. We engaged the whole system."

One member of the LEPC and community participant in CEREG commented, "There was a lot of pride in recognizing that we were doing something that is not commonly done in all areas . . . that our model could be used by other people to improve emergency response in their own areas and

TABLE 18.2
Selected Results from Safety Street Pre- and Post-Surveys

Opinion Towards Various Industries

	Percentage Favorable		Percentage Unfavorable	
	July 1994	March 1994	July 1994	March 1994
Banks	84	87	12	10
Public utilities	81	75	17	25
Hospitals	77	81	20	17
Chemical companies	62	61	31	35

Perspectives on Environmental Issues

	Percentage Major Problem		Percentage Minor Problem		Percentage No Problem	
	July 1994	March 1994	July 1994	March 1994	July 1994	March 1994
Potential for major leak or explosion at chemical plant	58	60	32	34	9	5
Disposal of hazardous waste	58	46	15	35	21	14
Air emissions from chemical plant	50	57	33	34	15	7
Automobile emissions	20	17	47	59	31	21

Source: Ryan, McGinn, Samples Research, Inc., August 1994.

management of chemicals and interaction with communities. Locally, we recognized that we were able to work together and could be very effective at doing it, and really enjoyed it once we got into it."[18]

At present, the CMA is in the process of supporting worst-case projects in other areas of the country. Can the Safety Street experience be replicated in other areas or for other industries? To the degree the process was successful, was it an artifact of strong personal leaders both within the chemical industry and from the public? How important was the geographical setting of the Kanawha Valley? To what degree was the interest of the public and the willingness of the industry to participate related to accidents that had occurred before and during the process?

18.7 REFERENCES

1. Erekson, O. H. and Johnson, P. C., "Responsible Care® and worst-case scenarios in the Kanawha Valley, USA," presented at the *Second International Conference on Building the Ethics of Business in a Global Economy,* November 1995.
2. Wood, A., 10 Years after Bhopal, *Chem. Week,* 155, 25, 1994.
3. Begley, R., After Bhopal: CMA 'war room,' new programs, and changed habits, *Chem. Week,* 155, 32, 1994.
4. Ward, K. Jr., 'Worst case' release is just the beginning, *The Charleston Gazette,* June 6, 1994.
5. Rayport, J. F. and Lodge, G. C., Responsible Care, Harvard Business School Case N9-391-135, 1991.
6. The Canadian Chemical Producers' Association, Codes of Practice Commitment Package, Ottawa, ON, 1992.

7. Chemical Manufacturers Association, Community Awareness and Emergency Response Code of Management Practices, Responsible Care®, 1989.
8. Chemical Manufacturers Association, Process Safety Code of Management Practices, Responsible Care®, 199.
9. Personal communication with C. Aldag, Miami University, Oxford, OH, October 27, 1994.
10. Chemical plants list leaks during the last five years, *Charleston Gazette,* June 3, 1994.
11. Correspondence with Richard M. Doyle, June 22, 1995.
12. Ollis, R., Managing our risk together: the Kanawha Valley worst case scenario project, 1994.
13. Interview with Pam Nixon, April 14, 1995.
14. Interview with Paul Hill, April 1995.
15. JBF Associates, Inc., KB Hazard Assessment Project: third-party results, June 3–4, 1994.
16. Interview with Mary Francis Bleidt, June 27, 1995.
17. Ryan and McGinn, Samples Research, Inc., Risk management and worst case scenarios: pre-and post-survey reports, August 1994.
18. Interview with Lillian Morris, July 26, 1995.
19. Memorandum from Paul Hill, May 25, 1994.
20. *Ammonia Risk management Plan Summary,* DuPont Belle plant.
21. Safeguards with anhydrous ammonia, *Layers of Protection: Our Commitment to Responsible Care,* Dupont.

PERMISSIONS FOR USE OF PUBLISHED MATERIAL

Figure 1.1	Reprinted from *Beyond the Limits* copyright ©1992 by Meadows, Meadows, and Randers. With permission from Chelsea Green Publishing Co., White River Junction, Vermont.
Figure 1.2	World Resources Institute. World Resources, 1994–95: A guide to the global environment. New York: Oxford University Press, 1994.
Figure 1.5	Reprinted from *Ecological Economics,* Volume number 12, Kurt R. Wetzel and John F. Wetzel, Sizing the earth: recognition of economic carrying capacity, pp. 13–21, Copyright 1995, with permission from Elsevier Science.
Figure 1.7	Allenby, Braden R., Achieving Sustainable Development through Industrial Ecology, *International Environmental Affairs,* 4(1), pp. 58–59, 1992.
Box 2.1	From Hett, J. M. and R. V. O'Neill. Systems Analysis of the Aleut Ecosystem. ARCTIC ANTHROPOLOGY, Vol. 11, No. 1 (1974) Copyright 1974. Reprinted by permission of The University of Wisconsin Press.
Figure 3.2	From Richard B. Howarth and Richard B. Norgaard, "Environmental Valuation under Sustainable Development," *American Economic Review,* Volume 82, no. 2, May 1992, pp. 473–477.
Figure 3.3	Reprinted from Milon, J. Walter, Environmental and Natural Resources in National Economic Accounts, in *Integrating Economic and Ecological Indicators,* J. Walter Milon and Jason F. Shogren, Copyright ©1995 by J. Walter Milon and Jason F. Shogren. Reproduced with permission of GREENWOOD PUBLISHING GROUP, INC., Westport, CT.
Figure 3.4	From Cobb, Clifford and Ted Halstead, *The Genuine Progress Indicator: Summary of Data and Methodology.* Reprinted with Permission, copyright 1995, Redefining Progress.
Table 3.2	Wealth Figures, 1990 Values from Passell, Peter, The Wealth of Nations: A 'Greener' Approach Turns List Upside Down, *New York Times,* September 19, 1995, Copyright ©1995 by The New York Times Co. Reprinted by Permission. Global Environmental Indicators, GNP in 1991, per 1990 Population: Adapted from Rodenburg, Eric; Dan Tunstall; and Frederik van Bolhuis, Environmental Indicators for Global Cooperation, Working Paper No. 11. Washington, D.C., The World Bank, 1995.
Table 3.3	From Repetto, Robert; William Magrath; Michael Wells; Christine Beer; and Fabrizio Rossini, Wasting Assets: *Natural Resources in the National Income Accounts.* Washington, D.C.: World Resources Institute, 1989.
Table 3.4	From Bartelmus, P., Stahmer, C., and van Tongeren, J. Integrated environmental and economic accounting—a framework for an SNA satellite system, in *Toward Improved Accounting for the Environment,* Lutz, E. The World Bank, Washington, D.C., 1993, p. 45.
Figure 4.3	From Barker, Jerry and David Tingey, *Air Pollution Effects on Biodiversity,* Van Nostrand Reinhold, New York, 1992.
Figure 5.1	Granted with permission from *Environmental Strategies for Industry,* Johan Schot and Kurt Fischer, © by Island Press, 1993. Published by Island Press, Washington, DC and Covelo, CA.
Figure 6.1	From ACCOUNTING INFORMATION SYSTEMS by Leitch/Davis, ©1983. Reprinted by permission of Prentice-Hall, Inc., Upper Saddle, NJ.
Figures 6.2a and 6.2b	From ACCOUNTING INFORMATION SYSTEMS by Leitch/Davis, ©1983. Reprinted by permission of Prentice-Hall, Inc., Upper Saddle, NJ.
Box 6.1, 6.3–6.9	Permission granted from *Greenwire,* a news service of *The National Journal.*
Table 7.2	Hart, Stuart, "A Natural Resource Based View of the Firm," *Academy of Management Review,* Vol. 20 (4), October 1995, pp. 986–1014.

Figure 7.5	From *Strategic Environmental Management: Using TQEM and ISO 14000 for Competitive Advantage,* Grace H. Wever, Copyright ©1996 by John Wiley & Sons, Inc. Reprinted by permission of John Wiley & Sons, Inc.
Figure 7.6	From *Strategic Environmental Management: Using TQEM and ISO 14000 for Competitive Advantage,* Grace H. Wever, Copyright ©1996 by Council of Great Lakes Industries. Reprinted by permission of John Wiley & Sons, Inc.
Reference 8.1	From *Everything for Sale: The Virtues and Limits of Markets,* Robert Kuttner, copyright ©1996 by Alfred A. Knopf, Inc. Reprinted with permission of Alfred A. Knopf Incorporated.
Figure 8.3	Reprinted from *Ecological Economics,* Vol. 14, Nick Johnstone, Trade Liberalization, Economic Specialization, and the Environment, pp. 165–173, Copyright 1995, with permission from Elsevier Science. Original source cited in Johnstone was K. Anderson and R. Blackhurst, eds., *The Greening of World Trade Issues.* London: Harvester Wheatsheaf, 1992, and World Resources Institute, *World Resources 1992–93,* New York: Oxford University Press.
Figure 8.4	Adapted from Peter Glick, *Water and Crisis: A Guide to the World's Fresh Water Resources,* Oxford University Press, 1993, as cited in Kenneth D. Frederick, Water as a Source of International Conflict, *Resources,* Spring 1996, pp. 9–12.
Figure 8.5	From Bryan Norton, Georgia Institute of Technology, as cited in Michael A. Toman, The Difficulty in Defining Sustainability, *Resources,* Winter 1992, p. 5.
Table 11.2	From *Whole Foods,* December, 1993.
Figure 12.1	With permission of Miami University Applied Technologies.
Figure 12.2	From Figure 1.4 in Pongracz, D.D. 1978. *Practical Viticulture,* Capetown: David Philip Publisher, p. 7.
Figure 12.3	From ECOLOGY AND FIELD BIOLOGY, 5th Edition, by Robert Leo Smith; Copyright ©1974, by Robert L. Smith. Reprinted by permission of Addison-Wesley Educational Publishers.
Figure 12.4	From Figure 7–17, from Nebel, Bernard J. 1990. *Environmental Science: The Way the World Works,* 3rd edition. Englewood Cliffs, N.J.: Prentice-Hall, p. 169.
Figure 12.5	Swift, M. J.; O. W. Heal; and J. M. Anderson. 1979. *Decomposition in Terrestrial Ecosystems.* Oxford, U.K., as cited in Freckman, Diana W. *Life in the Soil, Soil Biodiversity: Its Importance to Ecosystem Process.* Report of a Workshop Held at the National History Museum, London, England, August 30–September 1, 1994. Reprinted by permission of Blackwell Science, Ltd.
Figure 12.6	University of California Division of Agriculture and Natural Resources. 1982. *Grape Pest Management.* Oakland: UC DANR, p. 286. Regents of the University of California. Reprinted with permission.
Figure 12.7	University of California Division of Agriculture and Natural Resources. 1982. *Grape Pest Management.* Oakland: UC DANR, pp. 291–293. Regents of the University of California. Reprinted with permission.
Figure 12.8	With permission of Miami University Applied Technologies.
Table 14.1	*Ward's Automotive Yearbook* 1994, p. 234. Reprinted with permission of The Polk Company.
Figure 14.2	Reprinted with permission ©1992 AUTOMOTIVE ENGINEERING magazine—October 1992. Society of Automotive Engineers, Inc.

Permissions

Figure 15.1	Grand Trunk CN North America. Research by Patty Vegella, graphic by David Kordaiski: With permission of *The Detroit News,* May 24, 1995.
Figure 15.5	TRANSCAER: Transportation Community Awareness & Emergency Response, With permission of Chemical Manufacturers Association, Washington, D.C. 1995.
Figure 15.6	*Macmillan Visual Dictionary,* cited by Sean McDade, With permission of *The Detroit News,* May 26, 1995.
Figures 18.1 and 18.2	Reprinted with permission of National Institute for Chemical Studies.
Figure 18.3	Reprinted with permission of JBF Associates, Inc.
Figure 18.4	Adapted from Memorandum from Paul L. Hill, May 25, 1994. Reprinted with permission of National Institute for Chemical Studies.
Figures 18.5 and 18.6	With permission from DuPont Health, Environment, and Safety.
Figures 18.7a and 18.7b	Reprinted with permission of JBF Associates, Inc.

Author Index

A

Ackoff, R. L. 97, 104
Adams, M. 151, 165
Aldag, C. 3, 20, 39, 40, 146, 165, 174, 179
Allen, T. F. H. 50, 61
Allenby, B. R. 15, 20
Anderson, J. M. 256, 269
Arrow, K. 8, 10, 20
Auster, E. R. 84, 103

B

Banks, R. D. 119, 133
Barbier, E. B. 12
Barde, J. P. 144, 166
Barker, J. 78, 79, 80
Barrett, C. B. 174, 179
Barrett, G. W. 39
Bartelmus, P. 53, 57, 61, 62
Bartholomew, J. A. 85, 103
Beer, C. 55, 56, 61
Begley, R. 338, 339, 352
Behre, C. E. 65, 66, 80
Behrens, W. W., III 4, 14, 20
Bishop, R. 116, 133
Black, F. 111, 112, 132
Bolin, B. 8, 10, 20
Boockholdt, J. 119, 133
Booth, D. E. 215, 225
Bormann, F. H. 177, 179
Boulding, K. 88, 95, 96, 103, 104
Bounds, G. 151, 165
Branson, W. H. 8, 20
Brealey, R. 109, 132
Brennan, M. 110, 111, 132
Brennan, T. J. 46, 48, 61
Brooks, P. 125, 126, 127, 134
Brookshire, D. 115, 116, 133
Brown, J. B 12, 20, 27, 28, 40
Brown, T. C. 117, 133
Brundtland, G. H. 167
Buchholz, R. A. 88, 103

C

Cairncross, F. 310, 321
Caldwell, A. K. 8, 20
Callenbach, E. 101, 104
Callicot, J. 108, 132
Cameron, T. A. 116, 133
Capra, F. 101, 104
Carpenter, G. D. 153, 165
Carson, C. S. 57, 62
Case, D. R. 124, 133
Catton, W. R., Jr. 94, 104

Cernea, M. 16, 21
Chamberlin, J. H. 324, 335
Chaney, R. 216, 225
Chaprnka, R. 296, 302
Chavas, J. 116, 133
Chemical Manufacturers Association 158, 159, 166, 340, 353
Chew, D. 132
Chilton, K. 141, 165
Clawson, M. 114, 133
Clegg, S. R. 104
Cleveland, C. J. 54, 61
Cobb, C. W. 52, 54, 57, 58, 59, 60, 61, 62
Cobb, J. B., Jr. 11, 20, 26, 39, 52, 57, 61, 62
Colby, M. E. 93, 94, 103, 104
Cook, J. A. 252, 268
Cooper, C. 117
Cordray, S. M. 30, 31 40
Cornell, B. 148, 165
Costanza, R. 8, 9, 10, 14, 20 21, 39, 40, 42, 61, 85, 103, 110, 132
Council of Great Lakes Industries 155, 156, 165
Council on Environmental Quality 36, 37, 40, 79, 80
Covington, W. W. 61
Crocker, T. D. 42, 60
Cromack, K. 215, 217, 225
Cronon, W. 28, 30, 40, 67, 80
Cropper, M. L. 48, 61
Cross, F. 106, 114, 116, 117, 118, 132
Cruz, W. 26, 39
Cummings, R. 115, 133
Cushman, J. 72, 80

D

Daly, H. E. 9, 11, 16, 20, 24, 26, 39, 42, 49, 50, 61, 85, 103, 110, 132, 172, 173, 178
d'Arge, R. 116, 133
Dasgupta, P. 8, 10, 20
Davidson, L. 125, 126, 127, 134
Davis, K. 120, 121, 133
DeBano, L.F. 61
deCamino, R. 26, 39
Deming, W. E. 150, 151, 152, 165
Denison, W. 215, 217, 225
Desvouges, W. H. 118, 133
Devall, W. 93, 95, 104
Ditz, D. 119, 133
Drew, K. 84, 103
Drumwright, M. 308, 321
Dunlap, R. E. 94, 104

E

Egri, C. P. 93, 96, 104
Ehrenfeld, D. 117, 133

Ehrlich, A. H. 5, 10, 11, 20
Ehrlich, P. R. 5, 10, 11, 20
El-Ashry, M. T. 37, 40
Emery, F. E. 97, 104
Emmons, G. 139, 165
Environment Canada 72, 80
Environmental Liaison Centre 40
Erekson, O. H. 3, 20, 39, 40, 146, 165, 169, 174, 178, 179, 337, 352

F

Feigenbaum, A. V. 150, 165
Fischer, K. 103
Flader, S. L. 80
Flavin, C. 171, 178
Folke, C. 8, 9, 10, 14, 20, 21, 39, 40
Fortier, I. 94, 104
Franklin, J. F. 215, 217, 225
Freckman, D. W. 253, 256, 268
Frederick, R. 103, 104
Frederick, K. D. 173, 178
Freeman, A. M. 114, 132

G

Gale, R. P. 30, 31, 40
Gannon, J. 224, 225
Garber, P. 8, 20
Gellings, C. W. 324, 335
General Accounting Office 68, 69, 80
Gilbert, R. J. 324, 326, 335
Gladwin, T. N. 89, 93, 103, 104
Glick, P. 173
Gokcek, G. 148, 165
Goldman, L. 101, 104
Goodland, R. 9, 12, 16, 20
Goodspeed, L. 139, 165
Gorman, R. F. 169, 178
Gottfried, R. R. 58, 62
Gottlieb, E. 139, 165
Gray, R. 119, 131, 133
Greber, B. 214, 225
Green, R. 88, 103
Grossman, G. M. 8, 20, 175, 179
Grubb, M. 171, 178
Gugino, S. 250, 261, 265, 268
Gupta, P. 126, 134

H

Haas, P. M. 170, 178
Hale, G. J. 161, 166
Halliday, J. 250, 254, 268
Hall, R. M. 124, 133
Halstead, T. 54, 58, 59, 60, 61, 62
Hamlin, S. 265, 269
Hammer, M. 9, 20, 39, 40
Haneman, W. 116, 133
Hanson, M. E. 12, 20, 27, 28, 40

Hardin, G. 31, 33, 36, 39, 40, 118, 133
Hardy, C. 104
Harper, E. L. 295, 296, 302
Hart, S. 147, 148, 149, 165
Harte, J. 40
Hartman, C. L. 148, 165
Hartwick, J. M. 59, 62
Hawken, P. 99, 104, 143, 165
Heal, O. W. 256, 269
Heberlein, T. 116, 133
Hemenway, C. G. 161, 166
Henning, J. 114, 132
Henriques, D. 216, 225
Hett, J. M. 27, 40
Hicks, J. 50, 61
Hilborn, R. 30, 31, 40, 70, 80
Hildebrandt, C. C. 319, 321
Hillel, D. 34, 40
Hinrichs, D. 173, 178
Hoekstra, T. W. 50, 61
Hoffman, W. M. 96, 103, 104
Holdren, J. P. 10, 11, 20
Holling, C. S. 8, 10, 20
Homer-Dixon, T. F. 173, 178
Howarth, R. B. 42, 45, 46, 48, 60, 61
Hunt, B. C. 84, 103

I

Ingels, C. 260, 264, 265, 268, 269
International Chamber of Commerce 3, 20, 120, 122, 128, 133
Isaacs, W. 101, 104

J

Jaeger, W. K. 46, 48, 61
Jansson, A. 9, 20, 26, 39, 40
Jansson, B. 8, 10, 20, 26, 39
Janzen, D. C. 319, 321
Jarrett. H. 104
JBF Associates, Inc. 348, 349, 351, 353
Jiménez, J. 26, 39
Johnson, K. N. 214, 225
Johnson, P. C. 169, 178, 337, 352
Johnstone, N. 172, 178
Joskow, P. L. 326, 335
Juday, G. 215, 217, 225

K

Kahn, J. R. 142, 165
Kahneman, D. 115, 116, 133
Karr, J. R. 12, 20
Kaufmann, R. K. 54, 61
Kellert, S. R. 177, 179
Kelly, M. 148, 165
Kennally, J. J. 93, 104
Keohane, R. O. 170, 178
Kliewer, W. M. 252, 268

Klonsky, K. 260, 264, 265, 268, 269
Kluth, F. J. 159, 166
Knetsch, J. 114, 115, 116, 133
Koch, D. G. 319, 321
Koch, M. 171, 178
Koretz, G. 146, 165
Krause, T. 89, 93, 103, 104
Krehbiel, T. C. 169, 178
Kreuze, J. G. 124, 133
Krueger, A. B. 8, 20, 175, 179
Küng, H. 94, 95, 104
Kuta, C. C. 319, 321
Kuttner, R. 168, 178

L

Ledec, G. 12, 20
Leitch, R. 120, 121, 133
Leopold, A. 97, 104
Lesser, J. A. 46, 48, 61
Levin, S. 8, 10, 20
Levy, M. A. 170, 178
Lewin, R. 252, 268
Li, D. H. 119, 133
Lider, L. A. 252, 268
Lind, R. C. 48, 61
Lis, J. 141, 165
Liu, J. 115, 133
Liverman, D. M. 12, 20, 27, 28, 40
Livingston, P. 260, 264, 265, 268
Loebbecke, J. 123, 130, 133
Lodge, G. C. 339, 352
Loucks, O. L. 3, 20, 35, 39, 40, 146, 165, 169, 174, 178, 179
Lucas, G. 225
Ludwig, D. 30, 31, 40, 70, 80
Lutz, E. 61, 62
Lutz, R. 101, 104
Lyons, B. 148, 165

M

MacIntyre, A. 101, 104
MacKenzie, J. J. 37, 40
Magrath, W. 55, 56, 61
Maler, K. 8, 10, 20
Mäler, K-G 14, 21
Mankiw, N. G. 41, 44, 60
Marburg, S. 101, 104
Marglin, S. 110, 132
Markandya, A. 12, 47, 61
Marron, D. B. 326, 335
Martin, G. 218, 222, 223, 225
Martin, P. S. 29, 40
Martinez-Alier, J. 48, 61
Maser, C. 68, 80, 215, 217, 225
Matson, P. A. 5, 20
Maxey, M. 130, 134
McFarland, W. 123, 130, 133
McGivney, M. P. 118, 133

McKee, A. 215, 217, 225
McLean, R. A. N. 160, 161, 166
McPhail, A. 115, 133
Meadows, D. H. 4, 5, 14, 19, 20
Meadows, D. L. 4, 5, 14, 19, 20
Medhurst, J. 84, 103
Megan, R. 309, 321
Merideth, R. W., Jr. 12, 20, 27, 28, 40
Merton, R. 111, 132
Meyer, P. B. 294, 302
Milbrath, L. W. 87, 88, 93, 103
Milon, J. W. 52, 53, 60, 61
Morgan, R. 324, 326, 335
Munasinghe, M. 16, 21
Munson, A. 171, 178
Myers, S. 109, 132

N

Nadel, S. M. 324, 326, 335
Naess, A. 94, 104
Nash, J. A. 102, 104
National Public Radio 250, 268
National Research Council 129, 134
Nebel, B. J. 255, 269
Newell, G. E. 124, 133
Newell, S. J. 124, 133
Nord, W. 104
Nordhaus, W. 56, 62
Norgaard, R. B. 42, 45, 46, 47, 48, 50, 60, 61
NRDC 218, 225

O

Oates, W. E. 145, 146, 147, 165
Office of Technology Assessment 76, 80
Okorafor, A. 115, 133
Okore, A. 115, 133
Ollis, R. 343, 353
O'Neill, R. V. 27, 40
Ough, C. S. 263, 264, 269

P

Paddock, J. 111, 132
Palamides, J. 125, 126, 127, 134
Palmer, K. L. 146, 147, 165
Papendick, R. I. 257, 260, 268
Parikh, K. 144, 166
Parr, J. F. 257, 260, 268
Parrish, M. 215, 224, 225
Passell, P. 54, 55, 61
Pauchant, T. 94, 104
Pearce, D. W. 12, 47, 61
Pelline, J. 222, 225
Perrings, C. 8, 10, 20, 36, 40
Petry, E. S., Jr. 103, 104
Pimentel, D. 8, 10, 20
Pinfield, L. 93, 96, 104
Piper, T. 261, 269

Pongrácz, D. D. 252, 253, 268
Ponting, C. 28, 29, 30, 33, 34, 40
Porter, M. E. 146, 147, 165
Portney, P. R. 48, 61, 146, 147, 165
Postel, S. 7, 20
Powers, M. B. 151, 166
Puican, S. E. 334, 335

Q

Quelch, J. 308, 321

R

Raab, J. 331, 332, 335
Randers, J. 4, 5, 14, 19, 20
Ranganathan, J. 119, 133
Ranney, G. 151, 165
Rapport, D. J. 24, 25, 26, 39
Ratliff, R. 123, 130, 133
Ray, M. 126, 134
Rayport, J.F. 339, 352
Rees, W. 49, 61
Reganold, J. P. 257, 260, 268
Reid, M. W. 324, 326, 335
Regier, H. A. 24, 25, 26, 39
Repetto, R. 26, 39, 55, 56, 61
Rhodes, S. 161, 166
Ridker, R. 114, 132
Rigoglioso, M. 139, 165
Robert, R. 30, 40
Rodenburg, E. 54, 55, 61
Rolston, H., III 97, 104
Romm, J. J. 16, 21
Roodman, D. M. 173, 178, 324, 326, 335
Rosen, H. S. 45, 61, 171, 178
Rosenberg, R. 39
Rossini, F. 55, 56, 61
Rowe, J. 54, 58, 61
Rowe, R. 116, 133
Rubenstein, D. B. 39, 40, 118, 129, 133, 134
Ryan and McGinn, Samples Research, Inc. 351, 352, 353

S

Sachs, A. 6, 20
Safina, C. 73, 80
Saxe, D. 129, 134
Schabath, G. 301, 302
Schaltegger, S. 118, 130, 131, 133
Schmidheiny, S. 3, 20, 88, 89, 90, 103, 147, 148, 165
Scholes, M. 111, 112, 132
Schot, J. 103
Schultz, E. 216, 225
Schulze, W. 115, 116, 133
Schwartz, E. 110, 111, 132, 216, 225
Sedell, J. 215, 217, 225
Seller, C. 116, 133
Senge, P. 14, 20
Sessions, G. 93, 95, 104

Shapiro, A. 148, 165
Sharplin, A. 82, 83, 103
Sheridan, P. J. 130, 134
Shogren, J. F. 52, 60, 61
Shrivastava, P. 86, 93, 96, 99, 103, 104, 147, 165
Siegel, D. 111, 132
Simpson, R. D. 146, 165
Skow, J. 223, 225
Smart, B. 3, 20, 34, 38, 39, 40, 79, 80, 125, 133, 164, 166
Smith, J. 111, 132
Smith, R. 260, 264, 265, 268
Smith, R. L. 254, 269
Smith, V. K. 115, 118, 133
Socolow, R. H. 40
Solórzano, R. 26, 39
Solow, R. 45, 49, 55, 61
Stafford, E. R. 148, 165
Stahmer, C. 57, 62
Steer, A. 170, 171, 178
Sterling, J. 262, 269
Stinson, C. 118, 130, 131, 133
St. John, P. 301, 302
Stoft, S. 324, 326, 335
Stoll, J. 116, 133
Such, 296, 302
Sullivan, F. 171, 178
Swanson, F. 215, 217, 225
Swift, M. J. 256, 269

T

Thayer, M. 116, 133
Thomson, K. 171, 178
Thorpe, C. 24, 25, 26, 39
Tietenberg, T. 4, 20, 141, 165
Tingey, D. 78, 79, 80
Tippett, P. 154, 165
Tobin, J. 56, 62
Toman, M. A. 8, 20, 45, 61, 176, 179
Tosi, J. 26, 39
Tourte, L. 260, 264, 265, 268, 269
Tunali, O. 171, 178
Tunstall, D. 54, 55, 61
Turner, J. 250, 266, 267, 268
Turner, J. E. 24, 39
Twining, C. E. 67, 80

U

UN Food and Agriculture Organization 71, 80
U.S. Citizens Network Working Group on Ethics 95, 104
USDA Forest Service 66, 67, 80, 217, 225
US Geological Survey 74, 80

V

van Bolhuis, F. 54, 55, 61
van der Linde, C. 146, 147, 165
van Tongeren, J. 57, 62
Vasquez, A. 26, 39

Villalobos, C. 26, 39
Vitousek, P. M. 5, 20

W

Wackernagel, M. 49, 61
Waigner, L. 14, 21
Wainman, D. 123, 133
Wallace, W. 123, 130, 133
Walley, N. 148, 165
Walters, C. 30, 31, 40, 70, 80
Walters, M. 222, 225
Ward, K., Jr. 339, 352
Warner, C. 117
Watson, V. 26, 39
Wells, M. 55, 56, 61
Westra, L. S. 97, 104
Wetzel, J. F. 11, 20
Wetzel, K. R. 11, 20
Wever, G. H. 149, 150, 154, 155, 165, 166
Whitehead, B. 148, 165
Whittington, D. 115, 133
Wigglesworth, D. 122, 133
Williams, H. E. 84, 103
Williams, J.-O. 133

Willson, J. S. 160, 161, 166
Wilson, E. O. 177, 179
Winkler, A. J. 252, 268
Wolcott, D. R. 324, 326, 335
Wood, A. 155, 166, 338, 352
Woodward, R. 26, 39
Woodwell, G. M. 35, 40
World Commission on Environment and Development 8, 12, 20, 37, 38, 40, 86, 88, 103
World Industrial Conference on Environmental Management 36, 38, 40
World Resources Institute 6, 20, 66, 80
Wright, H. E., Jr. 40
Wu, F. 119, 133

Y

Yao, J. 308, 321
Yorks, L. 151, 165
Young, A. H. 57, 62

Z

Zerbe R. O. Jr. 46, 48, 61

Subject Index

A

ABC, see Activity based costing
Accounting, see Auditing and Reporting
Accounting Information Systems (AIS)
   definition, 119–122
   effective decision-making, 199–120
Activity based costing
   definition, 119, 125–128
   example, 126–127
   steps, 126
   value vs. nonvalue activities, 126
Adaptive ecosystem management, see Ecosystem, adaptive management
AFL-CIO Industrial Union Department, 159
Agriculture, see also Fetzer Vineyards, Walnut Acres
   crop rotation, 235–236
   fertilizer use, 236, see also soil
   pest control, 236, 255–257, 260–261
   subsidies and production techniques, 172
   weed control, 236, 261
AICPA, see American Institute of Certified Public Accountants
Air
   bubble approach, 78–79
   cleansing, 77–78
   permit trading, 79, see also Market-based incentives, tradeable permits
   pollution, 77–79
      control, 78–79
      emissions, 77
      greenhouse gases, 171
   quality, 35–36
   regeneration system, 77
   resources, 77–78
AIS, see Accounting Information Systems
Aldicarb oxime, 342
Allocative efficiency, see Economic concepts, economic efficiency
American Forest and Paper Association, 154
American Institute of Certified Public Accountants (AICPA)
   organization, 128, 130
   statement of auditing standards, 130
American Iron and Steel Institute, 285
American Plastics Council, 285
Anheuser-Busch, 153
Anhydrous ammonia, 347–348
Anthropocentrism, 86, 92–94, 108
API, see Ashland Chemical Company, Audit Priority Index
Appropriations rescission
   Emergency Two-year Salvage Timber Sale Program rider, 224
Arthur D. Little, 285

Ashland Chemical Company
   Audit Priority Index (API), 190
   auditing process, 188–197
   Compliance Audit Program, 188, 210
   environmental activities, 185–186
   Operations Auditing Group, 187–188
   organization chart, 186
   participation in Responsible Care®, 185–186
   preaudit questionnaire, 199–208
   response tracking, 209
   Simply the Best program, 185
   vision statement, 184–185
Assimilative capacity, 177
Attitudes, public, see Public opinion
AT&T, 153
Auditing and reporting, 31, 118–138, 170, 300, see also Green accounting and Ashland Chemical Company
   benefits, 122–133, 175
   CERES principles, 159
   corporate use of environmental audits, 123
   disincentives, 124
   external reporting, 128–129
   industry variation, 124–125
   plant specific, 124
   protocols, 122
   Satellite System for Integrated Environmental and Economic Accounting (SEEA), 57
   taxes, 130–131
   types, 120–122
Automotive industry, 283–284, see also David J. Joseph Company [check GM entries]
Automotive Recyclers Association, 286
Avenor, 154

B

Babbitt v. Sweet Home, 223
Balance, 69–72, 77
   science and economics, 32
BASF, 154
Bayer Rubber Corporation, 154
BCSD, see Business Charter for Sustainable Development
Bell Laboratories, 150
Ben & Jerry's Homemade, 159
Bhopal, India accident, 82, 85, 155, 338
Biocentrism, see Deep ecology
Biological diversity, 36, 64
   sustainable reproduction of species, 36
   importance to agriculture, 260–261
BMW of North America, 286
The Body Shop, 87
Boundaries, 17, 88, 168–169, 345
Bristol-Myers Squibb, 153

British Petroleum (BP), 125
British Standards Institution, BS 7750 system, 160
Brown-Forman, 250
Browning-Ferris Industries, 153–154
Bruntland, Gro Harlem, 37, 148, 167
Bruntland Report, 45
Business Charter for Sustainable Development (BCSD), 3, 85, 120, 128
Business decision-making, see also Environmental management systems
   accounting system, 119–120, 126–128
   advantages to environmental protection, 38–39, 146–149
   competitive advantage, 38, 86
   compliance, 85, 121–125, 326
   cradle-to-grave activities, 175–176, 287
   designing for the environment, 175, 284–287, see also Stewardship, product
   environmental pressures and response, 84–85, 103, 148
   internalizing costs, 37, 139–146
   partnerships with environmentalists, 3, 37–39, 101, 129–130, 147
   self-regulation, 146–149

C

CAER program, see Community Awareness and Emergency Response program
California Certified Organic Farmers (CCOF), 251, 263–265
California Department of Fish and Game, 223
California Department of Forestry, 222–224
California Forest Practices Act, 223
California Organic Foods Act, 263
California Redwood Association, annual timber shipments, 217
Call option, 111
Canadian Chemical Producers Association, 158, 339
Canadian National, 154, 293–294, 299
Canadian Petroleum Products Institute, 154
Carbon dioxide emissions, 10, 78, 172
Carbon monoxide emissions, 8
Carrying capacity, 4–7, 177
CCOF, see California Certified Organic Farmers
Center for Excellence in Environmental Engineering and Sciences, 286
CERCLA, see Comprehensive Environmental Response, Compensation, and Liability Act
CEREG, see Worst-case scenarios, Kanawha Valley, Community Emergency Response Evaluation Group
CERES, see Coalition for Environmentally Responsible Economies
CFCs, see chlorofluorocarbons
CGLI, see Council of Great Lakes Industries and TQEM
Chemical Emergency Preparedness program, 338
Chemical Manufacturers Association (CMA), 83, 100, 155, 158–159, 164, 175, 186, 298, 338–339
Chemical residuals, 34–35
Chemical Safety and Hazard Investigation Board, 339

Chernobyl, 88
Chlordane, 34
Chlorofluorocarbons (CFCs)
   commercial use, 34
   ozone, 34, 142
   terminating use, 34, 87, 139, 142
Chlorine, 296
Chromic acid, 272, 277
Chrysler Corporation, 283, 286
Ciba Specialty Chemical Corporation, 153
Cincinnati Gas & Electric Company
   company description, 327–331
   electricity load management, 324–326
      Demand-Side Management (DSM), 324–326, 328–329, 331, 334
      DSM collaborative, 324–326, 331–334
      Least-Cost Planning (LCP), 324
      Integrated Resource Planning (IRP), 327–329
   electric utility industry, 324–327
   regulatory setting, 326–327, 334
   stakeholder involvement, 332–334
   Total Resource Cost Test (TRC), 330–333
Circular flow diagram, 42–43
Clark Hill, 154
Clean Air Act (CAA) and Amendments, 35, 79, 141, 145, 276, 288, 326, 338, 341
   Risk Management Plan, 339
Clean Water Act (CWA), 35, 76, 118, 288
Climate Change, Framework-Convention on, 171
Clinton, President, 68, 131, 224–225
Club of Rome, see Growth, Limits to Growth
CMA, see Chemical Manufacturers Association
Coalition for Environmentally Responsible Economies (CERES)
   CERES principles, 159–160, 162–163
Coca-Cola, 153
Cod fishery, see Fisheries, Grand Banks
Colgate-Palmolive, 153, 312
Command-and-control, see Externalities
Commons, Tragedy of the, 31, 118
Common property goods, 31, 36, 70–71, 79, 140
Community Awareness and Emergency Response (CAER) program, 158, 339–340
Compliance, see also Auditing and Reporting, Business decision-making
Comprehensive Environmental Response, Compensation, and Liability Act (CERCLA), 36, 131, 288–289
Consumption
   nonrival, 44
   nonexcludable, 44, 171
Contingent valuation, see Valuation
Continuous quality improvement, 149–151, 155, 159
Coors Brewing, 153
Cornell-Shapiro stakeholder model, 148
Corporate vs. national income accounting, 119
Council of Great Lakes Industries (CGLI) and TQEM, 154–155, 164

Subject Index

D

David J. Joseph Company
    automobile industry and vehicle scrappage, 283–284
    company history and description, 279–281
    designing for the environment, 284–287
    disassembly, 284–287
    scrap business, 281–283, 287–288
DDT, see Dichloro diphenyl dichloroethane
Decision option analysis, 168–170
Declaration of the Business Council for Sustainable Development, 90–91
Deep ecology, 92–96
Demand side management (DSM), see Cincinnati Gas & Electric Company
Deming, W. Edwards, 150–153
Deregulation, see Cincinnati Gas & Electric Company and Regulation, command-and-control
Developing countries
    gap with developed countries, 88
    prospects for sustainable development, 175
Dialogue, see Sustainability, principles, information, dialogue, and expectations
Dichloro dipehnyl dichloroethane (DDT), 34–35, 256
Dieldrin, 34
Dioxin, 34
Discounted cash flow, see Valuation
Discounting
    choice of discount rates, 47–48, 109–110
        opportunity cost of capital, 47
        pure time preference, 48
        social discount rate, 48
    desirability, 46–47
    present value, 47
Dofasco, 154
Domino's Pizza Distribution Corporation, 159
Double dividend, 145
Dow Chemical Canada, 154
Dow Chemical Company 124, 148, 151, 153, 286, 300
Dow Elanco, 154
Downriver Mutual Aid Task Force, 297–298
DSM, see Demand side management
Duke Power, 153
DuPont Company
    Belle plant, 337, 347–348
    CFCs, 34, 87
    membership in GEMI, 153
    recycled products for automobiles, 286
    reporting, 125
    stockholders, 344
    TQM, 151

E

Earth Charter, 94–95
Earth Summit, 171
Eastman Kodak Canada, 154–155
Eastman Kodak Company, 151, 153, 154
Eco-efficiency, 96, 139
Eco-labeling, 310
Ecological functioning, 26–27, see also Resilience
Economic concepts
    economic efficiency, 16, 44, 140–141, 143
        dynamic efficiency, 143
        exchange efficiency, 44
        product-mix efficiency, 44
        production efficiency, 44
    efficiency versus equity, 45
    functions
        allocation, 42, 44–45
        distribution, 42, 45–46
        stabilization, 42, 49–50, see also Macroeconomic goals
    measures of well-being, 50–58, see also EDP, GDP, GPI, MEW, NCI, and PAI
    wealth per capita, 55
    systems and, 16
    measures of integrity, 25
Eco-resource security, 88–89, 102
Ecosystem
    adaptive management, 68
    health, 24
    integrity, 25–26
    productivity, 34
EDP, see Environmentally Adjusted Net Domestic Product
Efficiency, see Economic concepts, economic efficiency
Ehrlich-Holdren Identity, 10–11
Electricity load management, 324–326, see also Cincinnati Gas & Electric Company
EMAS, see European Union, Eco-Management and Audit Scheme
Emergency Planning and Community Right-to-Know Act (EPCRA), 272, 276, 278, 342
Endangered Species Act, 108, 117, 222–223
Energy consumption, 9, see also Cincinnati Gas & Electric Company
Environment
    economics and the environment, 8–10, 37, 92, 94
    environmental systems, 26
    global dimensions, 88
    international security, 173
    international trade, 170–173
    technology and the environment, 8–11
    treaties, 170
Environmental management systems, 154, 161–163, see also TQEM and Responsible CareÆ
    binding codes of practice, 161–162
    performance measurement, 161–162
    stakeholder dialogue, 161–162
    systemic nature, 161–162
Environment Protection Information Center (EPIC), 223
Environmental audit, 122–125, see also Auditing and Reporting
Environmental Research Institute of Michigan, 154
Environmental trust accounting, 134–138
Environmentally Adjusted Net Domestic Product (EDP), 57
EPA, see United States, Environmental Protection Agency
EPCRA, see Emergency Planning and Community Right-to-Know Act

EPIC, see Environment Protection Information Center
Equity, see Ethics
ERPG, see Worst-case scenarios, Emergency Response Planning Guidelines
Ethics
  and equity, see Sustainability, principles
  global integrity, 97–100, 103
  instrumental values, 89, 98
    commitment, 98, 100
    compatibility, 98–99
    holism, 98–99
    responsibility, 98–100
    scale, 98–99
    stewardship, 98, 100
  intergenerational equity, 8, 12, 16, 45–46, 48–49, 86–88, 110–111, 170, 176
  intragenerational equity, 16, 170
  intrinsic value of nature, see Value, intrinsic
  land ethic, 97
  premises for sustainable development, 87, 89, 92–93
  socio-geographic equity, see Ethics, intragenerational equity
  stewardship, 96, 98, 100
European Union Auto Recycling network, 286
European Union, European Community, 72
  Common Agricultural Policy, 172
  Eco-Management and Audit Scheme (EMAS), 160
Existence value, see Value
Externalities
  air emissions, 77
  costs to business, 37–38
  definition, 45, 140
  internalizing costs, 34, 45, 96, 140–146
    command-and-control, see Regulation
    legal remedies, 141, 148, see also Maxxam Group Inc., lawsuits to stop clear-cutting
    market-based incentives, see Market-based incentives
  relationship to environment, 17, 26
  quantifying, see Sustainability
Exxon, 79, 159

F

Falconbridge Limited, 154
FAO, see Food and Agriculture Organization
Farm Bill, see Food, Agriculture, Conservation, and Trade Act
FASB, see Financial Accounting Standards Board
Federal Emergency Planning and Community Right-to-Know Act, 338
Federal Railroad Administration, 292, 300, 302
Federal Reserve Bank of Chicago, 154
Federal Water Pollution Control Act, 288
Fertilizers, see Agriculture
Fetzer Vineyards, 12, 176
  Club Bonterra, 265
  conventional viticulture process, 257–260
  economic feasibility of Bonterra wines, 264–265
  growing conditions, 252–254
  history, 250–251
  marketing and sales of Bonterra wines, 266–268
  organic practices and flavor of wine, 249–250
  organic viticulture process, 255–257, 260–261
  product certification, 251, 263, 264
  soil structure, 252, 254–257, 261
  wine production, 262–263
  wine varieties, 250
Financial Accounting Standards, No. 5, 118, 129
Financial Accounting Standards Board, 118–119, 129
Fisheries, see also Maxxam Group Inc., impacts on fisheries
  Grand Banks, 71–72
  harvests, 70–72
  management, 72
    Magnuson Fisheries Management and Conservation Act, 73
  overfishing, 71–72
  sustaining, 169
Finland, 171
Food and Agriculture Organization (FAO), 71
Food, Agriculture, Conservation, and Trade Act (Farm Bill), 246, 263
Ford Motor Company, 151, 154, 283, 286
Ford Motor Company of Canada, 154
Forests
  California timber harvests, prices, shipments, and values, 220–221
  clear-cutting, 217, 222–225, 228–229
  Coast Redwoods, 216–218
  depletion, 28–30
  harvests, 65–68
  health, 65
  management and watersheds, 228–230
  mean annual volume increment (MAI), 226
  pacific northwest, 66–67
  relationship to fisheries, 222
  selective cutting, 214–215, 230
  stocks, 66–67
  tree growth, 225–227
Foron, 148
Frontier economics, 92
Future generations, needs, see Ethics, intergenerational equity

G

GDP, see Gross Domestic Product
GEAE, see General Electric Aircraft Engines
Geer vs. Connecticut, 117
GEMI, see Global Environmental Management Initiative
General Electric Aircraft Engines (GEAE)
  chemical use, 275–276
  company description, 272–275
  environmental cost containment, 276–277
  environmental waste costs, 271–272
General Motors Corporation (GM), 148, 154, 159, 283, 286
Genuine Progress Indicator (GPI), 57–60
Geon Corporation, 154

Subject Index

Georgia-Pacific Corporation, 153–154
Germany, 148, see also Procter & Gamble
Global Environmental Management Initiative (GEMI), 150, 153
Global concerns
   greenhouse gases, 171
   marketplace, 38
   water supply, 75
Global welfare curve, 11
Globalization, 27, 38, 147–148, 161, 264, 307, see also Procter & Gamble
   free trade, advantages and disadvantages, 171–173
   of commerce, see Sustainability, principles
GM, see General Motors Corporation
GNP, see Gross National Product
GPI, see Genuine Progress Indicator
Grand Trunk Western Railroad
   accidents and derailments, 291–297, 300–302
   company description, 293–294
   emergency procedures, 297–300
   hazardous materials, transportation, 294–297
   risk, 294–297
Great Lakes
   DDT and food chains, 35
   depletion of forests, 30
   disposal of production wastes, 35
   forests, 67–68
Green accounting, 128–129, see also Auditing and Reporting
Green consumers, see Marketing
Greenpeace, 148
Gross Domestic Product (GDP), 55–56, 59–60
   definition, 51–52
   shortcomings, 52–54, 58
Gross National Product (GNP), 51, 54–55
Growth
   economic growth, 3, 8, 94
      conflict with development, 49
      relationship to environmental regulation, 146–147
      relationship to international trade, 171–172
   exponential growth, 7
   Limits to Growth, 4–5, 19
   population growth, 4–5, 9–10
   steady-state, 49
   sustainable economic growth, 9, 49–50, 53–54
GTE North, 154
Gun-boat diplomacy, 72

H

Halliburton, 153
Hardin, Garrett, 31, 33, 36, 39–40
Hazardous Materials Transportation Uniform Safety Act, 294
Hazardous Substance Superfund, 130
Headwaters forest, 223–224, 228–229
Hedonic analysis, see Valuation
Henkel, 312
Hooker Chemical Company, 129, 134–138
Hudson Bay Company, 30
Hughes Electronics, 153
Hunter-gatherer system, 27–28
Hydrologic cycle, 74
Hydrologic modeling, 76

I

ICC, see International Chamber of Commerce
ICI of Canada, 293
Index of Biotic Integrity, 12
India, 123, 338
Indonesia, resource depletion, 55–56
Information, dialogue and expectations, see Sustainability, principles
Integrity of individuals, communities and nature, see Sustainability, principles
   health, 24–26
   measures, 25
Interfaith Center on Corporate Responsibility, 159
Internal reporting, 120–125
International Chamber of Commerce (ICC), 3, 38, 85, 122
International Joint Commission, 154
International Organization for Standardization (ISO), 159
International Paper Company, 147
Ireland, 171
ISO, see International Organization for Standardization
ISO 14000 standards, 159–161, 163–164, 170
ITT Automotive, 154

J

Japan, 150–151, 171
Japanese Union of Scientists and Engineers, 150
Johnson, S.C. and Son, Inc., 139
Johnson & Johnson, 153

K

Kanawha Valley, see Worst-case scenarios
KPMG Peat Marwick, 154

L

Landfill regulations, 288–289
Leaking Underground Storage Tank Trust Fund, 130
Life cycle
   analysis, 3, 149, 176
   inventory, 319
Limits to Growth, see Growth
Lindahl-Hicks income, 50
Linear-sink model, 14–15
Loblaw International Merchants, 87
Love Canal, 129, 337

M

Macroeconomic goals
   price stability, 8, 49
   full employment, 8, 49

economic growth, see Growth, economic growth
Malcolm Baldrige National Quality Award, 152, 155
Management Information System (MIS), 119
Marginal abatement cost, 142–143
Marginal damage cost, 142–143
Market-based incentives
 internalizing costs, 143–146, 309–310
 limitations, 168
 OECD countries, 144
 tradeable permits, 79, 144–146
 taxes, 86, 143–145
Marketing
 green consumers, 312
 green products, 38, 87, 284, 286
 product certification, 87, 160–161, 236–237, 263–264
Markets
 market failure, 42, 44–45, 140, 168
 price system, 44–45
 role of markets, 44
 supply and demand, 44
Maryland Automotive Reclamation Corporation, 285
Maxxam Group Inc.
 forest operations, 222–225
 history, 215
 Hurwitz, Charles, 213, 215–216
 impacts on employees, 219, 222
 impacts on fisheries, 222
 lawsuits to stop clear-cutting, 224–225
 Pacific Lumber Company (PALCO), 213–215
  forest management policy, 219–222
 takeover activities, 216
McDonald's, 87
Measure of Economic Welfare (MEW), 56
Measuring and adapting, see Sustainability, principles
Merck & Company, 153
Methyl isocyanate (MIC), 155, 338, 350
MEW, see Measure of Economic Welfare
MIC, see Methyl isocyanate
Michigan Department of Environmental Quality, 300
Michigan Department of State Police, 298, 300
Michigan Department of Transportation, 300
Michigan Railroads Association, 298
MIS, see Management Information System
Montreal Protocol, 34, 142
Motor Vehicles Manufacturing Association, 284

N

National Association of Risk Auditors, 124
National Audobon Society, 159
National income accounting, see Economic concepts, measures of well-being
National Institute for Chemical Studies, 342
National Institute of Standards and Technology, 152
National Organic Standards Board, 235, 247, 263
National Transportation Safety Board, 300, 302
National Wildlife Federation, 131, 159
Natural assets, see natural capital
Natural capital, 8–9, 28–31, 52, 56
 conversion, 33
 surplus, 29
Natural Capital Indicator (NCI), 54–55, see also Natural capital
Natural-resource-based view of the firm, 148–149
NCI, see Natural Capital Indicator
NDP, see Net Domestic Product
Neste, 125
Net Domestic Product (NDP), 55–56
New England Electric System, 79
New York Power Authority, 154
Norsk Hydro, 125
North Pacific Fishery Management Council, 72

O

Occidental Chemical Corporation, 154
Occidental Petroleum, 129, 134–138
OGBA, see Organic Growers and Buyers Association
Ohio Consumers' Counsel, 332
Oil Spill Liability Trust Fund, 130
Olin, 153
Ontario Hydro, 154
Option Pricing, see Valuation
Organic farming, see Fetzer Vineyards and Walnut Acres
 Organic Food Production Act, 245–246, 264
 Organic Growers and Buyers Association (OGBA), 245
Our Common Future, 8, 12, 37–38

P

Pacific Herring, economic value, 72
Pacific Lumber Company (PALCO), see Maxxam Group Inc.
PAI, see Pollution-Adjusted Economy Indicator
PALCO, see Pacific Lumber Company
Pareto Optimality, see Economic concepts, economic efficiency
PCBs, see Polychlorinated biphenols
PCSD, see President's Council on Sustainable Development
Pesticides, 259
Plan-Do-Check-Act cycle, 153, see also Deming
Plimsoll line, 49–50
Pollution
 by waste disposal, 32–33, 272, 277
 cost savings, 39, 149, 277
 Pollution Prevention Act, 276
 prevention, 149, 154, 277
 Rhine River, 88
 taxes, 86, 143–145
Pollution-Adjusted Economy Indicator (PAI), 54–55
Polychlorinated biphenols (PCBs), 34, 273, 287
Population, 10–11, 35, 45, 92
Positive sum outcome, see Win-win outcome
President's Council on Sustainable Development (PCSD), 131
Price Waterhouse, 123, 128
Process offsets, 146

Procter & Gamble (P&G), 39, 87, 176
  company description and history, 304–308
  environmental concern of consumers
    Germany, 309–310
    United States, 310–311
  environmental quality policy, 308
  fabric softener market, 311–317
  GEMI membership, 153
  globalization issues, 307–309
  packaging waste, 317–318
  recycling plastic waste, 318–321
Product offsets, 146
Property rights, 112–113, 140
Productive capacity, 5, 64, 146–147, 177
Public goods, 29–31, 44, 79, 112–114, 140, 171
  free rider problem, 171
  water as, 75
Public opinion, 158, 240–242, 311–312, 351–352
Public Utility Commission of Ohio (PUCO), 323, 327–331, 334–335
PUCO, see Public Utility Commission of Ohio
Put option, 111

R

RCRA, see Resource Conservation and Recovery Act and Amendments
Recycling, 284–286, 318–321
Redwood National Park, 218
Regenerative capacity, 64
Regulation, see Externalities
  command and control, 37, 39, 141–143, 146–147
Replacement cost, see Valuation
Resilience, see also Sustainability principles, system resilience
  ecological, 16, 29, 53, 154, 177
  economic, 16, 53, 177
  social, 16, 49, 53, 177
Resource Conservation and Recovery Act (RCRA) and Amendments, 35, 131, 141, 285, 288
Resources, see also Natural capital
  appropriation, 28–30
  asset liquidation, depletion, 3–4, 16, 23, 29–30, 52, see also Maxxam Group Inc.
  impoverishment, 63–64
  non-renewable, 3–4, 64
  overuse, 23, 31
  productivity, 28
  renewable, 4, 63–64, see also Maxxam Group Inc.
  stocks, 25, 30
  substitution of natural capital and human-made capital, 9, 29, 49
Responsible Care®, 158–159, 164, 185–187, 276, 339–340, 342, 350
  CAER code workshops, 342, 347
  Codes of Management Practice, 100, 158–159, 340
  Guiding Principles, 83, 158, 340
  origin, 158
Restoration cost, see Valuation
Revlon, eliminating CFCs, 87

Rhone-Poulenc Institute plant, 342, 344
Riker labs, 79, 87
Risk, 64–69
  assessment, see Sustainability, principles
  business-science synthesis, 129–130
  effects on financial statements, 130
  management plan, 339
  measuring, 64, 344–348, see also Worst-case scenarios
  transporting hazardous materials, 294–297
Roper Organization, 310–312

S

Safe minimum standard, 175–176
SARA, see Superfund, Amendments and Reauthorization Act
Save-the-Redwoods League, 215, 218
Scale and technology, see Sustainability, principles
Schmidheiny, Stephen, 90, 147–148
Scoping, see Boundaries
Scrap, see David J. Joseph Company
Sealed Air, Inc., 238
SEEA, see Auditing and Reporting, Satellite System for Integrated Environmental and Economic Accounting
Selling the right to pollute, see Market-based incentives, tradeable permits
Seventh Generation, 87
Sierra Club, 159, 215
Smart, Bruce, 39–40, 125, 164
SNA, see System of National Accounts
Social contract, 31, 39
Social Investment Forum, 159
Socio-economic security, 88–89
Soil
  ecology, 254
  erosion, 29, 33
  structure, 235–236, see also Fetzer, soil structure
Southern, 153
Spaceship economics, 92
Staggers Act, 296
Stakeholder interests, see Sustainability, principles
  community, see Worst-case scenarios, Kanawha Valley
Stewardship, product, 96, 149, 154, 159, 174, 304
Sulfur dioxide emissions, 8, 131, 145, 172
Sumarian society, 29
Sun Company, Inc., 159
Superfund, 36, 117, 128, 272
  Amendments and Reauthorization Act (SARA), 36, 141, 288, 339
Sustainability
  definition, 3–4, 8, 12–13, 17–19, 45
  foundational divisions, 12–13, 178
    coordination processes and measures, 13, 17–19
    ethics and values principles and measures, 13, 17–19, 174
    systems principles and measures, 12–18
  harvest and, 32
  principles

auditing and reporting, see also Auditing and Reporting
closing cycles and loops, 4, 16–17, 77, 83, 168, 287–289, 319–321
ethics and equity, 37, 83, 142, see also Ethics
generality of a systems view, 12, 14, 151, see also Systems
globalization of commerce,147, 170, 172–173, see also Globalization
information, dialogue, and expectations, 101, 147, 162, 170, 174–175, 337–338, 340, 350–351
integrity of individuals, communities, and nature, 17, 24, 26, 53, 149, 170
measuring and adapting, 64, 100, 155, 161–162, 170
quantifying externalities, 32, 140, 170 345–346, see also Externalities
risk assessment, 38, 64–65, 129, 294–297, 345–346, see also Risk
scale and technology, 12, 31, 42, 49–50, 98–99
stakeholder interests, 16, 91, 100, 147, 159, 162, 174–175
system resilience, 16, 24, 53, 177, see also Resilience
valuation, 51, 106–118, 170 see also Valuation
relationship to resource use, 64
strong versus weak sustainability, 49
worldview elements, 102
Sustainable development, see Sustainability
System of National Accounts (SNA), 56–57
Systems
closed, 16–17, 26–28, 151
complex feedback loops, 14–15, 26–28, 97
concepts, 3–4,13–17
emergent properties, 15
environmental management, 162
integrity, 26, 28, 97
limits and thresholds, 14–15
nonlinear interactions, 14–15
open, 17, 28, 42
view, generality of, see Sustainability, principles

T

Takings laws, 112–113
Technology
benign, 175
economic growth and development, 42
Tenneco, 153
3M Company, Pollution Prevention Pays, 39, 87
Total Quality Environmental Management (TQEM), 100, 149–150, 153–157
Total Quality Management (TQM), 100, 150–151, 299–300, see also DuPont Company and Grand Trunk Western Railroad
Toxic Substances Control Act, 141
TQEM, see Total Quality Environmental Management
TQEM Primer and Self-Assessment Matrix, 155–157
TQM, see Total Quality Management
TRANSCAER, see Transportation Community Awareness and Response program

Transportation Community Awareness and Response program (TRANSCAER), 298–300
Travel cost, see Valuation
Trust and consensus, 174–175, 337–338, 342, 349–350

U

UNEP, see United Nations Environment Program
Unified-cycling model, 14–15
Unilever, 312
Union Carbide, 82, 155, 338, 349
United Nations Commission on Environment and Development, 88, 90, see also *Our Common Future*
United Nations Environment Program (UNEP), 38
United Savings Association of Texas, 224
United States
  Census Bureau, 150
  Department of Agriculture
    Forest Service, 217
    livestock standards, 235
    Standard of Identity, 240
  Department of Commerce, 26
  Department of Justice, 141
  Environmental Protection Agency (EPA), 118, 141, 147, 184, 276, 287, 338, 350
  Fish and Wildlife Service, 222–223
  Food and Drug Administration, 240
  Navy, 151
  Supreme Court, 117, 223
University of California Division of Agricultural Sciences, 257, 264
Use value, see Value
U-shaped relationship, environmental quality and economic growth, 8–9, 172
Utility possibilities frontier, 45–46

V

Valdez Principles, see CERES Principles
Valuation, see also Sustainability
  comparison among methods, 116
  contingent valuation, 115–116
  discounted cash flow, 108–111
    Brennan and Schwartz, 110–111
    critique, 110–111
    example, 109–110
  externalities, 32
  hedonic analysis, 114
  indirect, 112–116
  option pricing, 111–112
  public policy aspects, 117–118
  replacement or restoration cost, 113–114
  travel cost, 114–115
  uses, 106
Value
  existence, 108, 118
  intrinsic, 95–97, 108, 117–118
  measuring, 106
  use, 106, 108, 114

Subject Index

Variability of natural systems, 32
Vehicle Recycling Development Center, 286

W

Walnut Acres
  Anderson, Bob, 232–233, 240–243, 247
  company organization and financial performance, 233–235
  Keene, Paul, 232–233, 242–243
  laws on organic farming, 245–247
  marketing, 241–243
  operations, 235–238
    food production, 237
    organic farming practices, 235–237
    order processing, 238
  product certification, 236–237, 245–257
  relationship to community, 240
  Whole Foods industry, 238–240
Waste
  agricultural production, 33
  disposal, 32, 272, 277, 287, 304
  industrial production, 32–33, 42
  water, 33
Water
  consumption, 74–75
  freshwater supplies, 73–76
  global supply, 75
  hydrologic system, 74, 79, see also hydrologic cycle and hydrologic modeling
  imported surface water, dependence on, 173
  pollution, 75–76
  quality, 73–76
  sustainable management, 75–76
  withdrawal, 74–76
WCED, see World Commission on Environment and Development
Wever, Grace, 149–150, 154
Whole Foods industry, 238–240
Wild and Scenic Rivers Act, 117
Wilderness Preservation Act, 117
Willingness to pay, see Valuation, contingent valuation
Win-win outcome, 175
WMX Technologies, 153
Wolves, 107
Woodbridge Group, 286
World Bank, 54, 123
  loans, 170–171
World Commission on Environment and Development (WCED), 8, 38, 86
World Health Organization, 75
World Industrial Conference on Environmental Management, 38
World Resources Institute, 146
Worst-case scenarios
  definition, 338–339, 341, 344–349
  Kanawha Valley, 337–338. 341–343, 352
    Community Emergency Response Evaluation Group (CEREG), 351
    Emergency Response Planning Guidelines (ERPG), 345–349
    Hazard Assessment Committee, 342–343
    Hill, Paul, 342, 344
    JBF Associates, Inc., 343, 350–351
    Knowles, Richard (Dick), 337–338, 342–344, 349–351
    Local Emergency Planning Committee (LEPC), 342–347, 350–351
    Nixon, Pam, 338, 343–344
    People Concerned about MIC, 338
    Responsible Care® guidelines, see Responsible Care®
    Risk Communications Committee, 342–343, 349
    Safety Street, 348–351
    worst-case process, 338–339, 342–352
      community stakeholders, 342, 348–351
  more probable release scenarios, 345–349

X

Xerox
  CGLI membership, 154
  environmental policy, 151
  Leadership Through Quality, 152

Z

Zero sum outcome, 175